MW00763131

Practical Controls

A Guide to Mechanical Systems

Practical Controls

A Guide to Mechanical Systems

Steven R. Calabrese

THE FAIRMONT PRESS, INC.
Lilburn, Georgia

MARCEL DEKKER, INC.
New York and Basel

Library of Congress Cataloging-in-Publication Data

Calabrese, Steve, 1965-.
Practical controls: a guide to mechanical systems/Steve Calabrese.
p. cm.
ISBN 0-88173-447-0 (print)
1. Automatic control. I. Title.

TJ213.C245 2003
629.8--dc21 2003048337

Published by The Fairmont Press, Inc.
700 Indian Trail, Lilburn, GA 30047
tel: 770-925-9388; fax: 770-381-9865
http://www.fairmontpress.com

Distributed by Marcel Dekker, Inc.
270 Madison Avenue, New York, NY 10016
tel: 212-696-9000; fax: 212-685-4540
http://www.dekker.com

Printed in the United States of America

10 9 8 7 6 5 4 3 2 1

0-88173-447-0 (The Fairmont Press, Inc.)
0-8247-4618-X (Marcel Dekker, Inc.)

Dedication

To Barbara and in Loving Memory of Richard,
For your knowledge, wisdom, and unconditional love

And to Connie and Anthony,
With undying commitment and adoration

TABLE OF CONTENTS

FORWARD

The title of this book, *Practical Controls*, sets the tone and style of the text within. The approach that it takes is one of practicality. To that end, the author attempts to describe the content in terms of "real world" practices and principles. The subject matter is purposefully short on theory, and often long on reality. The concepts covered in the following chapters stem from the practical experience gained by the author as an employee of a mechanical contracting company, one of which maintained control systems design, installation, and commissioning capabilities. The book's intent is to try to convey the practical methods of control as learned by the author throughout his years as a control systems designer working for a mechanical contractor. Although written from a mechanical contracting perspective, the book hopefully appeals to all corners of the HVAC industry, from consulting engineers to controls contractors.

Here is the place to define a suitable candidate for this book, what they should know going into it, and what they can expect to get out of it. This is not intended to be a college textbook. It is intended to be read by the HVAC professional; someone with at least some experience in the industry, perhaps knowing a little bit about most of the topics. The content herein assumes that the reader has a requisite knowledge of HVAC in general, of the fundamental concepts of mechanical systems and design, and at least an idea of how mechanical systems should operate. The content also assumes a prerequisite familiarity with the basics of electricity by the reader. The book expands on these presumptions, with the intentions of giving the reader a "nuts and bolts" explanation of the fundamental concepts of control. It will not make a control systems designer of the reader. However, it will give the experienced HVAC professional a well-rounded education on the practical methods of HVAC control.

The topic of controls in general is large and continually evolving. Whereas many of the basic design concepts of mechanical systems have been solidified generations ago, control system design seems to be ever-changing, in terms of what technological advances have to offer. Duct and pipe design and construction hasn't changed much throughout the ages. Many of the basic concepts and design criteria for air and water systems have been established and accepted by the industry, to the point of standardization. Yet the means to control these systems is in a state of continuous evolution, insofar as to what we have at our disposal in terms of tools and techniques. The basic concepts of control may be well established, but the means and methods are constantly changing, driven mainly by technology and ingenuity.

There is no possible way to cover every aspect of this broad subject in one single edition. To focus upon the intention of this writing, the author has taken the liberty of omitting certain topics that fall under the general subject of controls as they apply to the HVAC industry. Classical controls and control systems, such as pneumatics and electro-mechanical sequencing controls, are not covered herein, even though they are still in wide use even in this day and age. The material presented in this book deals strictly with electrical, electronic, and microprocessor-based controls and control systems. Also excluded for the most part are any detailed discussions on the internal controls of packaged equipment. Items such as packaged rooftop units, boilers, and self-contained chillers have most (if not all) of their controls components factory installed and wired, thus providing for a complete, factory furnished control system. Though in places the author will touch upon factory equipment controls, the material mainly focuses on those types of equipment and systems that must be fitted with engineered control systems.

The direction that this book takes, is that of laying down the basics, and building upon them. This is found to be true as the reader journeys from chapter to chapter. It is even more so the case within the chapters themselves, with each section of a chapter building upon the previous section. It is important that the reader start each chapter from the beginning, and read and comprehend each section before moving on to the next. Once read and understood, this book can from then on serve as a "reference manual" for the reader, perhaps to be utilized as a design tool and/or as a practical resource.

In attempting to succeed in the goal of this book, the author periodically provides insight into how mechanical systems are designed, and how they are designed to operate. Stopping short of full-blown discussions and descriptions of mechanical design concepts, the author will include enough on a given topic to illustrate not only how the particular equipment and system is controlled, but also why it's controlled the way it is.

The topic of controls and control systems in HVAC is an extensive one. By no means will this book cover all aspects of the topic, nor will it cover any single aspect to any great extent. However, it will delve into and discuss a broad array of different concepts falling under the general topic, and will thus serve as a good foundation to a further education in controls, should the reader choose to pursue it...

Chapter 1

INTRODUCTION

Welcome! Unless you are reading this as a form of torture or as a cure for insomnia, you are most likely a member of the **Heating, Ventilating, and Air Conditioning (HVAC)** industry. How you became that could have been any number of ways. You may have fallen into it by accident, or you may have trained for a career in this industry. Regardless of how you got here, you are now part of a special club, a nationwide, better yet, a worldwide network of engineers, designers, technicians, and installers, all (most!) of whom are dedicated to providing "personal comfort," through the proper and efficient application and utilization of mechanical equipment and systems.

You can call it what you want, but the HVAC industry is primarily a "mechanical" industry. The equipment and components that make up a typical HVAC system are mechanical in nature. Fans, pumps, dampers, valves, ductwork, and piping. Put 'em all together and you have yourself a **mechanical system**. The designers of these systems, the mechanics that install them, and the technicians that work on them, are all trained in the mechanical vein.

How these systems are controlled, however, might very well fall under a different discipline, and is the very subject of this writing. An HVAC system can be described as being a mechanical system *plus* the control system that is required to properly and efficiently operate it. A typical mechanical system consists of many subsystems. Each of these subsystems in itself must be controlled. A fan must be "told" to turn on and off. A damper must be told when to be open, and when to be closed. A valve must be told what position to assume; whether to allow flow through it or not, or maybe to allow partial flow through it. To control any of one these subsystems on its own requires at least a little bit of insight, as to what function the subsystem is serving. More importantly, though, is how all of these various subsystems should work together, so as to operate as a single system.

author's intention to describe the role of controls in me- ...ns. This includes: how the various mechanical systems ... subsystems should operate, how these systems should be designed to operate, and how to use "practical" controls methods to correctly control the operation of these systems. The first step to that end is to define what an HVAC system is. As stated earlier, an **HVAC system** can be thought of as **a mechanical system *plus* the associated controls and control system required to operate it**.

Mechanical Systems

In HVAC, mechanical systems are typically designed to perform heating, cooling, and ventilation of spaces requiring such types of environmental control. The complexity of these systems ranges from the simple to the sophisticated. A ducted exhaust fan, that is manually turned on and off, is an example of a simple mechanical system. The system is composed of the fan, and the associated distribution ductwork required to convey the air, from the space being exhausted, to the outdoors (Figure 1-1a).

As an example of a more complicated mechanical system, consider Figure 1-1b: a hot water piping/pumping system consisting of two hot water boilers, two hot water circulating pumps, and the required hot water distribution piping going out to miscellaneous hydronic (hot water) heating equipment.

In each of the above examples of mechanical systems, we notice two distinct components: the equipment, and the required mechanical means of connecting the equipment, to other equipment, and to the real world. In the simple example of the exhaust system, the exhaust fan is the equipment, and the ductwork is the mechanical means. In the more complex example, the equipment consists of the boilers, pumps, and the miscellaneous heaters. The mechanical means of connecting together all of the equipment, in some meaningful manner, is the hot water piping.

We can say that mechanical systems are typically made up of these two components: the equipment, and the mechanical means of connecting the equipment. In all but the simplest of mechanical systems, equipment alone does not make up the system. Unless designed and manufactured as completely "stand-alone," a piece of equipment does

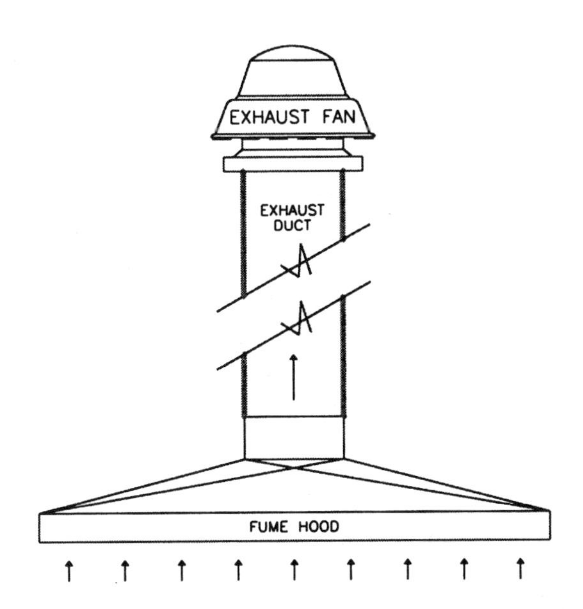

Figure 1-1a. Roof mounted exhaust fan ducted to a fume hood.

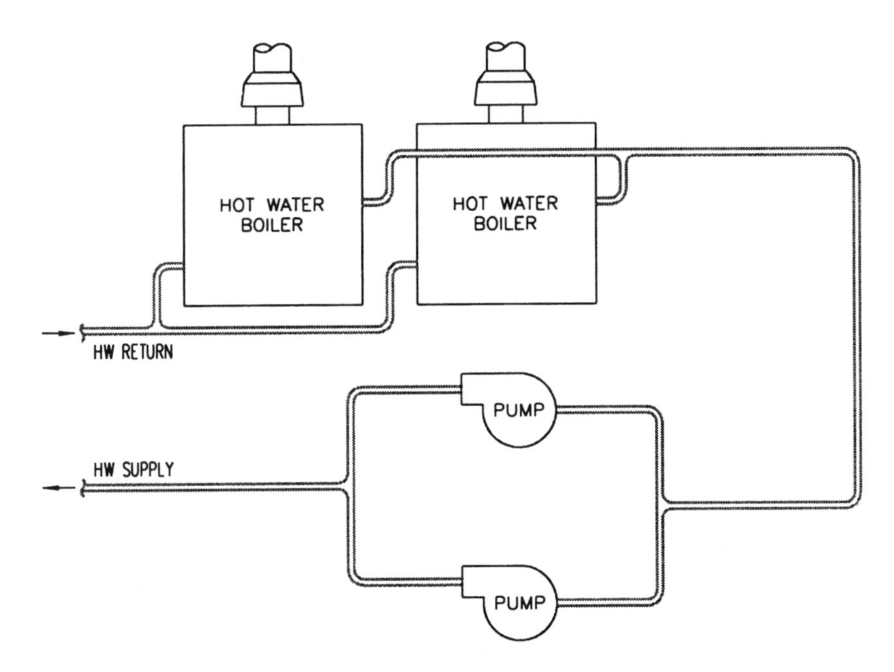

Figure 1-1b. Hot water piping/pumping system consisting of two boilers, two pumps, and hot water distribution piping.

not do much good by itself. An example of a stand-alone piece of equipment would be perhaps an electric heater that you just plug into a wall outlet, or an oscillating fan that you might buy for your basement. For the majority of heating, ventilating, and air conditioning applications that are encountered in our industry, we're talking systems. We, as specifiers and designers, are selecting equipment and designing the systems, integrating the equipment with properly designed mechanical distribution (ductwork and piping) systems that enable the equipment to function the way that it's intended to: as part of a system!

Okay, okay. Enough talk about equipment, ductwork, and piping already! Time to switch gears and talk about the other component of a typical HVAC system: the controls!

Controls and Control Systems

In our two examples of mechanical systems, we need some method of control. First and foremost, we need to have some idea, some inkling, of how the system should be controlled. The designing engineer of a mechanical system should have an idea of the system's method of control, as he is the one designing it. As the designer of the system, the engineer may choose to write a description of how he would like the system to operate. This description, commonly referred to as a **Sequence of Operation**, should describe in detail how each piece of equipment and each subsystem should operate, so that the system as a whole is properly functional.

For the exhaust system example, it's a no-brainer. The sequence of operation for this system may read as follows:

Fume hood exhaust fan EF-1 is manually operated by a user switch, located in the space being served by the fan.

For the hot water system example, however, the sequence of operation is not that simple and straightforward. For this example, it is extremely important that the engineer's intentions be communicated to the control systems designer. When should the system be in operation? Is it seasonal? How are the pumps controlled? Does one run, with the other serving as a backup? Should the backup pump automatically start upon failure of the primary pump? How are the boilers controlled? By a single controller? Is hot water setpoint to be "reset" as a function of outside air temperature?

The engineer's intentions for the operation of the system can be communicated in several different ways, or a combination of ways. He can choose to create a sequence of operation. He can specify certain controls or controls methods in his design. He can provide informative clues in his selection and sizing of equipment. And he can communicate his intentions verbally. For the hot water system example, the general sequence of operation may read like this:

> *Hot water system to be in operation whenever the outside air temperature is below 60 degrees F. (adj.). System enable/disable to be performed automatically, by an outside air temperature controller. When enabled, the pump selected as the "primary" pump is to run continuously. Selection of the primary pump is a manual procedure, performed by a primary pump selector switch. Upon failure of the primary pump, the backup pump is to automatically start. Boilers are to operate, via a common temperature controller, to maintain hot water temperature setpoint. Setpoint is to be a function of outside air temperature. As the outside air temperature increases, the hot water temperature setpoint is "reset" downwards. Miscellaneous unitary hot water heating equipment throughout the building is to operate via integral controls, to maintain desired comfort levels.*

With the operation of the mechanical system described as such, the control systems designer has a clear direction to follow, and can design his or her control system in accordance with the sequence. During the design phase, the designer may also choose to elaborate on this given sequence of operation, so as to include important operational details, temperature settings, equipment specific information, etc.

Now that the sequence of operation has been developed, the next step is to identify the various "points of control." For the first example, this is a relatively simple task. By looking at the sequence, we gather that the "point of control" is a user switch located in the space served by the exhaust fan. In this case, the control device is a switch, that is used to manually turn the exhaust fan on and off (Figure 1-2a).

For the second example, identifying the points of control is a more complicated task. Figure 1-2b shows the various points of control required for the hot water system. The first point that we can identify is the outside air temperature controller. This device is to allow the hot water system to operate when the outside air temperature is below 60, and disallow its operation otherwise. In essence, the device is nothing more than a temperature actuated switch that closes when the temperature drops below the setpoint of the device. When the switch is closed, the primary pump runs, and the boilers are enabled for operation.

The second point that we can identify from the sequence of operation is the primary pump selector switch. The switch is a manual control that determines which of the two pumps is to be the primary pump. Another point that is associated with the pumps, that is perhaps a bit more difficult to identify, has to do with determining primary pump failure. The sequence states that the backup pump is to automatically start upon failure of the primary pump. How do we determine that the primary pump has failed? We can look at a couple of different things. We can monitor water flow with a flow switch, or we can monitor the pump motor's current draw with a current sensing switch. Either device can alert us to a failure of the primary pump.

The final point of control that remains to be identified here, is that of hot water temperature control. The sequence mentions that the boilers are to be operated, by a common controller, to maintain hot water temperature setpoint. As such, we need to measure, or sense, the hot water supply temperature, common to both boilers, and also establish a means of controlling the operation of the boilers to maintain some setpoint. In simple terms, we are talking about installing a sensor in the common hot water supply piping, and transmitting the temperature "signal" to some central controller. At the controller, we have a means of establishing a setpoint. The controller can therefore calculate the difference in sensed temperature and setpoint, and stage the boilers accordingly, in an attempt to minimize this difference. This particular controller, as implied in the sequence, must also be able to reset the hot water temperature setpoint as a function of outside air temperature.

The next step in designing a control system for our given mechanical system is to begin selecting practical, "real-world" methods and controls to implement our sequence of operation. A mechanical system can be designed and a sequence of operation can be written in advance. On paper, and in theory, what is designed mechanically and what is written may be quite feasible. Yet in practice, what is being asked for the mechanical system to do by the sequence of operation may be impractical, inappropriate, or even impossible! This especially holds true for systems consisting of many subsystems. While each subsystem may be able to be

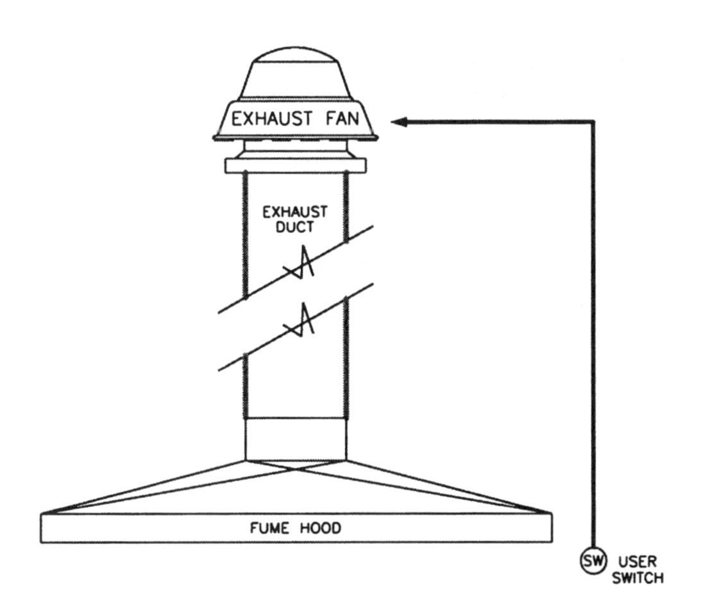

Figure 1-2a. Exhaust system with user switch as the "point of control."

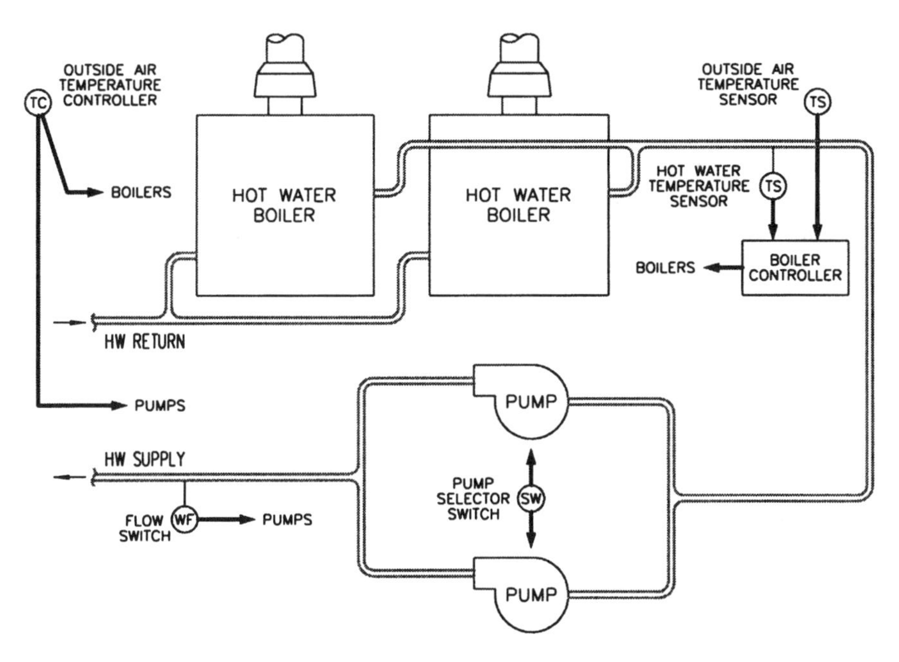

Figure 1-2b. Hot water system, conceptually showing all required points of control.

controlled adequately on its own, the specified mode of operation for each of these individual subsystems may be counteractive to overall system operation.

The upcoming chapters will discuss many of the common mechanical systems that are popular in this day and age, and will attempt to define "practical" methods of control for each, by defining basic rules, equipment requirements, rules of thumb, pros and cons, do's and don'ts, etc. Please read on, as we begin the next chapter, and attempt to give an overview on mechanical systems and equipment.

Chapter 2

Mechanical Systems and Equipment Overview

Before we begin taking a look at the various types of systems and the equipment that makes up these systems, now is probably a good time to draw the distinction between "**packaged**" equipment, and equipment that must be "**built up**." A packaged piece of equipment is one that is furnished with some amount of factory mounted and installed controls devices. These controls may partially or even completely allow the equipment to operate, without any other added controls. In our examples from the previous chapter, the hot water boilers are "packaged" pieces of equipment. Each boiler can operate via its factory installed control system, with virtually no other controls required. Just hook up the gas, water, and power, connect it to the piping/pumping system, and let it fire!

The exhaust fan from our first example is typical of a piece of equipment that must be "built up." The fan itself comes with no controls at all, and must be equipped with controls upon installation. An air handling unit that comes from the factory as nothing more than a fan, a hot water coil, and a mixing box, is another example of a piece of equipment that needs to be "built up." Controls required to make this air handling unit operate include, but are not limited to, a fan controller, a control valve for the hot water coil, a damper actuator for the mixing box, and some kind of a temperature controller.

It's interesting to note that just because a piece of equipment comes with factory controls, it doesn't mean that it can't be it can't be part of a built up "system." Either of the two boilers from our hot water system example have the ability to operate completely via their own factory installed controls. Yet in the example they are additionally controlled by a central boiler controller. A "hierarchy of control" exists here, and will be explored more deeply in the chapters to come.

With the distinction made between packaged equipment and built up equipment, let us now begin our discussion on the various types of mechanical systems. We can break down these systems into three major types: airside, waterside, and "miscellaneous." We'll talk about the airside first...

Airside Systems and Equipment

The topic of airside systems encompasses those systems and types of equipment that are primarily dealing with the movement and conditioning of air. As such, we will be taking a look at fan systems and air handling equipment, as well as zoning equipment. What follows is a brief description of some typical air systems and associated equipment.

Rooftop Units

Rooftop units, in the purest sense, are packaged air handlers designed to provide single zone heating and cooling. Heating is typically gas-fired, though it could be electric, and cooling is done by refrigeration, with the entire "refrigeration cycle" integral to the unit. They are packaged in the sense that virtually every control device required for unit operation, less a space thermostat, is factory furnished and mounted. Although fundamentally designed for single zone applications, they can be adapted for use in multizone applications as well.

Make-up Air Units

Make up air units, as the name implies, are air handling units designed to replace, or "make up" air that's being exhausted by some exhaust system. Typically designed for 100 percent outside air, they primarily operate to maintain a constant discharge air temperature. They can be bought as packaged units with gas-fired heating, or they can be built up systems with electric or steam heating coils. Although cooling isn't a concern in many make-up air applications, it can be added. The packaged make-up air unit manufacturer may be able to offer cooling as part of the "package." If not, a separate cooling coil would need to be added. This being the case, and for a built up unit requiring cooling as well, the cooling coil can be a chilled water coil, or a DX coil with a remote condensing unit.

Fan Coil Units

Fan coil units consist minimally of, you guessed it, a fan and a coil. A unit such as this might have another coil, and perhaps a mixing box, and would have to be built up, in the sense that it would not come with any factory installed controls. However, fan coil units can be purchased as packaged units also. The physical size of these units is relatively small; were talking about a unit that could fit in the space above a ceiling, in a wall soffit, or in a small closet. These are single zone systems, and operate to maintain zone comfort levels by heating, cooling, or both. The different configurations that are possible with fan coil units are too numerous to list here, and will be covered later.

Built Up Air Handling Units

Built up air handlers can be thought of as "the big brother" to built up fan coil units. Component-wise they are very similar, yet there is a distinction or two to be made between the two types of units. Size is one of them. More notably, whereas fan coil units are strictly for small, single zone applications, built up air handlers can serve in single zone and multi-zone applications, with virtually no limitation in physical size. It's not uncommon to see a built up air handler designed to deliver 40,000 cfm (cubic feet per minute) of air. With the physical size and the added complexity of air handlers comes the requirement for more sophisticated control strategies. That may be the biggest difference of all between fan coil units and air handlers.

RTU Zoning Systems and Stand-alone Zone Dampers

An RTU (rooftop unit) zoning system is a system of components designed to operate in conjunction with a single zone, constant volume rooftop unit, in order to provide multiple zoning. The space served by the rooftop unit is broken down into zones. The distribution ductwork from the rooftop unit serves each of these zones, via "networked" zone dampers (one per desired zone). Each zone gets to cast a "vote," on whether it wants the rooftop unit to be in a heating mode or a cooling mode. The thermostat that would normally be wired to the single zone rooftop unit is replaced by a zoning system control panel, which polls the zones and controls the unit accordingly. Those zones that are in the majority get what they need from the unit (heating or cooling), and the zone dampers modulate to allow the proper amounts of air into the zones in order to achieve and maintain zone setpoints. Those zones that are in the minority are "outta luck," so to speak, at least for the time being, and their zone dampers remain substantially closed until they become the majority, at which point the rooftop unit changes over to the opposite mode. The last component required in an RTU zoning system is a "bypass damper." This is a control damper that is ducted between the supply and return of the rooftop unit. As zone dampers close off, the static pressure in the supply duct tends to rise. In response to this increase in static pressure, the bypass damper modulates open in order to bypass air from the supply to the return of the unit, thereby maintaining suitable duct static pressure. Such types of zoning systems are sometimes referred to as "poor man's VAV."

Stand-alone zone dampers are another means of providing additional zoning to a single zone, constant volume rooftop unit. The unit may serve a row of offices, with it's controlling thermostat in the Big Guy's office, which is on the southwest corner of the building. On the northwest corner is the main conference room, also served by this rooftop unit. On a cold, sunny December afternoon, the boss's office may be feeling the effects of the early afternoon sun, and the thermostat may be requesting the rooftop unit to be in a cooling or a ventilation mode. On the other side of the building, a small meeting is taking place in the conference room. With its northwestern exposure, this room is not yet benefiting from the sun, and quite possibly could use some heat. By making the conference room a "subzone" and equipping it with a stand-alone zone damper and a room temperature controller to control it, we allow for some degree of additional control. While the subzone cannot command the rooftop unit to change its mode of operation from cooling to heating, it can at least minimize the amount of that cool air that's being delivered into the conference room from the rooftop unit, by closing down the zone damper. The term "stand-alone" pertains to the fact that the zone damper is not networked as part of a larger system, and thus operates in a "stand-alone" fashion, as per its temperature controller.

VAV and Fan Powered Boxes

VAV (Variable Air Volume) and fan powered boxes are zoning equipment, in the sense that they provide for individual zone control when connected to a VAV air handling system. In its simplest form, a VAV air handler operates to provide cold air. This cold air is delivered to all of the VAV boxes (and fan powered boxes) that are connected to the system. A VAV box is, in essence, a zone damper, with a few added features. For each zone served by a VAV box, there is a space temperature controller located in the zone and wired back to the box. The controller modulates the damper in the VAV box to allow the appropriate amount of cold air into the zone, in order to achieve and maintain zone temperature setpoint. A VAV box that is equipped with a heating coil also has the capability of providing heating for the zone, if required.

Fan powered boxes are VAV boxes that also have a fan, and most always have a heating coil. Depending upon the configuration of the fan powered box, the fan may run continuously, or may only run upon a request for heating by the zone controller. The configurations and operational descriptions of fan powered boxes rate as too complex to be fully discussed here, and will be covered in detail in the chapter devoted to them.

Reheat Coils

Reheat coils are heating coils that are applied as zoning equipment. A simple reheat system starts with a constant volume air handler operating to provide cold air. Zones are connected to the air handler (by ductwork), in the form of heating coils. For each zone served by a reheat coil, there is a temperature controller or thermostat located in the zone and wired back to the coil assembly. The coil can be electric, hot water, or even steam. With electric coils, the zone controller or thermostat controls the amount of electric heat generated by the coil in order to achieve and maintain zone temperature setpoint. With hydronic or steam coils, the zone controller is wired to a control valve that regulates the hot water or steam flow through the coil. The term "reheat" stems from the notion that the air being delivered from the air handler is cooled by the air handler, and then "reheated" as necessary by the reheat coil.

Exhaust Fans and Systems

Fans and systems that exhaust air from spaces are a substantial part of most HVAC systems, and play a big role in all of these applications, in terms of ventilation requirements from the exhaust equipment, and from other associated equipment as well. Exhaust fans normally take the form of one of three configurations. Roof mounted exhaust fans are physically mounted outdoors, typically on the roof of a facility, and are ducted "down and around" to their point(s) of exhaust. Sidewall exhausters are mounted in the wall, and may or may not be ducted. Ceiling and in-line fans are ducted; the fan itself resides inside the building, with ductwork to and from it, to the outdoors and from the point of exhaust. Exhaust requirements in HVAC applications range from fume hood exhaust, to toilet exhaust, to ventilation exhaust, to temperature and pressure controlled exhaust.

Waterside Systems and Equipment

The topic of waterside systems deals with those systems and types of equipment that are primarily dealing with the movement and conditioning of water. Hence, in this section we will be looking at pumping systems, boilers, and chillers. Read on for a brief description of some typical water systems and associated equipment.

Pumps and Pumping Systems

Pumping systems vary in complexity, from the basic, easy to understand systems, to those with highly sophisticated configurations. In its simplest form, a pumping system consists of a single pump that is either running or not running, depending upon some pre-established scheme of operation. Quite often, redundancy is built into pumping systems, by adding another pump. The two pumps are piped in parallel, and are either each sized for the full required pumping capacity (with only one running at any given time), or are each sized for half capacity (both run at the same time). With either method, backup is achieved, in that upon a failure of one of the two pumps, the remaining pump will begin, or will continue, to operate, while the failed pump is repaired or replaced.

More complicated pumping systems, such as three pump systems, variable speed pumping systems, and primary/secondary pumping systems, will be touched upon in the chapter devoted to this topic.

Boilers

Boilers fall under one of two categories, hot water or steam. These are packaged units, in the sense that everything required for the operation of the boiler is factory furnished and installed. Hot water boilers operate to maintain hot water supply temperature setpoint, whereas steam boilers operate to maintain steam supply pressure setpoint.

If a single boiler is to serve a hot water or steam system, there may be no other controls required for the boiler to operate, short of an outside air temperature controller to enable and disable its operation. For hot water boilers, a flow switch may be recommended by the boiler manufacturer, to be field installed in the supply piping and wired back to the boiler.

If multiple boilers are to serve a common hot water or steam system as a boiler "plant," it is often a good idea (application permitting) to control the operation of the boilers by a common controller, instead of the boilers' individual "on-board" controllers. This way, only one setpoint needs to be set for the system, rather than having to individually adjust setpoints for each boiler in the plant. Other benefits are reaped by using a common controller, yet the application at hand must allow for it. The chapter devoted to this topic covers this in greater detail.

Chillers

While chiller systems can come in an array of different types of configurations, the purpose of all chillers is the same; to chill water. A packaged air cooled chiller is one that is installed outdoors, and operates via an integral leaving water temperature controller, to maintain chilled water setpoint. Chilled water piping is run to and from the chiller, and into the building. An air cooled chiller may have one or more of its refrigeration components located indoors, and thus would be categorized as a "split system." In these types of systems, all chilled water piping is run indoors, and refrigeration piping is run between the indoor and outdoor components of the chiller. A water cooled chiller is one that is located indoors, yet requires condenser water piping from it out to some type of "heat-rejecting" equipment, such as a cooling tower. All types of chiller systems require proof of water flow for them to be allowed to operate, and this is typically accomplished by a flow switch or differential pressure switch that's field installed in the chilled water supply piping and wired back to the chiller's control system.

Like boiler plants, if multiple chillers are to serve a common chilled water system, then it is often recommended that the chillers be controlled by a common controller.

Miscellaneous Systems and Equipment

This section picks up the types of systems that were not categorized (in the previous sections) as either airside or waterside systems. These systems and equipment may be able to be categorized as either or as both, depending on their application, or perhaps as neither. Thus, we create a category for these systems and equipment to call their own...

Heat Exchangers

In general terms, a heat exchanger is a device that transfers heat from one medium to another. The gas-fired heating section of a rooftop unit consists of a heat exchanger, that heats up and transfers its heat to the air blowing over it. The heat exchangers that we will concern ourselves with are those that are not a part of a packaged piece of equipment. Common types of heat exchangers that we see in our industry are steam-to-water and water-to-water. A steam-to-water heat exchanger utilizes steam to produce hot water. A water-to-water heat exchanger may take in chilled water on its "primary" side, and produce chilled water on its "secondary" side. The purpose of this type of heat exchanger is to provide isolation between the two separate systems. Heat exchanger construction takes the form of "shell and tube," or "plate and frame." Shell and tube construction, normally suited for steam-to-water applications, consists of a container, or shell, of which the interior contains a network of piping. Steam enters the shell at the top, and leaves (as condensate) at the bottom. Water is pumped through the network of piping, and is heated up by the steam as it passes through. Plate and frame construction, suited for water-to-water applications, consists of a series of plates, stacked together and secured in place by a frame. When pressed together, the plates form two independent paths, each of which water can flow through. With two systems of water pumping through the heat exchanger, heat is exchanged between the two, via the heat conductive metal plates.

Humidifiers

In commercial HVAC, humidification is most always done by means of a fan system and a duct mounted humidifier. Humidifiers generally are in the form of "steam-utilizing" or "steam-generating." With either case, steam is introduced into the air stream of an air moving system, by means of a duct mounted dispersion tube. With steam-utilizing humidifiers, a separate steam-generating system must exist, such as a steam boiler. Steam-utilizing humidifiers consist of the dispersion tube and a steam control valve, and not much more. With steam-generating humidifiers, the steam used in the humidification process is produced by the humidifier itself. Steam-generating humidifiers are packaged equipment. Steam is generated by electric heating probes that are submersed in water. The steam is delivered from the humidifier "hotbox" to the duct mounted dispersion tube. The rate of steam production is governed by the humidifier's control system.

Unitary Heating Equipment

Equipment under this category includes unit type heaters that are powered by electricity, hot water, steam, or even natural gas (typical for unit heaters). Unit heaters are fan driven heaters that are normally hung from above, mostly found in spaces having "exposed construction." Cabinet heaters are normally fan driven as well, and sometimes attempt to "blend in" to their surroundings by hiding above a ceiling, or inside a wall soffit. Wall heaters are fan driven unitary devices that are surface mounted to a wall, or recessed in it. Radiators and baseboard heaters have no fan, and rely on the principles of radiation to provide warmth. These usually hide close to the ground, at the perimeters of their served spaces.

Computer Room A/C Systems

Computer room air conditioning units and systems are those that are specifically designed and manufactured to provide environmental control for rooms containing a lot of heat generating equipment (computer and telecommunication rooms). While heating isn't normally a concern with these applications, the packaged computer room A/C unit will be equipped with cooling capabilities, and possibly the ability to humidify and dehumidify as well. These units are designed for demanding applications. As such, they are equipped with microprocessor-based control systems, in order to provide precision environmental control. In addition, redundancy is often built into systems utilizing this type of equipment, by having more than a single "fully sized" unit serving the same space. The selection of the "primary" unit, and the start-up of the "backup" unit if needed, is also often done with an outboard microprocessor-based supervisory control system.

Water Cooled Systems

Refrigeration equipment that is water cooled, as opposed to being air cooled, will have all components that make up the refrigeration cycle integral to the equipment, with the equipment being physically located indoors. This gets all of the moving parts inside, instead of needing to have one or more of the components of the refrigeration cycle, for each piece of equipment, located outdoors. Heat rejection must be performed by a "condenser water system," which pumps condenser water through each piece of equipment and up to a cooling tower. The refrigeration equipment (chillers, water cooled A/C units, heat pumps, or what have you) rejects heat to the condenser water. The water is pumped up to the cooling tower, where it gives off its heat to the outdoors and returns to the equipment, for another "go-around." The chapter on water cooled systems will go into detail on condenser water systems themselves, and associated equipment (cooling towers, pumps, etc.), and will also touch on the typical equipment served by a condenser water system. The chapter will also discuss how condenser water systems can serve heating equipment as well as cooling equipment, wherein the heat is not rejected to, but extracted from, the condenser water system.

Chapter 3

Introduction To Controls: Methods of Control

With an overview of the types of systems that we will be looking at and a brief description of them under our belts, we will switch gears now and talk about, what else, controls! These next few chapters delve into the various controls and control methods that are in use in this day and age. This current chapter introduces the reader to the fundamental methods of control, thus providing the basis or "laying the carpet" for the following chapters on sensors and controllers, and end devices. The final chapter before "Intermission" covers a few of the more popular control schemes that are in common use today, providing a nice bridge into the subsequent chapters on mechanical systems, equipment, and control. It is important to remind the reader at this time that the later chapters build upon the principles and concepts introduced in these next few chapters. The reader should come away from these chapters with at a least a solid fundamental understanding of controls and their function, and thus should be well equipped to forge ahead into the chapters that follow. With no further ado then, let us begin our journey...

Control Methods

In HVAC systems, we typically run across three types of control methods, each of which we will take a good look at. **Two-position control, staged control**, and **proportional control** are all methods of controlling mechanical equipment in HVAC systems, each with its place of proper application, each with its benefits when applied properly, and with its potential drawbacks when misapplied.

To illustrate, in simple concepts, the difference between the three above-mentioned methods of control, consider an example of a light bulb. The light bulb is the controlled device, and you are the controller. Being the controller, you have the ability to establish a preference, a "**setpoint**" so to speak. You can prefer to have the light on, to have it off, or even perhaps to have it glow at some intermediate level. Also as the controller, you have the power to control the bulb to your preference. The closer that the status of the bulb is to your preference, the better the overall control.

With two-position control of the light bulb, only two positions, or two states are achievable. The bulb can be on or off, with no in-between. If your preference, as the controller, is only for either one of these two states, then two-position control is adequate. However, if your preference is more demanding, and you require intermediate levels of light from the bulb, then two-position control doesn't cut it.

Staged control breaks up the control of the "end device" into parts, from zero percent to one hundred percent. In the case of our light bulb, zero percent is the light being off, and one hundred percent is the light being on. If we break up the control of the light bulb into, let's say, three stages, then we can basically have the bulb assume any one of three different states, or levels of light. With no stages called for by the controller, the light is off. If the controller calls for one stage, then the light illuminates at approximately 33 percent. A call for two stages will result in the light burning at 66 percent, and a call for three stages will result in the light burning at full power. As the controller, you have three levels of light at your disposal, to try to satisfy your preference.

What if your preference is in the middle of two stages? In other words, what if your preference is, say, 50 percent? If you had more stages of control at your disposal, you could at least get closer to your preference. You might conclude from this that the more stages of control at its disposal, the better chance the controller has at achieving and maintaining its setpoint. If we could "extrapolate" the number of stages to infinity, we would no longer be limited to "discreet" stages, or levels, of light. We would be able to assume any level that we want, as driven by our preference, in the whole range of zero to one hundred percent. This type of control, of which the controller can command the end device to be in any position throughout its whole range of operation, is the fundamental concept of proportional control.

This chapter will attempt to define each method of control, yet will not go into too much detail as to the application (and misapplication) of each. This, alas, will be covered more completely in the upcoming chapters on mechanical equipment.

Two-Position Control

Two-position control of a device or component is that which the component can assume only one of two positions, or "states." An exhaust fan that is on, or off. A damper that is open, or closed. A three-way control valve that is positioned to allow full flow through a hot water coil, or positioned to completely bypass the flow around the coil. All of these are examples of two-position control.

How is the two-position device controlled? In the case of the exhaust fan, it can be as simple as an ON-OFF switch. Or perhaps by a thermostat, that turns the fan on when the temperature in the space served by the fan exceeds the setpoint of the thermostat, and turns the fan off once the temperature falls back below setpoint. For the damper, perhaps it is an outside air intake damper that's part of a ducted fan system. The damper is open whenever the fan runs, and is closed whenever the fan is off. For the control valve, it may be controlling the flow through the hot water coil of a small fan coil unit. A thermostat in the space served by the fan coil unit controls the operation of the valve, positioning it to allow full flow through the coil if the space temperature is below setpoint, and bypassing the flow around the coil if the space temperature is above setpoint.

Staged Control

Staged control can be thought of as "incremental control" of a process. The process is broken down into increments, or stages, with each stage being able to be individually controlled. For a simple example of staged control, consider the case of an electric heating coil installed in a branch duct of some upstream fan system. Without any regard to what the upstream system is doing, we can analyze the operation of the duct heater. Let's assume that the heater has four stages of control. We can therefore individually step through these stages of heat to produce 25%, 50%, 75%, and 100% of the full capacity of the heater. In other words, if we turn two of the four stages on, the heater will produce 50% of the full amount of heat that it can produce. Add a stage, and we're at 75%. Drop out two stages, and we're down to 25%. You get the picture.

Of course, if we really wanted to, we can turn all four stages on or off at the same time. This would be, in essence, two-position control. In fact, two-position control is a form of staged control, with the number of stages being one.

So why would we want to break down a process such as this (i.e., heating), into increments of control, and not just control it as "all or nothing?" Without going into any great detail here as to the reason, we can make a general statement that, for processes, the more stages of control we have, the better the overall control of the process. In HVAC applications, this almost always translates to better and more consistent comfort control.

In our simple case of the electric duct heater, we have some insight as to why this might be the case. Consider that the duct heater is able to impart a 40-degree temperature increase to the air passing through it. After the air passes through it, it is delivered to a space, let's say, a conference room. In the conference room is a wall mounted temperature controller, that controls the heater. If we have only one stage of control, then we are controlling the heater as "all or nothing." Hence, upon a drop in space temperature below the setpoint of the temperature controller, the heater turns on, at full capacity, and raises the temperature of the supply air, the air being delivered into the space by the ceiling diffusers, by 40 degrees. That's quite a difference in supply air temperature all at once! And though the temperature in the room will (quickly) rise back above setpoint, this is not a very precise way of controlling and maintaining the temperature in that room.

On the other hand, if we had four stages of control at our disposal, then upon a drop in space temperature below the setpoint of our space temperature controller, we would only engage one stage of electric heat, and thus bring up the air being delivered into the space by 10 degrees. The room temperature will rise back above setpoint, stay right where it is, or continue to drop. If the load, or "heating requirement," in the space is such that the temperature continues to drop, then more stages of electric heat are brought on. The further the temperature drops below setpoint, the more stages of heat are energized. From a temperature control standpoint, this is a much more precise method of control than the "all or nothing" method.

Proportional Control

To start off this section, we consider a classic example of proportional control, one that perhaps most of us are quite familiar with: Cruise Control! The intent of cruise control is to automatically maintain a fixed speed, without having to use the accelerator pedal. The first step is to establish setpoint by getting up to the desired speed and then pressing a button to "lock in" the setpoint. Once setpoint is established, the accelerator automatically positions itself to try and maintain the "speed setpoint." The accelerator can assume varying positions to accomplish this. If a hill is encountered, the accelerator will increase to compensate for the added "load" on the automobile. Likewise, when traveling downhill, the accelerator will "lighten up," as there is less pedal required in this situation. The speed of the automobile is being proportionally controlled; the accelerator is "modulated" in an effort to maintain a fixed, constant speed.

Proportional control, or "**modulating control**," as it is alternately referred to, has as much to do with the controller as it does with the controlled device. An end device that can assume any position within its entire range of operation can be considered to be a device that can be proportionally controlled. A proportional controller is required to operate the end device. Consider the control valve example that we discussed in the section covering two-position control. Only now, think of the valve in terms of proportional control.

Instead of a simple on-off thermostat in the space controlling the valve, we will be dealing with a proportional temperature controller. The space temperature setpoint is set at the controller. A proportional control signal is generated at the controller, which can be sent out to the control valve. For the sake of simplicity, let us generalize things and say that the control signal can be any continuous value between zero and ten, inclusive. This signal is a function of the difference in space temperature from the setpoint of the controller. If the space temperature is exactly at setpoint, the control signal is 50 percent of its entire range, or for this example, 5. As the space temperature falls below the setpoint of the space temperature controller, the control signal increases in magnitude, toward 10. Likewise, as the space temperature rises above setpoint, the control signal decreases in magnitude, toward 0.

Now suppose that the control valve actuator can accept a control signal, and position the valve in accordance with the signal. A value of

0 received by the valve would result in the valve being in a position to allow no flow at all through the coil. A value of 10 received by the valve would result in the valve being in a position to allow full flow through the coil. Any continuous value between 0 and 10 can be received by the valve; the valve assumes the position that corresponds to the signal received. Thus, the valve can assume any position, from fully closed to the coil, to fully open to the coil. We can have the valve travel through its entire range of operation, by sending it a signal that varies from 0 to 10.

To complete the picture, let's think about how the temperature controller can operate the valve. If the space temperature is at setpoint, then the signal sent out to the control valve is 5, and the valve is positioned to allow "half" the water to flow through the coil, and "half" of it to bypass the coil. If the temperature in the space falls from setpoint, then the signal increases toward 10, and the control valve repositions to allow more flow through the coil. If the space temperature rises above setpoint, then the control signal decreases toward 0, and the control valve repositions to allow less flow through the coil.

The temperature of the air passing through the coil, of course, is in direct proportion to the amount of hot water flowing through the coil. The end result is that the temperature of the air delivered into the space is proportional to the deviation in space temperature from setpoint. In simple terms, the colder it is in the space, the hotter the air is being delivered into the space. This is what is meant by proportional control.

The "**throttling range**" is the range in temperature, above and below setpoint, that it takes for the controller to output the 0 to 10 control signal. Suppose that the setpoint that we are trying to maintain in the space is 70 degrees, and our throttling range is 4 degrees. The throttling range is centered about the setpoint, so it's 68 to 72 degrees. Earlier we stated that if the space temperature is at setpoint, then the output of our controller, that which is sent to the control valve, is 5, and the valve is at mid-position. As the space temperature falls, the control signal sent to the valve increases, toward 10, and will reach 10 at 68 degrees. Likewise, as the space temperature rises, the control signal sent to the valve decreases, toward 0, reaching 0 at a space temperature of 72 degrees. This means that the valve will travel its entire range of movement, from fully open to the coil to fully closed to the coil, through the range in space temperature of 68 to 72 degrees. Now, provided that the heating capacity of the coil is properly sized for the heating needs of the space, the controller will control, or maintain, the space temperature within this range of 68 to 72 degrees.

We came across an interesting concept in the past few paragraphs. We stated that if the space temperature is at setpoint, then our control valve is at mid-position. In more general terms, we can say that our desired condition (setpoint) corresponds to "half-travel" of the controlled device. We can look at this another way. If we can conclude, from the need for the control valve to be at mid-position, that the space heating requirement is at "half load," we can also draw the conclusion that the only time the controller is actually controlling to maintain setpoint is during "half load" conditions. Simply put, the only time the space temperature is maintained at setpoint is when the space is loaded to "half capacity." During light load conditions, the temperature is maintained somewhere above setpoint, and the valve is substantially closed to the coil. And during heavier loads, the temperature is maintained below setpoint, with the valve being substantially open to the coil.

Throttling range defines our **control band**; if the heating system is sized for the load, the controller will always control to some point within the control band, though will only control to actual setpoint at a need for half the total system heating capacity. If the load is less than or greater than the heating capacity, then the controller will still control within the control band, but to a point other than the actual setpoint. The point at which the controller controls to at any given moment is called the "**control point**," and the difference between the desired condition (setpoint) and the control point is called "**offset**" or "**error**" (see Figure 3-1).

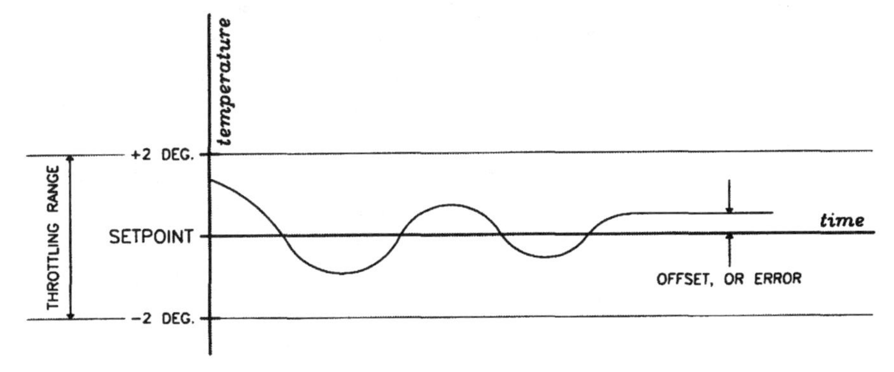

Figure 3-1. Graph illustrating a proportional control process. After the process has stabilized, the temperature is controlled to within the throttling range, yet not precisely and continuously to setpoint. The difference or "offset" in temperature from setpoint is defined as the "control point."

Wait a minute! You're telling me that, even with proportional control, I can't maintain an exact setpoint, but am only guaranteed to be able to control within a range? You gotta be kidding me? How do I improve upon that? Well, one thing you can do is simply decrease the throttling range. So in this example let's decrease it to 2 degrees. Now our throttling range is 69 to 71 degrees, and we should at least be able to control to within ±1 degree of setpoint, right? Maybe. There's a reason for throttling range, and it has to do with the dynamics of a process. Generally, the more dynamic a controlled variable is, i.e., the more quickly a change in the controlled variable registers a change at the controller, the larger the throttling range must be, for stability. For our example of space temperature control, a small throttling range is probably acceptable. Yet other applications may demand a larger throttling range. Reducing throttling range in an effort to reduce offset can cause the control process to become unstable. Instability is represented by a controlled device that is always in motion, or "hunting." If stable control with very small offsets is a requirement, and can't be achieved with proportional control methods, other options are available...

P+I and PID Control

P+I (Proportional Plus Integral) control is proportional control with another "dimension" added to it. P+I control is microprocessor-based. Not to say that we need a DDC system to perform this type of control. But we do need an electronic controller with some "smarts'.

The "I" stands for "integral." For those of you that recognize this as a mathematical term used in calculus (yikes!), my condolences. We won't explore the origin of this term and how it mathematically relates to our process at hand (whew!). We'll stick to simple terms. With P+I control, offset is measured over time, and minimized. A good P+I control loop will operate in a narrow band close to setpoint. The entire throttling range will not be used. P+I loops perform well with processes that don't experience large and rapid changes in load. They do not perform well when setpoints are dynamic (i.e., changing or being changed a lot).

PID (Proportional Plus Integral Plus Derivative) control is a modification to P+I control. Yep, another calculus term. What PID adds is a "predictive" element to the control response. Whereas the "I" in

PID asks "How far am I from setpoint?", the "D" asks "How quickly am I approaching setpoint?"

PID is a precision control strategy with origins in the process control industry, and is not normally needed in typical HVAC applications. Its proper application is labor intensive and time consuming; the control loop parameters associated with PID control require that the process be monitored and the parameters "tweaked," to ensure optimum performance of the control loop. One should not automatically use PID control for all proportional control processes in an HVAC system, even if it is at their disposal. Its application should be selective; only use it "if all else fails." In other words, when P or P+I just doesn't cut it.

Floating Control

The last control method that we will talk about is one that can almost be considered a form a proportional control, just not "true" proportional control. **Floating control**, or "**tri-state control**," as it is sometimes referred to, consists of a controller that can issue a "drive open" or a "drive closed" command to an appropriate end device. If our end device is that control valve that we keep talking about, a floating space temperature controller can command the valve to drive one way or the other, depending upon a bunch of things. Basically, the setpoint is set via the controller, and a "neutral zone" or "**null band**" is established, centered about the setpoint. If all is well and the temperature in the space is within the null band, then no commands are issued to the valve, and it "stays put" in the position that it's in. If the temperature strays out of the null band, then the controller issues the appropriate command to the valve, to drive it in one direction or another, in an attempt to bring the space temperature back within the null band. For as long as the space temperature remains out of the null band, the command is issued, and the valve drives. When the temperature comes back within the band, the command is released, and the valve "stops in its tracks." Figure 3-2 graphically illustrates the concept of floating control.

Of course, there are some problems with *this* method of control also (you're kidding!). Floating control has its place in today's HVAC applications. Its relatively cheap, as far as controllers and end devices go, as compared with proportional control. But many applica-

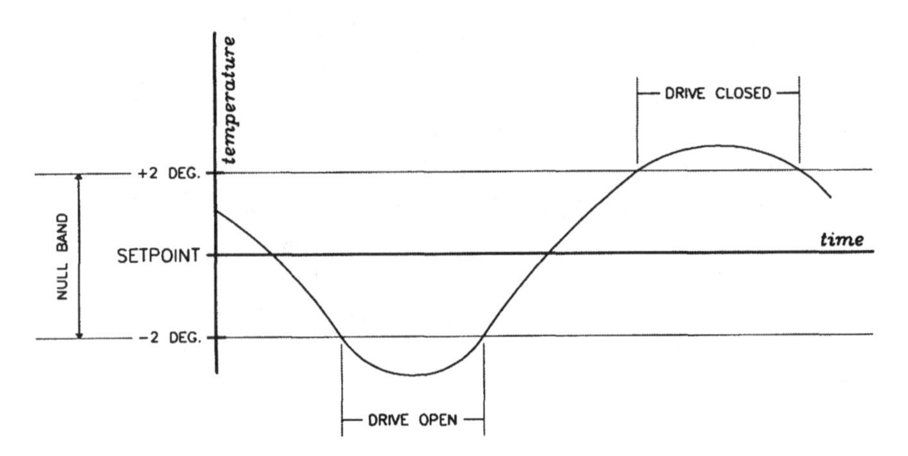

Figure 3-2. Graph illustrating a heating process implemented with floating control. When the temperature is within the null band, no action is taken upon the end device. When the temperature strays out of the null band, the controller takes the appropriate action on the end device (hot water valve), to bring the temperature back within the null band.

tions prohibit the use of floating control. For stable control, the controller must register the change in the controlled variable very quickly. In our wonderful example, we see that if the space temperature drops out of the null band, the controller will command the valve to "drive open" to allow increasing amounts of hot water to flow through the coil. With an increase in hot water flowing through the coil comes an increase in the temperature of the air delivered into the space. But by the time the space temperature controller registers the resulting increase in space temperature, the valve likely will have already driven fully open, and will have been sitting there for a while already. Now the space temperature rises into the null band, and keeps on going, basically overshooting the setpoint, and traveling right out the other end of the band. Now the controller issues the command to drive the valve closed, and the valve responds. The valve again will likely drive all the way, and be sitting there for a while, before the space temperature ever falls back within the null band. And so on...

What, if anything, can be done about this? Well let's just say that in the first place, space temperature control is not a very good application for floating control. One thing we see is that the "speed" of the end device might have something to do with control. How fast the device moves from its open position to its closed position upon receiving a command from the controller has an effect on how well we can control to within the null band. With this particular example, we could try a slower moving end device, a device better matched to the process. Really, though, floating control just doesn't function very well when thermodynamic lag (fancy term meaning slow moving temperature control process) exists, and if used in such an application, as we see in this example, control actually tends to approach "two-position."

Chapter 4

SENSORS AND CONTROLLERS

We've already been talking quite a bit about temperature controllers, without really taking the time to fully define these devices and their basic functions. Well, let's do that right now! We begin this section by talking about one of the most fundamental, and one of the most important, components of an HVAC control system: The sensor!

SENSORS

Contrary to popular belief, a sensor does not have the capability to control! Maybe it's just a matter of terminology, and the improper use of it, but quite often the term sensor is referred to in terms of "controlling the operation" of something. No, in the truest sense of the word (no pun intended), the only thing a sensor can do is "sense." As if that isn't an important enough task on its own! A sensor's main function is to measure a "controlled variable" in an accurate and continuous manner. In the HVAC industry, we are typically measuring and controlling temperature, pressure, and humidity (Figure 4-1).

Sensors take on many forms. It is not the purpose of this section to explore all of the different types of temperature, pres-

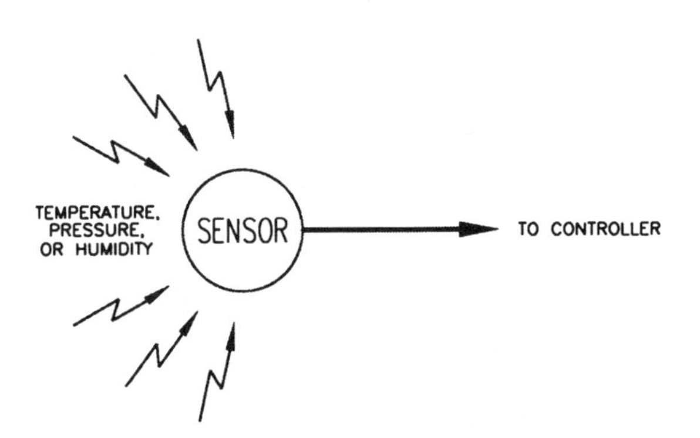

Figure 4-1. The All-important Sensor.

sure, and humidity sensors. Rather, it is meant to briefly touch upon the definition of the term, and more so to demonstrate its function and its place in a typical control system. As we stated earlier, a sensor on its own can't really do much of anything. The information measured by the sensor must be gathered and processed, so that appropriate actions can be taken on the controlled variable. Which leads us to the next section: Controllers!

Controllers: A General Description

Remember back in the last chapter, when we were talking about controlling a light bulb? If we think about this particular process of control, we can pinpoint some of the components that make up the process (Figure 4-2). The "controlled variable" was the level of light produced by the bulb. The "sensor," that which measured the level of light, was our eyes. And the controller? Us! We as the controller were able to receive data from the sensor (our eyes), process it, establish a preference, or "setpoint," and act upon the controlled variable in an effort to bring it closer to our liking, or to our setpoint.

With the help of this analogy, we can begin to define the basic functions of a "controller." The controller gathers information on the sensed variable, processes it, compares it with a setpoint, and acts upon the controlled variable in the appropriate manner, in an effort to bring the controlled variable closer to the controller's "desired" value. Figure 4-3 illustrates the functions of a controller, in block diagram form.

In HVAC, a controller can be anything from a self-contained temperature switch, to an input/output pair on a DDC controller. Often, the sensor is an integral part of the controller, as is the case with a simple thermostat or dehumidistat. With either device, the rise in the sensed variable (temperature or humidity) above setpoint causes a contact, or a switch, to close. The sensor and the switch are two parts of the same device. In other instances, the sensor is a separate device, and must be wired to the controller. This is most often the case with electronic and microprocessor-based controllers.

We see from the definition that the controller must be able to perform three distinct functions. The first is that the controller must be able to gather and process the information measured by the sensor. The second is that a preference, or setpoint, must be able to be established via

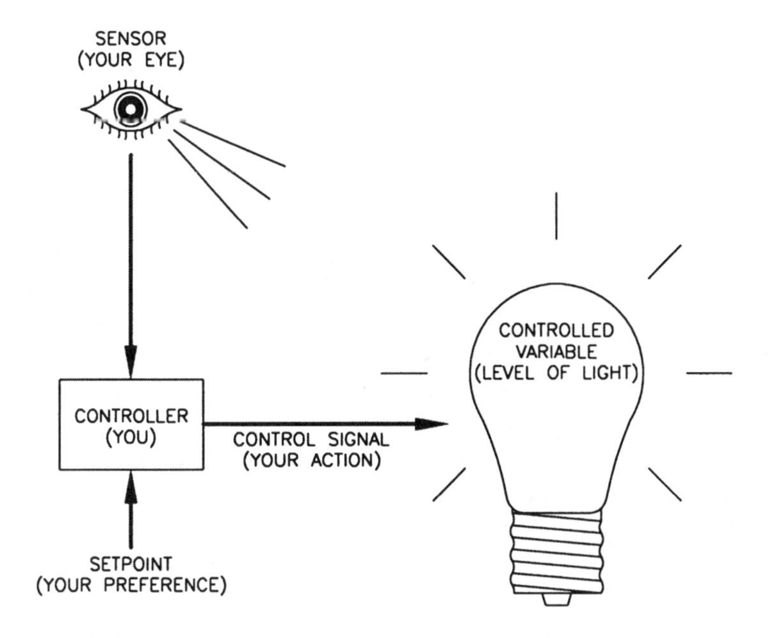

Figure 4-2. Analogy of a control process.

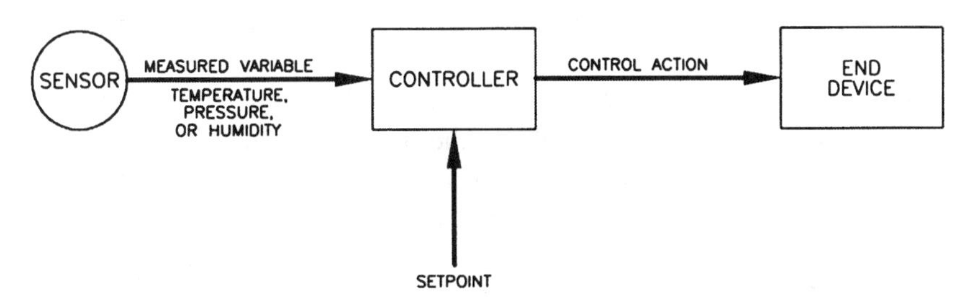

Figure 4-3. Block diagram of a controller, gathering information from a sensor, accepting a setpoint, and acting upon an end device.

the controller. And the third is that the controller must be able to act upon the controlled variable, in a manner beneficial to the controlled process. The first two functions are common to all controllers. The third function, however, varies greatly among controllers. How the controlled variable is acted upon is categorized in the same way that we had earlier categorized the various types of control. The following sections define the different types of controllers, and briefly discuss some of the characteristics of each type.

Single-stage or Two-position Controllers

When we think of what is meant by a two-position controller, we should think of a controller that can affect the position, or state, of a two-state device. In simple terms, a two-position controller is basically an automatic switch. As we had mentioned before, the sensor portion of the controller can be an integral part of the controller, or can be a separate device. For the sake of the following discussion on the various types of two-position controllers, we will make the generalization that all of the following examples of two-position controllers have sensing methods that are an integral part of the controller. Figure 4-4 shows a simple single-stage or two-position controller.

Let's start with the basic **temperature actuated switch**. This type of controller measures the temperature of some medium (air, water), compares the sensed temperature with an established setpoint, and operates its output, or its switch, as a function of these two parameters. If a temperature switch is used in a heating application, we might want the switch to close, or make, if the sensed temperature drops below setpoint, to turn on a source of heating, and then open, or break, once the temperature rose back above setpoint, to thus turn off the heating source. If we wanted to use a temperature switch in a cooling application, we would require that the action of the switch would be to make on a temperature rise, and break on a temperature fall, from setpoint.

What we are basically talking about here is something that we are all likely familiar with. **The common thermostat**! The term "thermostat" has been broadened over the years, to cover many different types of temperature controllers, from single-stage space temperature controllers, to proportional duct mounted temperature controllers. Whether it's right

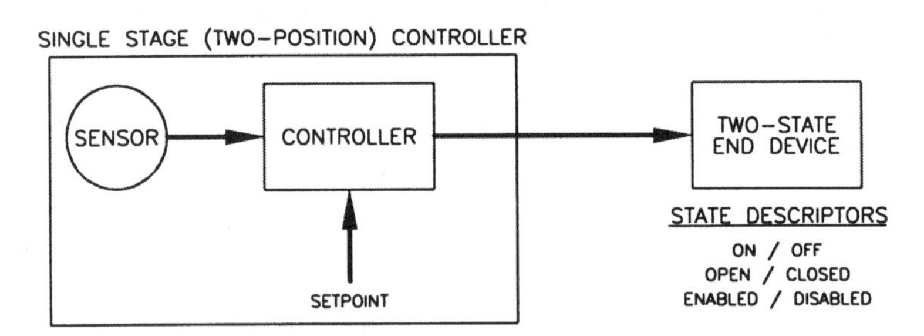

Figure 4-4. Single-stage controller with integral sensor, controlling an end device. State descriptors define possible "states" of the end device.

or wrong to generalize the meaning of this term to include proportional controllers is probably a matter of opinion. In the purest sense, the term refers to a single-stage or multistage temperature controller that is located in a space. For the remainder of this book, we will reserve the term "thermostat" for this type of controller, and refer to all other types of temperature controllers, as temperature controllers.

The temperature actuated switch of a typical thermostat is either a mercury filled glass bulb, or a "snap-acting" bimetal. The mercury filled bulb sits on the end of a coiled up strip of metal. As temperature changes, the coil expands and contracts. The bulb "rotates" as a result of this action. The mercury in the bulb, which is conductive, either "makes or breaks" an electrical connection within the bulb. The snap-acting type temperature switch consists of two dissimilar metal strips fused together. The strips react to temperature differently, expanding and contracting at different rates of change. As temperature changes, the bimetal deflects, as a result of the dissimilar properties of the two metals. This deflection translates into movement of a mechanical switch.

A single-stage thermostat can be selected to perform a heating function, or a cooling function, but not both. For a thermostat to be able to automatically perform both heating and cooling, it would technically have to have two independent stages, a heating stage and a cooling stage. A thermostat with a single stage can be used for both heating and cooling, however, provided that is has a "manual changeover switch." This switch allows the thermostat to perform either a heating function or a cooling function from the same temperature actuated switch, yet just not at the same time. In other words, a single-stage thermostat with a manual changeover switch can be set to either call for operation upon a rise in temperature, or upon a fall in temperature, but does not "simultaneously" have the ability to call for either mode of operation. The changeover switch has the labeling "HEAT-OFF-COOL." A thermostat with automatic changeover capabilities is one that can call for either mode of operation, without having to prohibit calls for the opposite mode of operation. This type of thermostat will have a system switch labeled "HEAT-AUTO-COOL-OFF," and will need to have at least one stage devoted to heating, and one devoted to cooling. Multistage thermostats will be covered later on, in great detail!

Another type of single-stage temperature controller, that technically has the "means of temperature sensing" as an integral component, is the **remote bulb temperature controller**. The sensor consists of a gas filled "bulb" at the end of a long capillary. The bulb is inserted into a duct, immersed in a pipe, hung outdoors, or whatever. The controller is mounted nearby, the distance being limited by the length of the capillary. Capillary lengths of up to 20 feet or so are commonly available. The gas in the bulb expands and contracts as a function of the temperature surrounding the bulb. This action translates into a movement of a mechanical switch back at the controller.

Other types of temperature controllers, such as those that do have remote sensing capabilities, are also widely used in the HVAC industry. With the evolution of electronic and DDC controls, the use of solid state sensors and electronic controllers has grown tremendously over the generations. Whereas a generation or two ago we would probably have no choice but to go with a remote bulb temperature controller for air temperature control in a duct, nowadays we may be more inclined to opt for electronic temperature control, utilizing a variable resistance type temperature sensor (thermistor), and an electronic, or even microprocessor-based temperature controller. The purpose again, of this writing, is not necessarily to explore the current technologies of control that we have at our disposal. Rather, it is meant to establish the fundamental concepts of control, by defining the traditional methods of control and types of controllers, and then applying these concepts to the control of mechanical systems. As technology changes and evolves, the technical aspects of control may change as well. However, the basic concepts of control will tend to remain, as these concepts are not necessarily time and technology sensitive. In other words, what we do in the future will be the same as what we do now. Technology will simplify our efforts, and improve our end results. Whew! That was an earful! With that said, let's turn our attention of other types of two-position controllers.

Humidity controllers that are two-position in nature are commonly referred to as **humidistats** and **dehumidistats**. A humidity controller employed as a humidistat is a switch that makes upon a decrease in relative humidity from setpoint, normally to turn on a humidifier. Employed as a dehumidistat, the controller's switch makes upon an increase in relative humidity from setpoint, to start some type of dehumidification process.

Enthalpy controllers look at not one but two conditions; temperature *and* relative humidity. The controller's function is to look at these two variables, and determine whether the combination of these two, or the enthalpy, is above or below the controller's setpoint. The setpoint of an enthalpy controller is generally not in any particular units (i.e., degrees and/or percent R.H.). Rather it is a setting that is derived from the psychrometric chart (if you don't know, please don't ask!). The setpoint can be adjusted so that the controller's output makes on relatively low enthalpies (conservative setting), or on high enthalpies (aggressive setting). Enthalpy controllers are most often used to determine whether the outside air is suitable for air conditioning purposes. To determine this, we must not only look at the temperature of the outside air, but also its humidity. An enthalpy controller set conservatively would only allow outside air to be used for "free cooling" if the temperature and humidity are relatively low. One that is set aggressively would permit the use of outside air starting at a higher combination of the two.

Flow switches are employed to determine the status of flow in air and water moving systems. An airflow switch can be as simple as a duct mounted "sail switch," with the sail basically *blowin' in the wind*. Seriously though, if there is no airflow, then the sail switch is open, and when airflow is established, the air pushes against the sail and causes the switch to close. Water flow switches basically follow the same principle. The switch consists of a paddle, which is installed in the water piping. When water flow is established, the water pushes against the paddle, and the switch closes.

Air pressure switches are used in many different ways. Common to all types of air pressure switches is the presence of "high" and "low" pressure ports. A difference measured between the ports can cause the switch to change state. Air pressure switches are typically used as another, more robust, means of determining flow in an air moving system. The high pressure port is piped into the duct, and the low pressure port is typically left alone. The pressure switch measures the difference in pressure between the interior of the duct, and the "ambient." No difference in pressure signifies no air movement. When the system fan is engaged, the pressure in the duct increases, and the air pressure switch makes. Another common use of an air pressure switch is to monitor the filter section of an air system. The two pressure ports are piped into the duct, before and after the filter section. A substantial difference in pressure as measured by the controller, when the fan is running, would cause the switch to close, indicating that the filters are dirty and in need of maintenance.

Liquid pressure switches are functionally equivalent to air pressure switches, yet are constructed quite differently. Common uses include another means of determining flow in a water moving system. In this application, the pressure ports are piped into the water piping, on either side of the main pump. With the pump off, there is no difference in pressure between the suction and discharge sides of the pump. When the pump is engaged, the difference in pressure across the pump is registered by the controller, and its switch makes, thus indicating that there is flow in the system.

Steam pressure switches are controllers that make or break upon an increase in steam pressure above the setting of the controller. The common use of such a controller is the "operating control" of a simple steam boiler. The controller monitors the pressure of the steam being generated by the boiler, firing the boiler when the pressure drops below setpoint, and turning it off when the pressure rises above it.

Float switches and solid state **water detection switches** are controllers that are used to determine if a level of liquid in a container (tank, vessel, drain pan, etc.) has exceeded or dropped below a certain point. A float switch might be used to turn on a pump to feed water into a tank whose water level has dropped below some critical point. A water detector is typically used to monitor a drain pan for an excessive water level. If the water level rises above the "setpoint" of the controller, the controller's switch opens to interrupt power to the equipment that's causing the problem.

Aquastats are used to monitor the temperature of water flowing though a pipe. Although they can be "immersed" in the pipe, the more common (simpler) means of mounting is to strap the sensing bulb to the outside of the pipe. Even though this method of mounting would appear to be somewhat less precise than actually getting the sensing element into the flow of the water, if installed correctly (a big if!), there is little difference in the precision offered between the two methods. Furthermore, aquastats are typically used in applications where precision is not required. For instance, to tell the difference of whether there's hot or cold water flowing through the pipe at any given time.

Current sensing switches provide a means of monitoring the electrical current being delivered to a piece of equipment. Commonly used with fans and pumps, they provide an alternate means of proving flow in an air or water moving system. The controller is installed so as to monitor current on one of the power leads to the fan or pump motor. With no power delivered to the motor, the current switch detects no current, indicating that the fan or pump is off. When power is applied to the motor, the current sensed by the controller rises above its setpoint, and the controller's switch closes, indicating that the fan or pump is running. Although this is an indirect way of proving flow—we are not directly measuring air or water flow—it has become an acceptable, and a relatively simple, method of verifying proper air or water system operation.

Light sensing switches (photocells) are devices that sense the level of ambient light and make or break when the light level drops below some fixed value. Typically used in outdoor lighting applications (when it gets dark out, turn on the parking lot lights!), they do have their uses in HVAC applications. A photocell, for instance, can be located in a general office space, that can override the heating and air conditioning equipment serving that space into an occupied mode. Thus, if someone comes in after hours to do some work, they flip on the lights and the A/C automatically comes on.

Occupancy sensing switches detect physical movement. Applications in HVAC would be similar to those of light sensing switches. Occupancy sensors can be mounted in several locations in a general office space. If any of the sensors detects movement during an unoccupied mode, the heating and air conditioning equipment comes to life.

With all of the controllers that we have talked about here, there is one thing common to all (well, most) that we have neglected to discuss, and that is the concept of "**differential**." In order to explain what is meant by this term, it is likely most easily understood in terms of temperature control (though is of equal importance in pressure and humidity control applications). Consider that we have an electric unit heater hanging from the ceiling of a small mechanical equipment room. The electric heater consists of a fan and a section of electric heat. When engaged, the heater heats up and the fan runs to circulate warm air in the space. The unit heater is controlled by a single-stage space heating thermostat, set at 65 degrees. If the space temperature is above the setpoint of the thermostat, then the heater is off. If the space temperature falls below setpoint, the thermostat makes, the heater turns on, and the space temperature begins to rise. How much the temperature must rise, above setpoint, in order for the thermostat's switch to open and cease operation of the heater, is referred to as differential.

Why is differential needed? Once people understand what it is, this is the question that most likely follows. To understand the need for differential, let's assume that the differential in our above example is zero.

When the space temperature drops below setpoint, the heater turns on. Almost immediately the space temperature begins to increase. With a differential of zero, as soon as the space temperature increases above 65, the heater turns off, and the space temperature begins its descent. The temperature falls back below 65, and the heater re-energizes. And so on... What we are seeing here is rapid on-off cycling of the heater to precisely maintain setpoint.

Now assume that we incorporate a differential of 2 degrees into the above temperature control process (refer to Figure 4-5). When the space temperature drops below setpoint, the heater turns on. The space temperature must now increase above setpoint and through the differential, before the thermostat commands the heater to turn off. So the heater comes on at 65, but doesn't turn off until 67. And won't come on again until 65. And so on...

By incorporating differential, we have reduced the cycling of the controlled equipment over time. With no differential in our example, the heater may turn on and off 50 times in one hour. But by establishing a differential of 2 degrees, we may have actually reduced that to 10 times. The larger the differential, the less cycling, the less equipment wear, and generally the more stable control of the process.

Of course, what do we give up by incorporating differential? Tight temperature setpoint control! With no differential, the space temperature in our equipment room doesn't have a chance to stray too much from the

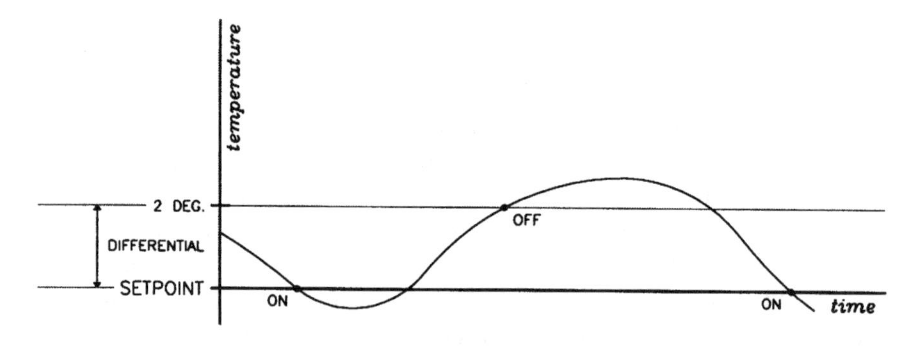

Figure 4-5. Graph illustrating a heating process implemented with two-position control. When the temperature drops to setpoint, heating is engaged. After a short time, the temperature begins an upswing. Heating remains engaged until the temperature rises through the differential.

65-degree setpoint. There may be a little bit of overshoot, but probably not much. However, with a differential of two degrees, we are allowing the space temperature to be in a range of 65 degrees up to 67 degrees. For this application, this won't normally be a concern. Yet for an application demanding tighter setpoint control, this may be an issue. There is basically a trade-off or compromise between tight setpoint control and acceptable equipment operation. This is an inherent characteristic of two-position control.

One last thing about differential (enough, already!). In our example, the differential is positioned above the setpoint. Depending upon the type of controller, the differential may be positioned below the setpoint, or may even be centered about the setpoint. If, in our example, the differential is below the setpoint, then the heater turns on at "setpoint minus differential" (63 degrees), and turns off at setpoint (65 degrees). If the differential is centered about the setpoint, then the heater turns on at "setpoint minus half the differential" (64 degrees), and turns off at "setpoint plus half the differential" (66 degrees).

Multistage Controllers and Thermostats

Staged controllers, or **sequencers**, measure the sensed variable, and offer as their output not one stage, but several stages of control. For a perfect example of this, we can go back to our original discussion on staged control, and consider the duct heater example. In the example, remember that the duct heater had four possible modes, or increments, of operation: 25%, 50%, 75%, and 100% of its full heating capacity. To control the heater, we can pretty much gather that we should use a four-stage temperature controller. For this example, let's imagine that we're using an electronic temperature controller, with a remote temperature sensor. The sensor gets to be in the conditioned space (lucky dog!), and continuously provides space temperature information back to the controller, which is installed back at the duct heater. The sensor in this example has the added feature of a small temperature setpoint adjustment, right at the sensor. This adjustment simply allows the occupant to make a change in temperature setpoint at the sensor, rather than having to do it back at the controller. This type of control scheme is depicted in Figure 4-6.

Control is pretty straightforward. We establish a setpoint, say, 72 degrees (room temperature!). If the temperature in the space drops below

72, the first stage of heating is engaged. If the space temperature continues to drop, more stages of heating are engaged. In other words, the further the drop in space temperature from setpoint, the more stages of heat are engaged. As temperature rises back toward setpoint, the stages of heating are disengaged, in the reverse order that they were engaged. Refer to Figure 4-7 for a graphic illustration of a multistage control process.

As we did with single-stage controllers, we need to talk about differential here as well. We also need to talk about a term called "**offset**." It is important to distinguish and differentiate the use of this term from the context that it's used in when discussing proportional control. In other words, the term "offset" (as it pertains to proportional control) has a different meaning when pertaining to staged control.

Since we beat to death the topic of differential earlier in this chapter, we don't need to go all the way down that path again (thank goodness!). Suffice it to say that each stage of control will have a differential value associated with it, so that there is a small difference in temperature between when the stage is activated, and when it is de-activated.

Offset, on the other hand, is something that we (unfortunately) have to beat around a little bit. Offset refers to the difference in temperature, from setpoint, that the second stage (and subsequent stages) will be activated. For instance, assume that the first of heating is engaged right at setpoint. So if the space temperature falls to 72 degrees, the first stage kicks on. The temperature must drop below setpoint, however, for stage two to come on. The controller knows not whether the space needs additional heating unless, or until, there is a marked deviation in space

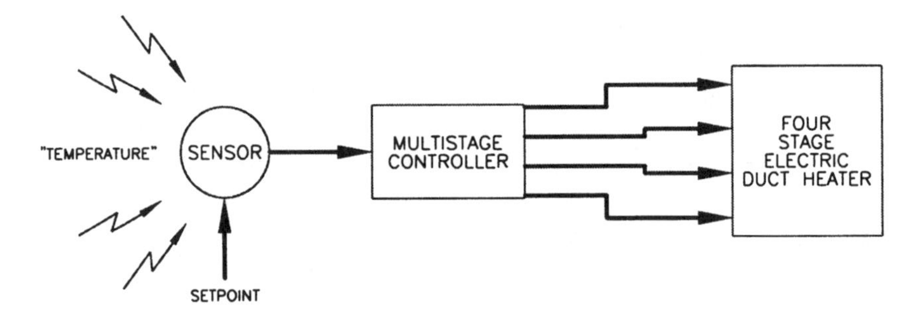

Figure 4-6. Multistage temperature controller with remote (space) sensor and setpoint adjustment, controlling a four-stage electric duct heater.

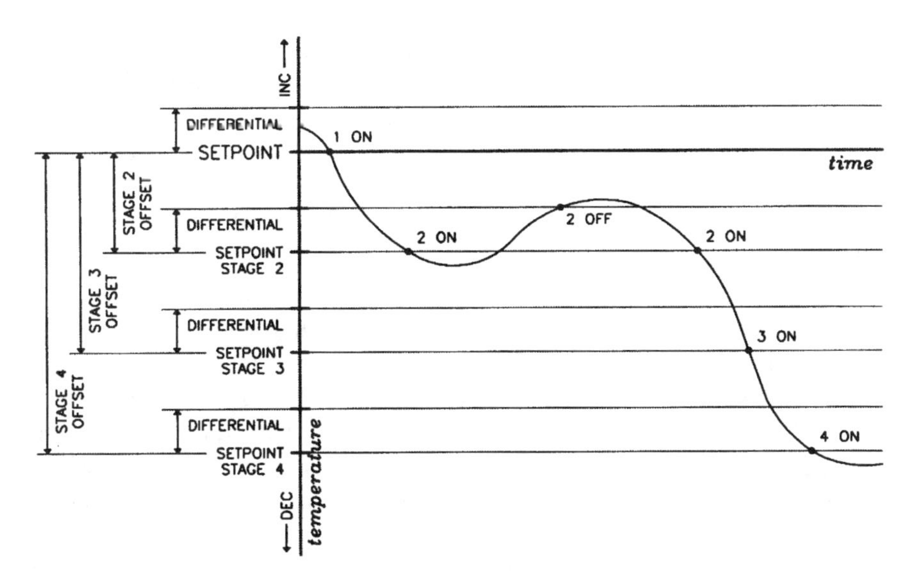

Figure 4-7. Graph illustrating a multistage (heating) temperature control process. At time t = 0, no stages are on and the temperature is falling. Stage 1 setpoint is reached, yet the temperature continues to fall. Stage 2 setpoint is reached, and the temperature begins an upswing, through the differential of stage 2. The temperature begins another downswing, and due to an increased load, falls through stage 2, 3, and 4 setpoints.

temperature from setpoint. Only then does it know to engage second stage heating. The same holds true for subsequent stages of control. Each subsequent stage of control must have an increasing amount of offset from the initial setpoint. This is an inherent characteristic of staged control. In our example, we will give stage two an offset of two degrees, stage three an offset of four degrees, and stage four an offset of six degrees.

Thus, the space temperature must drop to 70 for second stage heating, to 68 for third stage heating, and to 66 for fourth stage heating. In essence, these values are the actual setpoints for the subsequent stages of heat. Another point that we can conclude from this analysis, is that the only time we are really controlling the space temperature to maintain the setpoint that we have set via the controller, is when we have little or no space heating load. Ideally the heater is sized to handle the maximum possible load in the space. On the worst case day, all four stages of the heater may be needed. Yet we know that if we are calling for that fourth stage of heat, our operating setpoint is actually 66 degrees! That's cold for a "conditioned" space.

What can be done about the above situation? Well, it's not the heater's fault. If sized appropriately, the heater has the capability of heating the space to 72 degrees. It's the controller's fault. In this instance, the setpoint can be raised to 78 degrees, so that the fourth stage setpoint is 72. But then if nobody went and changed the setpoint back down, we would tend to have a problem of overheating on subsequent, lighter load days. What really needs to be done, at least in this instance, is that the offsets need to be reduced (and the differentials as well). Instead of spreading out the stages of control by 2 degrees each, the offsets can be reduced so as to spread the stages out by only a degree, or even a half a degree. Of course, by doing this, we are increasing the cycling rate of the stages, and while this may be acceptable for the application at hand (electric heat), in other applications this may be a real issue, in terms of equipment wear and stable control. Another issue has to do with the setup of the controller. Proper adjustment of offset and differential values is critical for acceptable multistage control. The more stages of control, the more critical these adjustments are, and the more room there is for error.

Multistage controllers and sequencers can take on the form of humidity and pressure controllers as well. We will take a look at applications requiring multiple stage operation when we start looking at the different types of mechanical systems.

Multistage thermostats are those of which have at least two stages of control, whether it be two heating, two cooling, one heating and one cooling, two heating and two cooling, or any other possible combination you can think of. Though the number of stages of either mode is generally limited to two, three-stage heating and cooling thermostats are out there. They're just not very popular. The types of thermostats that we will base our discussion on here are those that are generally applied to rooftop units and packaged equipment (which isn't to say that these types of thermostats can't be applied elsewhere).

Electromechanical thermostats typically use mercury type switches for temperature control. A simple single-stage heating/single-stage cooling thermostat will have two of these switches, one for either stage. When we have a thermostat that can call for both heating or cooling, we must either only allow for one mode of operation at a time and prohibit calls for the opposite mode (remember our discussion on "manual changeover), or we must incorporate what we call a "**deadband.**" A deadband is defined as a "neutral zone" between heating and cooling setpoints. In this neutral zone, no heating or cooling request is made of the equipment by the thermostat. If we establish a cooling setpoint at the thermostat of 72 degrees, and we allow for a 2-degree deadband, then our heating setpoint can be no higher than 70 degrees. This is primarily done so that the equipment doesn't bang back and forth between heating and cooling for every little temperature swing. For this reason, most thermostats will not allow a deadband of any smaller than two degrees.

One of the consequences of this "perpetual difference in heating and cooling setpoints," is the requirement for seasonal readjustment of the thermostat. Whaaat? You heard me. Depending upon your personal comfort range, you may be forced to "ride the levers" in order to keep yourself comfortable throughout changes in outdoor conditions. In the summertime, for instance, the thermostat is controlling the piece of equipment mostly between the cooling and ventilation modes. It follows that in the wintertime the thermostat is controlling the equipment mostly between the heating and ventilation modes. So in the summertime, we are actually controlling to maintain the cooling setpoint, and in the wintertime, we are controlling to maintain heating setpoint. If our deadband is 2 degrees, and we set the thermostat to maintain a "cooling setpoint" of 72 degrees in the summer, then when the heating season comes around, the thermostat will most likely begin working off its "heating setpoint," which is 2 degrees lower than the cooling setpoint, or 70 degrees. If your "comfort zone" is less than 2 degrees (you'd be surprised at what a difference 2 degrees makes with a lot of people), and the cooling setpoint is set to be in that zone, then the heating setpoint is "outta yer zone." The result of this? You go to the thermostat and turn up the heating setpoint to 72 degrees, which forces the cooling setpoint up to 74 degrees. Now come summertime, you have the same predicament, only opposite. The thermostat now controls to a cooling setpoint, of 74 degrees, and its too hot! So you go back and "ride those setpoint levers!" All of this isn't really a big deal, and this is probably an overdramatization. But it's a good way to nail down the concept of deadband.

A common component of thermostats, in addition to the setpoint adjusters, is a set of switches. "**Fan**" and "**System**" switches offer the user some control over how the equipment is to be operated. The Fan switch will normally have two positions, and will be labeled "ON-AUTO." The switch allows the occupant to select between continuous and intermittent fan operation. Intermittent operation means that the fan turns on upon calls for heating and cooling, and is off otherwise. The system switch can take on many forms; we will restrict our discussion to a system switch labeled as "HEAT-AUTO-COOL-OFF." With the system switch in the "OFF" position, thermostatic control of the equipment is entirely disabled. With the switch in the "HEAT" position, calls for heat only are allowed, with calls for cooling prohibited. The opposite is true when the switch is in the "COOL" position. With the switch in the "AUTO" position, calls for either mode of operation are allowed.

Traditionally, these switches are not physically part of the thermostat itself, but are contained on what is called a "**switching subbase**." The thermostat and the subbase are two separate components that are made to fit together and function as one device. The primary reason for the separation of the two components has to with the different configurations offered with these switches, and the numerous combinations possible. A consumer can select a standard one-heat/one-cool thermostat, and depending upon the application and/or the preference, can select from a variety of different switching configurations. Figure 4-8 shows what a typical electromechanical thermostat/subbase might look like.

Electronic thermostats are thermostats that sense space temperature electronically, via an integral thermistor (electronic thermostats usually have remote sensing capabilities as well). Its output stages are not mercury bulbs or electromechanical switches. They are electronic

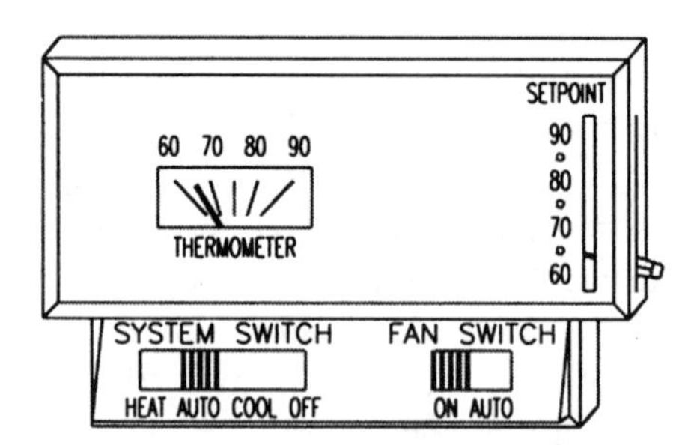

Figure 4-8. Generic depiction of a conventional electromechanical thermostat, with fan and system switches, built-in thermometer, and a single lever for setpoint adjustment.

switches or relays that are mounted on the electronic printed circuit board internal to the thermostat. Setpoint is established via a front mounted user interface, normally taking the form of electronic push buttons. Temperature and setpoint can be viewed at the front mounted digital display of the thermostat.

Programmable thermostats are electronic thermostats that have an on-board time clock function. Just because a thermostat is electronic doesn't necessarily mean that it's programmable. Programmable thermostats allow for programming of occupied and unoccupied modes and setpoints. In occupied modes, the heating and cooling setpoints are set for occupied comfort levels. The fan runs continuously or intermittently, as determined by the position of the fan switch. In unoccupied modes, the heating and cooling setpoints are "spread out," with heating usually set at around 60 degrees, and cooling set at 80 degrees or so. With nobody occupying the space, comfort control isn't generally required. The equipment just needs to operate to keep the space within a reasonable temperature range. During this mode, the fan operates intermittently, regardless of the position of the fan switch.

Programmable thermostats are often equipped with a "time-of-day" output, which is in the form of a relay contact. This output can be used in a number of different ways, of which will be discussed in the upcoming chapters.

Most programmable thermostats have the capability of remote sensing. Instead of locating the thermostat in the space served, the programmable thermostat can be mounted in an alternate location, and a remote temperature sensor (thermistor) can occupy the space and be wired back to the thermostat. The thermostat's integral sensor is disabled, and the remote sensor transmits temperature information from the space back to the thermostat. This is an attractive feature, especially if the customer is concerned about occupants "playing around" or tampering with the thermostat settings. Get the actual thermostat out of the space and that concern goes away!

Another use for remote sensing is temperature "averaging" of several areas within the same zone. A programmable thermostat controlling a rooftop unit that serves a large open office area is an application suited for averaging. The office area is a single zone, served by a single piece of HVAC equipment, yet is large enough that there may be differences in temperature throughout the space. With such an application, where do you locate the thermostat? Remote sensors can be mounted in "strategic locations" throughout the space, all wired back to the same thermostat. The thermostat "calculates" the average temperature of all the sensors, and uses the resulting value to perform temperature control.

The use of multiple remote sensors with programmable thermostats is *typically* limited to discreet series-parallel combinations of "N squared." Huh? Okay, first off, the terms "series" and "parallel" refer to the way the sensors must be wired together and to the thermostat. Refer to Figure 4-9 for sensor wiring configurations, as they are discussed here. "N squared" refers to the number of sensors, where N is a number from 1 to infinity (not really!). The result of the formula is the number of sensors permitted in an averaging network. For N equaling 1, 2, 3, and 4, the result of "N squared" is 1, 4, 9, and 16, respectively. This means that series-parallel combinations of these discreet quantities of sensors are permitted, and others prohibited. In plain English, we can average 4 zones, but not 2 or 3. We can average 9, but not 8 or 7. Or 6 or 5. And so it goes. This has to do with the sensors and how they must be wired together, its explanation being beyond the intended scope of this section. Suffice it to say that if your application at hand requires you to average, say, 3 zones, you may have to get creative. You must have 4 sensors wired in, yet with only 3 zones required to be averaged, you can simply place 2 of the sensors, side-by-side, in the same zone. Gives more "weight" to this zone, in terms of voting power or "say-so," but in an imperfect world, who's to argue?

Averaging using remote sensors has its limitations, and should not be considered "the ultimate solution" to all single zone temperature control problems. Still, it does have its uses, and if applied properly, does offer some advantage over "single point sensing."

Proportional and Floating Controllers

Proportional controllers take on the same form as many of the two-position and staged controllers that we've talked about thus far, the difference being that their output is in the form of a proportional or modulating control signal. Proportional temperature, pressure, and humidity controllers are available for many of the common HVAC applications demanding more "precise" control than what is achieved with other means.

Proportional temperature control, traditionally done with an elec-

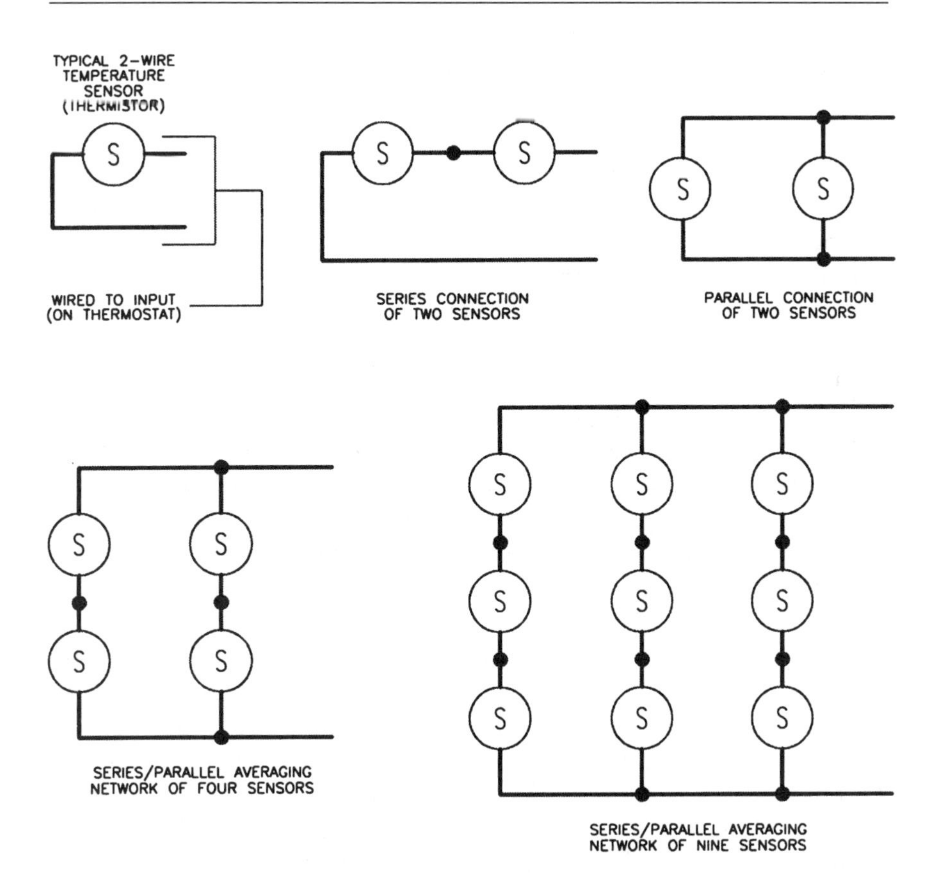

Figure 4-9. Sensor wiring configurations.

tromechanical "variable resistance" type of temperature controller, is nowadays more often done electronically, with the output signal taking the form of a DC voltage, instead of a varying resistance. Pressure and humidity applications calling for proportional control are virtually always done electronically. Whereas with electronic proportional temperature control, the sensed variable can be measured with a simple "passive" electronic device (thermistor), for proportional pressure and humidity applications, the sensed variable must be measured with "active" electronics. A pressure or humidity "transmitter" measures the controlled variable, converts it to a signal, amplifies it, and transmits it to the controller.

We have already discussed a proportional temperature control application. To even things up, let's look at a brief example of both a pressure and a humidity application, "proportionally speaking."

A typical example of proportional pressure control would be space static pressure control (Figure 4-10). The space pressure transmitter/controller monitors static pressure in the space. A setpoint is established at the controller, and the controller outputs a proportional control signal, as a function of the difference in space static pressure from setpoint. The controller can send this signal to a variable frequency drive (VFD), which has the capability of varying the speed of a fan delivering fresh air into the space. The controller can proportionally control the fan speed in an effort to achieve and maintain desired space static pressure setpoint.

For proportional humidity control, the classic example is that of a duct mounted, steam-utilizing humidifier. A space (or duct mounted) humidity transmitter/controller monitors humidity, and a setpoint is established. The controller outputs its proportional signal, and sends it to the humidifier. The humidifier control valve modulates in accordance with the control signal received, and humidity setpoint is (hopefully!) maintained.

Floating controllers are typically found as either temperature or pressure controllers. Remember we said that, for decent floating control, the controller must register a change in the controlled variable very quickly. Floating temperature control works pretty good in an application where we can quickly sense the changes in temperature caused by our control process. Like if we are trying to maintain a stable discharge air temperature in a duct, downstream of a chilled water coil, by controlling the amount of flow through the coil. We can use a floating type

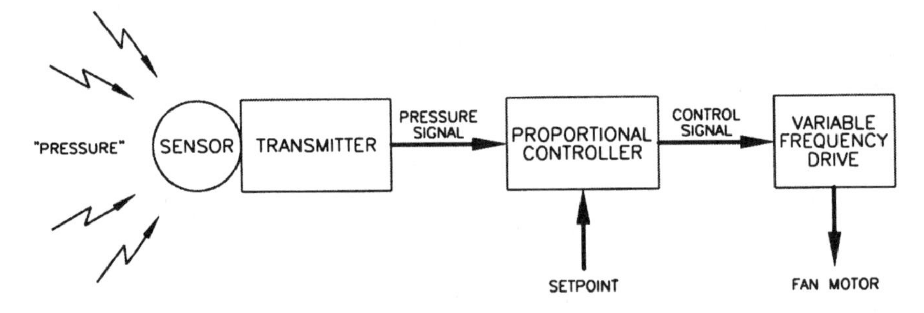

Figure 4-10. Proportional pressure controller with remote (space) sensor/transmitter regulating the power output of a VFD, thereby controlling the speed of a fresh air fan in order to maintain space pressure setpoint.

temperature controller, and a floating type control valve. The temperature of the discharge air changes quickly, with changes in the amount of flow through the chilled water coil. We can conclude that this could be an acceptable application for floating control, because the controller, located in the discharge air duct, can quickly register a change in the process.

Floating control is undoubtedly an acceptable method of control for many types of pressure applications, and that is where you find that "floating control is alive and well" in the 21st century! Pressure control is not subject to thermodynamic lag (as temperature control is). Normally, a change in the controlled variable in a pressure control process registers very quickly at the pressure controller. Think about the example of bringing in variable amounts of fresh air as a means of maintaining a constant space static pressure. Unlike in a temperature control application, the location of the pressure controller is less critical, even arbitrary in some instances, to the control process. A change, say an increase, in the amount of fresh air being introduced into the space should almost instantly register at the controller, regardless of where the controller is located (as long as it's in the space trying to be maintained, of course).

Microprocessor-based floating control is another reason for this type of control's continuing popularity. With a "smart" controller operating the end device, additional "intuitive" control can be built into the process. For instance, with a smart controller, we may be able to limit the drive rate of the end device by the controller. Instead of blindly driving the device continuously in one direction if the sensed condition strays out of the null band, a smart controller can drive it for a few seconds, wait a bit, check the condition, and see if the condition is converging or diverging. The controller can then take the next appropriate step toward the "ultimate goal" of bringing the condition back into the null band and keeping it there!

DDC Controllers

In order to better understand and appreciate the trend toward electronic and microprocessor-based control systems, we need to at least briefly touch on **Direct Digital Control (DDC)**. A "DDC" controller is essentially an input/output device, that can be programmed or configured to perform an array of control functions. Whether operating as a "stand-alone" controller, or networked together with other controllers as part of a full-blown Building Automation System (BAS), the same general principles apply. For this topic, we will limit our discussion to a single DDC controller, and define some of the typical components and functions found within such a controller. Figure 4-11 illustrates, in general block diagram form, some of the features and functions of a digital (DDC) controller.

As we had said, a DDC controller is basically an input/output device. As inputs, we can feed the controller with the switch type "outputs" of two-position controllers, and/or the proportional type signals generated by sensors and transmitters. A two-state input on a DDC con-

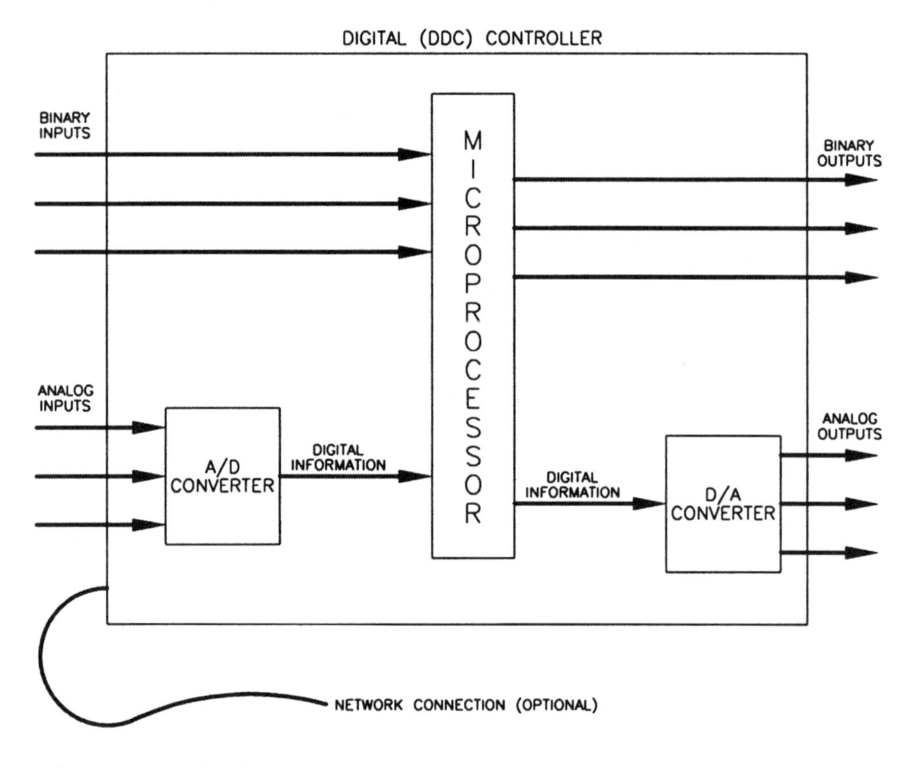

Figure 4-11. Block diagram of a digital controller, showing inputs, outputs, and internal components. Network connection is required if the controller is to be networked with other controllers as part of a Building Automation System. Access to the microprocessor is required to program the controller, and also to change various operating parameters.

troller is referred to as a "binary" or "digital" input, and a proportional input is referred to as an "analog" input. So the switching output of a two-position temperature controller, for instance, can serve as a binary input to a DDC controller, and the proportional signal of a humidity transmitter can serve as an analog input to the controller.

Now, the outputs of the DDC controller are categorized in the same manner as the inputs; we have binary (digital) outputs and analog outputs. Binary outputs are two-state outputs, in the form of an electronic switch or relay contact, that can affect the state of a two-state device. Likewise, analog outputs are proportional control signals that can control the operation of a proportional end device.

The term "digital" in DDC refers not to the digital type inputs and outputs found on the controller. There is a potential misinterpretation here of the term "digital" as it relates to Direct Digital Control. A digital (binary) input or output is simply a two-state "point." The term digital stems from what is meant in digital electronics by the ability of a piece of electronic information, or "bit," to assume one of two states, or "digits."

With DDC or microprocessor-based controllers and control systems, a microprocessor exists, virtually between the inputs and the outputs of the system. In order for the "brain" to process information from the inputs, it must be converted to "digital information" (0s and 1s), and in order for the brain to convey its decisions to the real world, it must be converted back to "analog information." This is essentially true for the proportional type, or analog, inputs and outputs. Analog information is converted to digital information, by means of an analog to digital (A/D) converter. The brain processes the digital information, and makes decisions that ultimately result in actions taken upon real world devices. However, the decisions must be turned into analog values before presented to the controller's outputs. This is done with a digital to analog (D/A) converter.

The true beauty of DDC is that most of the control strategies are implemented via software, and the flexibility offered with the software is virtually boundless, limited only by the programmer's imagination, his knowledge of control systems, and his abilities as a programmer. As an extremely simple example of DDC control, look at a simple input/output pair as it applies to proportional temperature control of a space. A temperature sensor resides in the space. The sensor is wired to an analog input of the controller. Space temperature setpoint is established through the software. An analog output is programmed to control a proportional end device, in preferential accordance with the space temperature input and the programmed setpoint.

Not much different than a run-of-the-mill temperature controller, directly controlling the operation of the end device, huh? Well, maybe not, if you just look at this one individual scheme of control. But when you start adding complexity to a control system, and this is just one small facet of the entire sequence of operation, then you can begin to see how DDC can really shine. With a large, non microprocessor-based control system, a substantial amount of engineering has to be done up front. The sequence of operation would have to be developed as the first step. The system would then be designed to conform to the sequence. Once built, it would be very difficult to implement changes to the sequence without hardware modifications. With a DDC system of equal size, the only thing that really needs to be done up front is to take into account all of the potentially required input and output points. The sequence can be formulated over time, revised, added to, and customized, without ever having to change the hardware.

One final note regarding DDC. The term DDC does not *necessarily* refer to a fully networked control system. A better term to use in describing a fully networked control system is the term "Building Automation System." DDC controllers can very well operate in "stand-alone" situations. A control system specified to be "non-DDC" can be interpreted to mean that the system is not to be networked. However, the control systems designer can choose to utilize DDC controllers to implement the various systems that make up the entire project. Moreover, the complexity of the control strategies called for may even make the use of non-DDC controls prohibitive. In other words, the requisite control schemes that must accompany the more complex types of mechanical systems may only be achievable by means of digital controllers. So even though the engineer is not asking for Direct Digital Control, he may approve the use of DDC controllers operating in a "stand-alone," non-networked mode.

Safety Devices

The very last types of controllers that we need to talk about (I promise!), are safety devices. These are two-position controllers whose primary purpose is to interrupt a process or the operation of equipment upon sensing an unacceptable condition. They are usually of the "manual reset" variety, meaning that if the condition they are monitoring becomes unacceptable, they break or "trip," and will remain tripped until their reset button is pressed. Of course, if the reset button is pushed while the condition being monitored by the device is still unacceptable, the device trips again, as soon as the reset button is released. Figure 4-12 illustrates a "chain" of safety devices wired together for a typical application.

Low limit temperature controllers, or **freezestats**, are manual reset temperature controllers that serve the purpose of equipment shutdown upon detection of excessively low temperatures by the device. The "sensor" is a long capillary or element (with no bulb). Typical application dictates that the element be draped across the leaving air side of an air handling system's hot or chilled water coil. The controller monitors the air temperature off the coil, and trips if a small segment of the capillary registers a temperature below the setting of the device. The controller is primarily used in air handling systems, to offer freeze protection of the air handler's water coil(s). Freezestats should also be used to protect any downstream water coils. This being the case, if the air handler itself does not have a water coil, the freezestat's sensing element should "cross-sec-

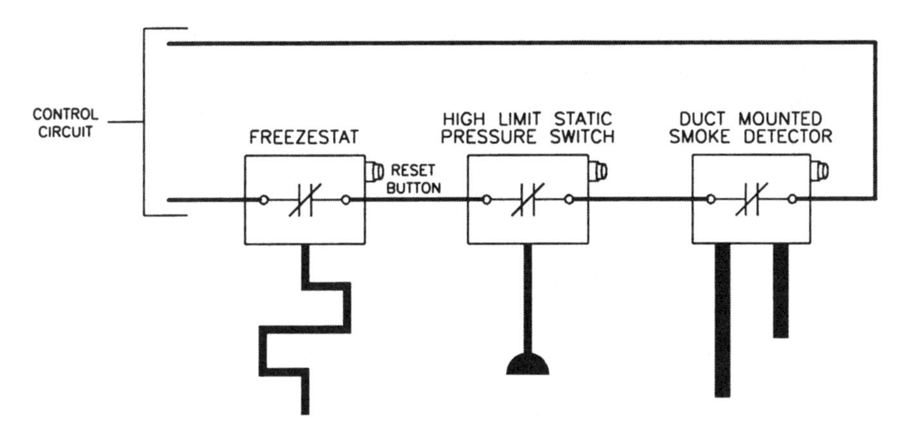

Figure 4-12. Series connection of multiple safety devices, as may be utilized in a built up air handler application. A trip of any device breaks the control circuit, to affect a complete shutdown of the unit. The tripped device must be manually reset, to restore the unit to normal operation.

tion" the duct shortly upstream or downstream of the air handler's supply fan.

High limit static pressure switches are used in VAV air handling systems. The high pressure port of the controller is piped into the supply duct, downstream of the supply fan. The controller monitors the supply duct for excessive pressure, and shuts down air the air handler if tripped. These are manual reset devices.

High limit humidity controllers are used in conjunction with duct mounted humidification systems. The controller is duct mounted, downstream of the humidifier dispersion tube. This type of controller is normally not a manual reset device. If the humidity in the discharge of the air moving system rises above the setting of the high limit controller, the humidifier is disabled. Once the level drops back down, the humidification process resumes. Although "cycling" on the high limit controller may be an indicator that there is a problem with the humidifier (or the air handling system), there may be times when this mode of operation is acceptable. This may be the case when the supply air is at its "saturation point" (i.e., can hold no more moisture), yet the space mounted humidity controller is calling for more humidity.

Refrigeration pressure switches are normally "equipment mounted controllers" that are part of a packaged refrigeration machine. These controllers simply monitor the pressure in the appropriate refrigeration lines, and shut down the compressor(s) upon tripping. These are most always manual reset devices.

Boiler high limit temperature and pressure controllers are "equipment mounted controllers" that are part of a packaged hot water or steam boiler. In either case, the controller serves to shut down the boiler if excessive conditions are reached. These are always manual reset devices (for good reason!).

High limit thermal cutout switches are typically found as part of electric heating equipment, such as duct heaters and unitary heating equipment. They are used, as the name implies, to "cut out" power to the heating equipment upon detection of excessive heat. They can be ordered with the equipment as either manual reset or "automatic reset." If ordered as automatic reset, a malfunction or abnormal operation of the associated mechanical system that causes the temperature of the heating equipment to rise above its cutout would result in the heating equipment cycling on its high limit. If ordered as manual reset, a system malfunction would cause the heater to "trip out" on its high limit, thus disabling heater operation altogether, until the reset button is pressed.

Duct mounted smoke detectors are manual reset devices that monitor ductwork for the presence of smoke, and shut down the air moving equipment if detected. These are generally code-mandated, depending upon the municipality and the code that it follows. Practically speaking, though, the use of smoke detectors just makes good sense. How they are applied, and what purpose they are serving, is of continual debate. The detector is supposed to detect smoke that is recirculated from the space, and turn off the blower so as not to "feed the fire." If the air handling unit protected by the smoke detector is equipped with an economizer section (and no exhaust section), the unit may recirculate little or no air from the space during an economizer cycle. A duct mounted smoke detector, whether mounted in the supply or return duct, will thus have much less of a chance to detect smoke generated in the space served, during economizer operation. A duct mounted smoke detector mounted in the supply duct of a 100 percent outside air system will only serve the purpose of shutting down the system if the air handler itself caught fire. With all that is being said here, installation of a duct mounted smoke detector is still "good practice," and is generally required for any air handling system delivering over 2,000 cfm of air. Two smoke detectors may be required for larger systems, one in the supply and one in the return, this depending on the size of the unit, code, engineer's specification, etc.

Chapter 5

END DEVICES

End devices are devices that play a critical part in the control of mechanical systems. The controllers in the previous section measure and process a sensed variable, make a decision as to what type of action should be implemented, and act upon this decision by producing an appropriate output, or command. If nothing is around to receive the commands, though, then no actions are taken, and nothing happens. The controllers need something to receive their commands, and to carry them out. This section explores the various end devices that we encounter in every day life in the HVAC biz.

RELAYS AND CONTACTORS

A relay is basically an electrically operated switch. Rated for a specific coil voltage, the relay's switch, or "contacts" change state when the proper voltage is applied to the "coil" of the relay. So a relay typically consists of contacts and a coil. The coil is capable of accepting a voltage, and the contacts change state when the voltage is applied. A relay with a single switch or set of contacts is referred to as a single pole relay. A relay with two independent sets of contacts is referred to as a double pole or two-pole relay. Three and four-pole relays are also commonly available. Figure 5-1 illustrates a simple single pole relay.

With a relay, we have the ability to switch an electrical circuit in and out, remotely, via another, completely isolated electrical circuit. Without getting "too much" into the construction of a relay and its functions and applications, consider the following example: a remote bulb temperature controller monitors outside air temperature, and makes when the temperature drops below 40 degrees. Upon this, an electrical circuit is completed to the coil of a relay. The relay has four isolated contacts (four-pole relay). Now, four separate circuits of electric radiant baseboard heating are required to be enabled whenever the outside air

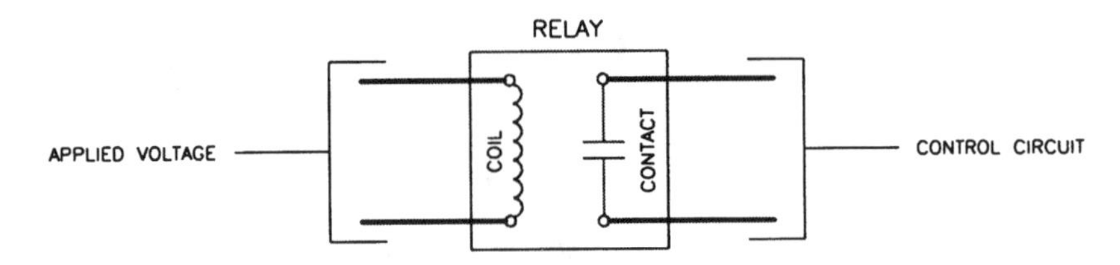

Figure 5-1. Conceptual diagram of a relay. Voltage applied to the coil of the relay causes a "change of state" of the contact. The normally open contact closes upon energizing the coil. The closed contact completes a separate circuit, by allowing current to flow.

temperature is below the setting of the outside air temperature controller. The controller can't do this on its own, because the four heating circuits are independent of each other and must remain isolated. However, the relay can do it, with its four separate switches, each switch handling a circuit of electric heating. The temperature controller, with one switch, controls the relay, with four switches, which in turn takes control of the four electric heating circuits.

Relays are normally rated for low power "pilot duty" applications, and are used more so as a "building block" in the implementation of control "logic" circuits. Relay logic, or ladder logic, is a method of using relays to accomplish some function or arrive at some outcome, given certain required conditions. By connecting contacts of different relays in series and in parallel, in some meaningful manner, and controlling the coils of the relays as a function of some condition, we can perform "operational logic." With the result of this logic, we can control some other end devices in their intended manner. Ladder logic is a tool in the controls designer's tool belt, and has been since the beginning. Yet nowadays many applications requiring any extent of logic can be implemented with a programmable digital controller with inputs and outputs, the ladder logic being replaced by programming. Nevertheless, relays and relay logic will continue to be a simple, cost-effective means of control in the modern world.

Contactors are essentially relays whose contacts can handle much more power. The previous example of the four-pole relay operating the electric baseboard heating would more likely be done with a contactor. For the sake of putting forth some order of magnitude here, let's briefly discuss something called "current rating." Current is "electrical flow." When a switch is closed to complete an electrical circuit, current flows through the switch and on to the device or equipment requiring electrical power. The more power a device requires, the more current flows to it, and thus through the switch. Current is in units of amperes, or more simply, amps. Whereas the contacts of a typical control relay might be rated to handle up to 10 amps of current, contactors are available with contact ratings of anywhere from 10 amps, up to 100 amps or more! In our electrical baseboard example, if the current draw of the individual baseboard heating circuits was more than what we could handle with a simple relay, we would be inclined to use a contactor with contacts of a suitable rating. Contactors are generally offered with one, two, three, and four poles.

STARTERS

The purpose of a starter is twofold. In general terms, starters provide for a means of manually or automatically switching power to a motor, and at the same time protect the motor from "burning out." In HVAC this is most always a fan or a pump motor. A basic starter consists of a contactor and an overload block. The contactor's coil is energized and de-energized to control the delivery of power to the motor. The overload block monitors the current draw of the motor, and if it draws more than its supposed to, the overload mechanism breaks the control circuit to the contactor, thus dropping power to the motor. The starter's overload block must be manually reset, in order to restore the starter to normal operation. A block diagram and description of a typical starter is shown in Figure 5-2.

The basic components of a starter (contactor and overload) are normally housed in a common enclosure. Other components that can make up a starter are control transformers, pilot devices, and auxiliary contacts. A control transformer is normally required to provide the power for the coil and the control circuit, instead of using the higher voltage power that's being delivered to the motor. Pilot devices are switches and lights that can be installed in the cover of the starter's enclosure. A pilot light can give the user an indication of the status of the starter. "START/STOP" switches give the user the ability to manually operate the motor, right from the starter. "HAND-OFF-AUTO" switches give the user the additional feature of being able to automatically control the motor remotely, perhaps from a stand-alone temperature controller,

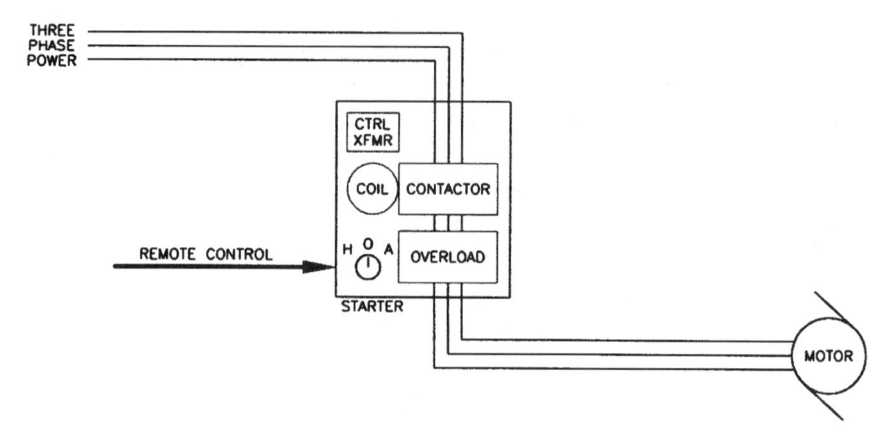

Figure 5-2. Three-phase motor starter shown with the requisite coil, contactor, and overload block. Also shown is a control transformer, which is often required, and a cover mounted "Hand-Off-Auto" switch. In the "Auto" position, starter operation is automatic, via remote control.

or from a full-blown control system. Auxiliary contacts are normally open or normally closed contacts that change state when the starter is energized, and basically can be used to provide information about the status of the starter, and therefore of the associated motor.

The requirement for overload protection of electrical motors is one dictated by the National Electrical Code (NEC). All motors must have some means of protection from overload occurrences. For small, fractional horsepower, single-phase motors, this protection is usually built into the windings of the motor itself. For larger, three-phase motors, though, this feature is (generally) not available. Hence the need for the starter! If a three-phase motor were equipped with internal overload protection, it would not need a full-blown starter; a contactor would suffice. We can draw a general conclusion from this and say that, for all three-phase motors, a starter (contactor plus overload) is required, and for most single-phase motors, a simple switch or contactor is sufficient to control the motor.

It is important to understand the need for starters on a typical HVAC project. We have stated that starters are generally required for all three-phase motors. When these motors are part of a piece of packaged equipment, the starter function (contactor and overload protection) will be built into the control system of the package. A rooftop unit will have at least two components utilizing three-phase motors as part of its package: the supply fan, and the compressor(s). This equipment does not need to be additionally equipped with starters for these components. It's all part of the package! Same goes with a chiller or a condensing unit. The motors making up these types of equipment are factory furnished with all of the control devices and overload protection that they need, and do not require additional, field mounted starters.

In summary, starters provide a means of switching power to a fan or pump motor, and protect the motor from overload occurrence. They minimally consist of a contactor and an overload block, yet are normally equipped with more appurtenances, such as control transformers, pilot lights and switches, and auxiliary contacts. Overload protection of an electrical motor is code-mandated. Single-phase motors usually have their means of overload protection built into them, thus eliminating the requirement for a full-blown starter. Three-phase motors are generally not internally protected, and thus require starters, for control and overload protection. Three-phase motor operated equipment purchased as "stand-alone," such as exhaust fans and pumps, require field installed starters. Motors purchased as part of packaged equipment will come equipped with starters, as part of the package.

Dampers and Actuators

Control dampers allow or prohibit the passage of air from one place or another (two-position control), or regulate the amount of air passed through them (proportional control). The dampers themselves take on many shapes, sizes, and configurations. The three configurations that we will limit our discussion to are butterfly, parallel blade, and opposed blade.

Butterfly dampers are cylindrical in shape. The damper consists of a disc, radially mounted on a shaft, inside a small segment of cylindrical duct. The shaft protrudes through the surface of the duct, to be accessible from outside the duct. A radial motion imposed on the shaft results in a rotation of the disc inside the duct. Varying the rotation of the disc results in variations in airflow through the damper assembly. Butterfly dampers are limited in size (up to 16" in diameter or so), and are typically used in zoning applications.

Parallel blade dampers are rectangular "windows," with strips or "blades" of metal filling the area of the window. The strips are generally just a few inches wide, and can be as long as the application dictates (within reason). A damper with an area of 24" high by 30" long may be constructed with four 6" wide blades, each 30" in length. Each blade has a shaft, bisecting the length of the blade. The blades are mounted within the framework of the damper assembly, and the shafts serve as the "point of rotation" for each blade. Finally, rod and joint assemblies connect the blades together, so that a movement of one blade translates into an equal movement of all blades of the damper assembly.

With the damper closed, the blades are situated to allow no flow through the framework; as the blades are rotated, more and more air is allowed to pass through the assembly. With parallel blade dampers, the blades all rotate in the same direction. This rotation in the damper blades, from fully closed to fully open, is not directly proportional to the amount of air passing through them. The position of the damper blades to the amount of air passing through a damper, in other words, is not a linear relationship. Simply put, the majority of the air capable of passing through a parallel blade damper does so at only "half the travel" of the damper blades. The travel of the blades, from half open, to full open, only translates into a small amount of additional air passing through the damper. Parallel blade dampers are suitable for most two-position applications.

Opposed blade dampers, on the other hand, offer a more linear relationship between the position of the blades and the amount of air being able to pass through them. The construction of the opposed blade damper is much the same as that of the parallel blade damper, except for the method in which the blades are linked together. With parallel blade dampers, the blades are connected so that they all rotate in the same direction. With opposed blade dampers, each blade rotates in the opposite direction as the two blades adjacent to it. With a "four blade" damper, then, as the top blade rotates clockwise, the blade below it rotates counterclockwise, the blade below that clockwise, and the bottom blade counterclockwise. Opposed blade dampers are suitable for proportional control applications, much more so then parallel blade dampers.

Now that we've talked about the common types of control dampers, we must talk about "actuation." The **damper actuator** gives the control damper "character." Without an actuator, the control damper is nothing; the actuator makes the damper what it is (see Figure 5-3)!

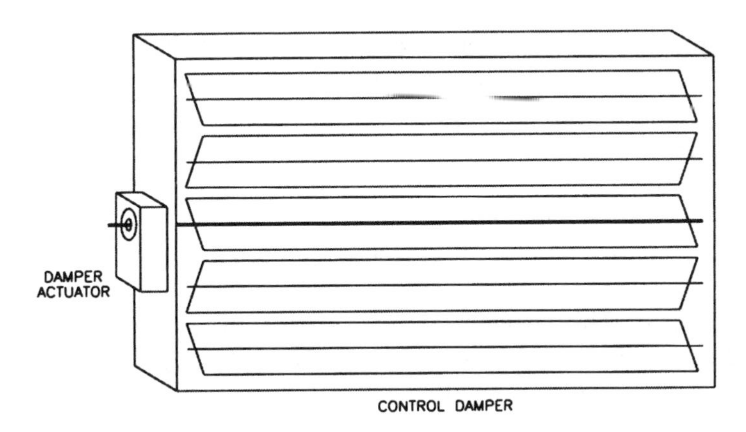

Figure 5-3. Opposed blade damper with direct-mount damper actuator.

Okay, enough of the silly references. Seriously, the damper itself can't do much on its own. An actuator gives life to the damper. The actuator connects to the damper in such a way that it can stroke the damper from fully closed to fully open. Prior to the introduction and subsequent popularity boom of "direct-mount" damper actuators, linking an actuator to a control damper was, at best, a cumbersome task. Ball joints, control rods, and crank arms were all required to link the actuator to the damper assembly. To tie all of this junk together in order for the actuator to properly stroke the damper throughout its entire range of motion is a bit of a trick traditionally reserved for the start-up technician. With the advent of direct-mount actuators came the substantial simplification of this task. A direct-mount actuator can be mounted directly to the control shaft of the damper assembly in a matter of minutes (don't quote me on that!). This of course is provided that the damper's control shaft is readily accessible, which isn't always the case, even nowadays.

Actuators are constructed to different "strength" ratings. **Torque** is the term that refers to the amount of rotational strength that a damper actuator has. For most applications, the torque rating directly translates into the amount of damper area, in square feet, that the actuator can adequately operate.

A major consideration to be pondered when specifying and selecting a damper actuator, is whether there is a need for the actuator to "**spring return**" upon removal of power from the actuator. This is a "fail-safe" feature, required for many damper applications. For example, the outside air intake damper of an air handler most often requires spring return operation. Hence, upon a loss of power to the air handler, or a trip by an air handler's safety device, the damper actuator loses power, and returns the damper to its fully closed position, by means of a wound up spring inside the actuator.

With the requirement of spring return comes a reduction in the torque offered by similar size actuators. What did he say? Well, what I tried to say was, you can get a lot more torque out of a much smaller package with a non-spring return actuator, as compared with a spring return actuator. This is partly because, with spring return actuators, the spring makes up a good portion of the actuator's physical size. Makes sense. So what if your torque requirements are out of the range of available spring return actuators, yet you need spring return operation for your application at hand? Double up! You should be able to have multiple actuators control the same damper assembly, at least in theory. You may have to get creative, though, perhaps having to extend the control shaft, to fit the two actuators on the same shaft, or grab the damper assembly from either side.

Two-position, modulating, **normally open, normally closed**. All of these terms mean nothing to a control damper until a damper actuator is mounted to it. With respect to the "normal" position of a damper, this relates to what position the damper is in, when equipped with a spring return actuator, when the actuator is unpowered. For example, the outside air intake damper from the "paragraph before last" would be classified as "normally closed"; the damper is forced closed by the actuator's spring upon removal of power from the actuator. A damper assembly equipped with a non-spring return actuator has no normal position. It is incorrect to refer to a non-spring return damper as normally open or normally closed. Sure, it may have a position that it *should* be in when its associated system is not in operation, but in reality, if power is removed from the damper actuator, the damper itself just stops right where it is. If this is not acceptable for the application, the damper really ought to be equipped with a spring return actuator. Oh, incidentally, spring return actuators are more costly than non-spring return actuators (that spring costs money!).

Control Valves

Well, it's a good thing that we talked about dampers and actuators before this section. Control valves are the "waterside" equivalent to control dampers, in that they control the passage of water (or steam) from one place to another. Many of the terms that we covered in the previous chapter, in regard to actuation, apply here as well. However, that might be where the similarity between control dampers and control valves ends.

As we did with dampers, let us talk about the mechanics of control valves first, and put off discussions on actuation until further on.

Valve Bodies

The first thing that we'll tackle is the topic of "valve bodies." A control valve assembly is basically made up of two components, the valve body, and the valve actuator (oh, yeah, and the linkage between the two!). Anyway, the valve body is a fairly complex entity in itself; forget about the actuator! Valve bodies can assume many different shapes and sizes, and can be either "two-way," or "three-way." A two-way valve body simply has two ports; the steam or water enters one port, and leaves via the other port. A three-way valve body has an additional port. This extra port brings about an immense increase in complexity and confusion, when it comes to application, installation, and control.

Why is this so? To answer that, consider first that a valve with three ports can have a couple of different flow configurations. Forget about steam for a while; three-way valves don't have much place in steam applications. So water can flow into one port of a three-way valve, and out of the other two ports. Or it can flow into two of the ports, and out of one port. A valve body that is designed as "one in and two out" is called a "diverting" valve body. And a valve body that is designed as "two in and one out" is called a "mixing" valve. This is standard terminology that is used with all three-way valve bodies.

The first potential point of confusion directly relates to these terms, and how the control valve is ultimately supposed to operate. For the purpose of establishing some standard means of referring to the ports of a three-way valve, let us call the common port, the port that handles all the flow, all the time, port "AB." The two remaining ports we will call port "A" and port "B." With a diverting valve, Flow is into the AB port, and (in some way) out of the other two. The term "diverting" *implies* that "full flow" is either out one port or the other. This would seem to imply "two-position" control. Now with a mixing valve, flow is into the A and B ports, and out of the AB port. The term "mixing" *implies* that flow can be into both ports at the same time, in differing portions, and mixing together. This seems to imply "proportional" control. One can mistakenly conclude that diverting valves are used solely in two-position applications, and mixing valves are used solely in proportional applications.

The truth of the matter is the terms diverting and mixing have nothing to do with whether the valve is controlled as two-position or proportional. It is strictly a convention, or a terminology, used to describe how the valve body has been designed to accept and deliver flow. We can put a two-position actuator on a mixing valve body, and we can put a proportional actuator on a diverting valve body. Do not be misled into thinking that the way a three-way valve body is described (diverting or mixing) has anything to do with how it is to be controlled.

Okay, now that we've cleared that up, what else can confuse us, in regard to three-way valve bodies, and the application thereof? Well, first off, what is the functional difference between the two configurations of three-way valves? When should either be applied? What governs the decision to go with diverting over mixing, and vice versa? How should the valve be ported (connected to the piping system), in order to be "properly" operated by the actuator and the associated control system? In a spring return application, which port is "normally open" and which is "normally closed?" These are just a few of the many questions that can arise when applying three-way valves. Perhaps the most obvious question altogether is "Why even bother with a three-way valve; why not just use a two-way valve?"

The following sections on the different types of valve bodies, and valve actuation, should go a long way in addressing most of these questions. For that last question, though, we need to think about pressure and flow in a typical water moving system. For this, draw in your mind a picture of a hot water pumping/piping system. A steam-to-hot water heat exchanger produces the hot water for our application. The hot water is pumped from the heat exchanger, out to a few dozen radiators, each with a local control valve, and each controlled by a temperature controller in the space served by the individual radiator. Assume first that all of these control valves are two-way valves. As space temperatures increase and control valves close off, the pressure in the piping system rises (not a particularly good thing on its own). In addition, the hot water pump, running at constant speed and pumping against closed control valves, begins to "deadhead." The term simply describes a pump trying to pump water that has nowhere to go. This is not good for the pump, and can eventually lead to mechanical failure of the pump. For this type of application, it would be wise to either install a system level pressure regulated "bypass" between the supply and return mains, or at least fit the project with a few strategically located three-way valves (for bypass at the zone level).

Suppose the pump was controlled by a variable frequency drive (VFD)? This is a device that allows the pump motor to run at variable speeds. In an application such as this, the VFD would operate to maintain a constant pressure in the system piping. As pressure builds in the piping, the VFD would operate the pump at a slower speed, thereby keeping the system pressure at an acceptable level. In this case, we would want *all* of the valves to be two-way valves. We want the pressure in the system to be affected by the valves, so that the VFD can react to this by slowing down the pump. Three-way valves would defeat the purpose of the VFD, which is to conserve energy (and save money!)

So far, so good? Good! Time to move away from this whole discussion and talk about the common types of valve bodies that we use in control applications. Specifically, we will be taking a look at **globe valves, ball valves, butterfly valves**, and **zone valves**.

The **globe valve** is the traditional control valve. Long before anyone got the idea of putting an actuator on a ball valve, the globe valve was fully adapted to electric actuation. Even these days, with less costly alternatives at their disposal, many in the business still prefer the globe valve as their control valve of choice. The body of a globe valve has a stem protruding through the top of it. Lateral (up and down) motion of the stem, and thus of the plug assembly inside the body, translates into variable flow through the valve body. An actuator is mechanically linked to the valve body, and operates the stem, moving it up and down, as illustrated in Figure 5-4.

Three-way globe valves are primarily offered as mixing valves, though can also be found as diverting valves. The fact that certain valve body manufacturers do not offer diverting globe valves in smaller sizes is generally a reason why many (most?) applications calling for three-way globe valves are implemented using mixing valves. The difference in physical construction between the two configurations is beyond the scope of this discussion. Suffice it to say, though, that the two cannot be interchanged. In most cases, the valve body manufacturer states that a diverting globe valve is not to be used in a mixing configuration, and vice versa. This has everything to do with how the valve bodies are

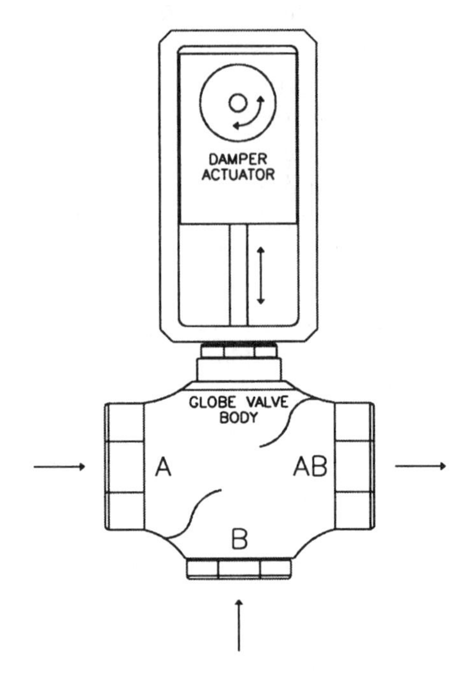

Figure 5-4. Three-way mixing globe valve body with linkage and actuator. Rotation of the actuator imparts a lateral motion to the valve stem, varying flow from ports A and B to port AB.

constructed; using a three-way globe valve body in a configuration other than it was designed for results in a shortened life span of the valve body.

Globe valves are generally available in the range of 1/2" to 6". They are suitable for two-position and modulating control of both water and steam, for applications falling within this size range.

Ball valves have come into their own as a low cost alternative to the globe valve, at least for applications requiring smaller size valves. Inside the body of the ball valve is... a ball... with a hole through the center of it. And a stem fused to the top of it, that protrudes through the body of the valve. Rotation of the stem, and thus of the ball with the hole in it, translates into variable flow through the valve body. An actuator is mechanically linked to the valve body, and operates the stem, turning it back and forth.

Three-way ball valves are diverting by the nature of their design, though they can be used in a mixing configuration. Certain manufacturers may throw in a disclaimer in their literature, saying that the use of their three-way ball valve in a mixing application can lead to premature wear of the valve.

Ball valves are available in the range of 1/2" to 2", and are generally limited to use with two-position and modulating water applications.

Given the same size ball and globe valve, the standard ball valve can handle quite a bit more flow through it than the globe valve. Application-wise, this basically translates into being able to use a smaller ball valve than a globe valve, for a given required flow capacity. While this sounds like a good deal (and it is when applied correctly), it does create a potential predicament, one that we will explore when we talk about valve sizing.

Butterfly valves are the waterside equivalent to butterfly dampers. Since we discussed the construction of the butterfly damper, we need not go into this too deeply for the valve. Picture that butterfly damper, except instead of ductwork, think in terms of piping. Good. Now think of two of those, turned 90 degrees with respect to each other, installed in a piping "tee," and you have your three-way butterfly valve.

Three-way butterfly valves can be used in either mixing or diverting applications; there's not much to be concerned with here, as far as porting goes.

Butterfly valves are normally available in the range of 2" up to 12" (or larger, if custom-built). They obviously take over where globe valves leave off, size-wise, but are also a cost-effective alternative to globe valves in the 2" to 6" range, if conditions are correct (I won't touch that one!).

Zone valves are small "globe type" or "ball type" valves suited for small, single zone applications. The reason that the two above terms are in "parentheses" is because they are not true globe or ball valves, but are a variation of one or the other, or a "hybrid" of the two types. Whereas most of the valve types that we covered prior consisted of valve bodies and separate, medium-cost actuation, zone valves offer a low cost alternative, for the range of 1/2" to 1" or so. Purchased separately as a valve body and a compatible actuator, or purchased as a manufactured assembly, zone valves are typically utilized in zoning applications, and can be two-way or three-way, two-position or proportional, spring return or non-spring return, for water or for steam.

Valve Sizing

Valve sizing is rocket science! No, it's not. But it is pretty critical. It's not critical in that it's difficult to perform. It's actually rather simple. It's critical in that the individual making the valve selections must know what he is doing and how to go about it. For those not familiar with standard pipe and valve sizes, here they are: 1/2", 3/4", 1", 1-1/4", 1-1/2", 2", 3", 4", 5", 6", 8", 10", 12".

The first thing that needs to be considered when sizing a control valve is how it will be controlled. More specifically, is it a two-position application, or is it a proportional application? For a two-position application, all you need to know is the line size. The valve size should be the same. That's it! Yeah, theoretically, we can get away with using a valve that's perhaps a size smaller, and save a few bucks. But that savings is usually eaten up in fittings and the extra labor required for installation.

For proportional applications, sizing the control valve is more involved. An improperly sized control valve can cause an array of systematic problems. An undersized valve can result in insufficient capacity at full load conditions. With an oversized valve, stable control may not be able to be achieved, as exhibited by the control valve constantly "hunting." To better understand the importance of "right-sizing" a control valve, we need to discuss three terms, and how they are related by a simple equation. We likely know of or have heard of the first two; flow rate (gallons per minute, or gpm) and pressure (P). The third term is the valve coefficient, or Cv. The Cv value is a flow characteristic of the valve, and is directly proportional to the size of the valve and its ability to pass fluid through it. More specifically, Cv is defined as the flow rate (gpm) through the valve at a pressure drop (P) across the valve of 1 psi (pounds per square inch). The equation that relates these three values is as follows:

$$P = (gpm/Cv)^2$$

The pressure drop across a proportional control valve when fully open, for optimal control, should be equal to the pressure drop across the coil that it's serving. Sometimes the coil's pressure drop is not known or readily available. However, for typical HVAC applications, the pressure drop across a hot or chilled water coil is usually less than 4 psi. Hence, a popular and commonly used rule of thumb exists that if the coil pressure drop is not known, select the valve so that the pressure drop across it when fully open is in the range of 3-5 psi.

Generally then, all that really needs to be known is the flow rate. The above equation can be manipulated, fixing the pressure drop at 4 psi and inputting the gpm, to come up with a suitable valve Cv rating. Valves come in discreet sizes though, and a valve may not be available for the Cv rating that you calculate. So then you select a valve with a Cv that's close to what you've come up with, and plug it into the formula, solving for pressure. If your result is between 3 and 5 psi, then you're done!

What if your result is not within the rule of thumb? What if it's too low? Or too high? Now it becomes a judgment call. You're pretty much stuck with these discreet valve sizes. You may be able to go to another manufacturer and find a valve body that better fits the rule. But what if you have a dozen valves to size and select? Chances are that with any manufacturer's product offering, you will end up with some valves that fall within the rule, and some that don't. And the last thing you may want to do is "mix and match" products from different manufacturers, simply for the sake of having all the valves fall within the rule.

If selection of a valve, using this approach, does not fall within the 3-5 psi rule of thumb, then it's time for some insight. Does the system pump have plenty of power, even to overcome the pressure drop of a valve selected as "kinda high?" Is precise control of critical concern, where oversizing may be less desirable than undersizing? How much is the valve selection undersized, and if the next larger size were selected, would the "deviation" from the 3-5 psi rule increase or decrease? Which valve should be selected, if the deviation is close to being equal?

These are all concerns that typically must be dealt with on a job-to-job basis. Realistically speaking, the body of a proportionally controlled "globe" valve will almost always be a size or two less than the pipe size. This is another rule of thumb that serves as a "check" when making the valve selection, and helps in determining the "appropriate" selection. Rocket science it ain't, but we do need to be aware and make a concerted effort when sizing control valves.

The above paragraph pointed out a useful rule of thumb regarding "globe" valves. This is not *necessarily* the case, however, with ball valves. As we made mention of earlier in this section, a standard ball valve can handle more flow through it than a globe valve of the same size. This is a general statement regarding ball valves and their "traditional" construction. Ball valves were not originally made to be automatically controlled. When actuation was first applied to ball valves, it was done so with existing valves. The port of a ball valve, i.e., the hole in the ball, was rather large with respect to the valve. This translated to a rather large Cv value for the valve. In essence, for a globe valve sized for a proportional control application as a 1" valve, a ball valve sized for the same application would end up as 3/4" or even 1/2". With valve manufacturers telling you to avoid using a control valve that's less than half the size of the pipe connecting to either side of it, a ball valve sized for the application at hand may very well fall short of this recommendation. And upsizing the valve to satisfy this rule of thumb puts you out of the 3-5 psi rule of thumb, thus resulting in an oversized valve. What can you do? Go with a globe valve!

Actually, the above discussion is all academic, for two things work in favor of going with ball valves on a project. One is the increase in electronic and microprocessor-based control. With current technologies, it is possible to build more intuitive control into our control schemes. A ball valve that is selected as slightly oversized will be less prone to unstable control when P+I and "smart" control is employed.

The other thing that ball valves have going for them, is that their manufacturer's have realized some of these "potential" problems in sizing, at least as being a concern among those that size valves. So what they have done is they have expanded their offering. A ball valve can now be purchased with differing port sizes (not to mention specialized configurations), given the same connection size. In other words, a 1/2" valve body may be offered in several different port sizes, and thus several different Cv values. Same for a 3/4" and 1" valve body. And even for the rest of the range of ball valve sizes. This gives the valve selector some flexibility, and opportunity, to apply ball valves in proportional control applications. Where in the past a required Cv of 10 would mean a 1/2" valve, now a valve with that same Cv value might be available in a 3/4" size, or even a 1" size! Great news for those of whom like to satisfy "all" of those recommendations and rules of thumb that we have talked about.

It is important to note that control valve sizing as described herein applies to sizing valves for coils, and doesn't necessarily extend to other control valve applications. Rules that apply in sizing control valves for applications other than coil control are not included here, and must be found elsewhere. Coils make up the majority of control valve applications in our business, and so the material presented here should go a long way in control valve sizing and selection. Yet there are other control valve applications out there, and the designer must be weary of other guidelines that govern valve sizing, for these less common applications. Such applications include hot or chilled water system (pressure) bypass control, hot water system (temperature) reset control, and cooling tower condenser water (temperature) bypass control, just to name a few.

To wrap up the whole topic of valve sizing, we have to at least make mention of steam valve sizing. Okay. For two-position control, the steam valve should be line size, and for proportional control, the valve will be one or two sizes smaller than line size. Enough said? Well, there is a formula for sizing proportional steam valves as there is for sizing proportional water valves. It's more complicated, though. Generally, tables are available that will allow you to find the required Cv for a steam application, given that the steam pressure and the steam capacity are known. This is the simplest way to size proportional steam valves; find the required Cv from the table, and then find a valve whose Cv rating most closely matches. For steam, it's most recommended to "err on the high side"; if the required Cv falls between two choices, pick the larger of the two.

VALVE ACTUATORS

Finally, we can talk about actuation of the control valve. Just as the damper actuator gives life to the damper, so does the **control valve actuator** to the valve. The linkage connecting the actuator to the control valve has evolved almost to a point of inconsequence, as technologies and construction methods have simplified the task. For that reason, we gloss over this topic and move onward to talk about actuator strength and how it acts upon the valve.

The concept of torque, as it applies to damper actuation, does not particularly apply to valve actuation, at least not with globe valves. The term used to classify the strength that an actuator can impart to a control valve is "**close-off rating**." This refers to how tightly a valve actuator can shut off flow through the port(s) of the valve, and is in units of pressure. The higher the close-off rating, the less chance of a fluid "leaking through" when the valve is fully closed. Like with damper actuators, if close-off rating is an issue, a valve body may be "doubled up," or equipped with two actuators (if linkage methods permit).

As with dampers, the need for spring return must be considered when applying control valves. Many valve applications require spring return operation, especially when there is a potential for "freeze" conditions. As such, the hot and chilled water coils of a large air handler may be fitted with spring return control valves. For instance, it is customary to have the hot water valve spring open to its coil upon removal of power from the control valve actuator.

When equipped with spring return actuation, the control valve can now take on the term of "normally open" or "normally closed." The two-way valve itself is termed as one or the other. A three-way valve's ports (specifically the "A" and "B" ports) take on these terms, when controlled by a spring return valve actuator. It is incorrect to refer to a two-way valve, or the A and B ports of a three-way valve, using these terms, unless the valve is equipped with spring return actuation.

So with a spring return three-way control valve, which port is normally open, and which is normally closed? That is a question of the ages. It used to be of great importance that the installer know, up front, which port was which, for that was dictated by the motor itself. If the installer mixed up the porting, all that could be done was to buy a motor of "the opposite type," or re-pipe it correctly.

These days, motors and linkages are generally offered to be more flexible and more forgiving (thank goodness!). You may only need to convey to the installer which port is common, and allow the installer the liberty of piping the other two ports in the configuration easiest for him. Then upon start-up, spring return operation can be validated; if correct, great, and if not, adjust the linkage and/or the motor so that it is.

Control valves are one of the most (if not the most) critical components of many, many HVAC systems. It's no wonder then, that there is so much to be said of them, and so much to misunderstand. This section has hopefully given a broad overview of control valves, the terminologies associated with them, and application tips and tidbits. There's a lot to know about our friend the control valve, and a little knowledge goes a long way in understanding its role in today's HVAC systems.

Variable Frequency Drives

We've already brought up the term and discussed the function of variable frequency drives (VFDs) a couple of times now, in some of the examples that we've considered, without formally describing the VFD. Now's the time! Refer to Figure 5-5 for a block diagram and description of a typical VFD.

A VFD is an electronic piece of equipment. Its function is to vary the speed of an electric motor, as a function of some control signal received by it. It is therefore classified as an "end device" and covered in this section. The VFD will vary the power delivered to a motor, typically a three-phase motor. More specifically, the VFD varies the "frequency" of the electrical power delivered to the motor. Standard power in the United States is delivered from the utility at 60 cycles per second, or 60 hertz (Hz). If the cycles per second were able to be reduced, so too would the power be to the equipment consuming the electricity.

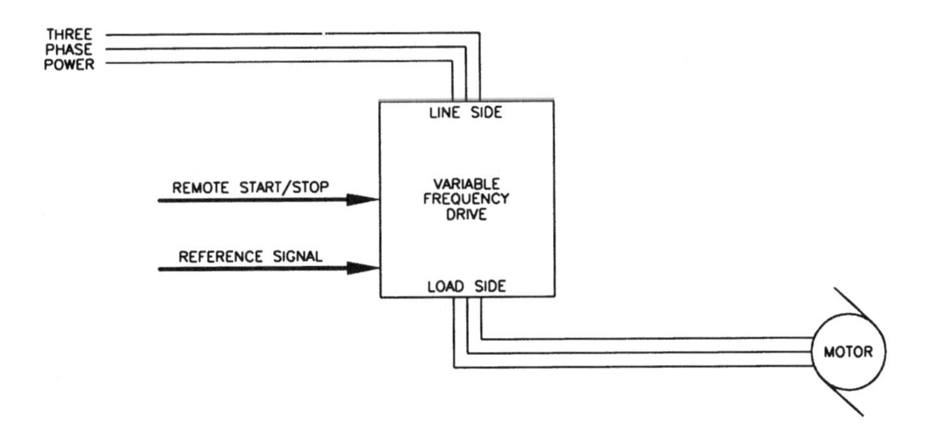

Figure 5-5. Three-phase variable frequency drive. Three-phase (60 Hz) power is connected to the input (line) side, and the motor is connected to the output (load) side. Start/stop can be done locally or remotely. Reference signal is typically the proportional output of some controller. Motor speed is a function of the frequency of the power applied to it.

So a VFD can vary the frequency from 0 to 60 Hz, as a function of a control signal received by it. The variance in frequency translates directly into a variance in speed of the connected motor. Thus, sometimes VFDs are referred to as variable speed drives (VSDs).

In HVAC applications, VFDs are primarily applied to fans and pumps. As stated, the VFD can accept a signal from a proportional controller, and translate that signal to a variable fan or pump speed. The examples that we have covered in previous sections were that of a fresh air fan maintaining pressure in a space, and of a hot water pump maintaining pressure in a piping system. Both of these applications are pressure applications. Temperature applications exist as well, for both fan and pump systems. A fan may be used to cool down condenser water in a cooling tower application; varying the speed to the fan offers more precise temperature control of the process. Likewise, a pump may be used to draw water from one hot water piping loop, and variably inject it into another loop, in an effort to maintain precise temperature in the second loop.

A VFD controlling a fan or pump motor may have a "minimum allowable point of operation," that is actually dictated by the requirements of the motor. A motor manufacturer may state that their motor should not be fed a power supply with frequencies of any less than 10 or 15 Hz. This more or less translates to around 16-25 percent of the full speed of the motor. Hence, a VFD controlling a motor may be limited as to "how low it can go," the parameter being set through the VFD. Motors specifically designed and constructed for operation with a VFD will typically have much less of an operating restriction, or perhaps none at all, where the VFD can operate the motor all the way down to 0 Hz.

VFDs can be purchased with a "bypass." This comes in the form of a manual user switch that allows the motor to be controlled by the VFD, or allows motor operation to completely bypass the VFD electronics. It is implemented with a series of contactors; when the switch is turned from the VFD to the bypass position, contactors on either side (line and load) of the VFD open and isolate the VFD from the power supply and from the motor. Another contactor closes to connect the power supply directly to the motor. This is offered as a factory-installed option, and it sounds like a pretty good feature. If the VFD electronics break down, you can still get motor operation, albeit at full speed, by going to the bypass mode. Caution must be exercised though. If that switch is thrown with no regard to the rest of the system being served by the VFD/motor combination, trouble can be had! For a fan system, loss of fan speed control can result in a trip of the high limit duct static pressure controller, or worse yet (were there no high limit controller), in ductwork actually coming apart at the seams! For a pumping system, loss of pump speed control could result in potentially excessive line pressures, which can cause an array of piping system problems.

It's important to note that, for a VFD application, a motor starter need not be furnished for the motor! The VFD will have all of the functions of a starter built into the package. This includes the requisite overload protection. The VFD manufacturer will likely even offer (as an option) a line side disconnect and input fuses.

In addition to offering precise control, the VFD is an energy saver. To run a fan or pump motor only as fast as needed at any point in time is efficient and energy conserving. First cost and reliability of these devices have historically been "bones of contention." Yet with prices (and physical size) continuously coming down, and reliability going up, technology has taken this one time luxurious (and potentially problematic) alternative to the status-quo in many of the more demanding HVAC applications that abound.

Chapter 6

Common Control Schemes

Before we move ahead and begin taking a more "in depth" look at the various mechanical systems that were listed out back in Chapter 2, it's probably a good idea to take a little time to describe some of the popular control "schemes" that are applied to many of these mechanical systems. For each of the following sections, we will demonstrate the control scheme as it applies to a particular type of system, for the sake of explanation. Be it known, though, that most of these control schemes can apply to many of the types of systems out there. With an explanation of these schemes under our collective belts, we need only refer to the defining terms when discussing systems in the subsequent chapters, rather than taking out more time then to explain the schemes. With no further commotion, let's jump right into it!

Reset

The definition of reset control is "automatically adjusting a setpoint as a function of some changing condition." Its most common application is with hot water systems. For our example, consider a steam-to-hot water heat exchanger. A steam valve, prior to the heat exchanger, proportionally controls the rate of steam entering the "primary" side of the heat exchanger, in order to maintain the temperature setpoint of the hot water on the "secondary" side. In straight temperature control applications, the hot water "supply side" temperature is sensed, and a temperature controller, with an established setpoint, processes this information, compares it to setpoint, and sends a proportional control signal out to the steam valve, the signal being a function of the difference in hot water temperature from setpoint. So the steam valve is modulated to maintain the pre-established constant hot water temperature setpoint.

With reset control, we need to measure an additional variable. In this example, we may want to "reset" the hot water temperature setpoint as a function of the temperature outdoors. More specifically, we would want to "lower" the setpoint as the outside air temperature rises, and "raise" the setpoint as the outside air temperature falls.

A reset controller is a controller that can perform this function. Instead of one sensor, the controller has two sensors associated with it. Older electromechanical type controllers employ the use of two remote bulbs, for sensing purposes. The electronic descendant of the reset controller uses two thermistors, wired back to the central controller. Of course, reset control can be performed with a DDC controller as well.

A "reset schedule," or table of values, needs to be decided upon, before putting the reset controller to work. The schedule defines the operation of the controller throughout its whole range of control. A typical schedule for the current example would look something like this:

O.A Temperature:	–10	60
How Water Setpoint:	180	130

The schedule states that, at an outside air temperature of –10 degrees, the hot water temperature setpoint is to be 180 degrees, and at an outside air temperature of 60 degrees, the hot water temperature setpoint is to be 130 degrees. This is a linear relationship; for the range of –10 to 60 degrees outside air temperature, the hot water temperature setpoint assumes a corresponding value in the range of 180 to 130 degrees. The reset schedule is sometimes illustrated as a graph, as shown in Figure 6-1; the graph can quickly show what the hot water temperature setpoint will be for any corresponding outside air temperature.

Once the reset schedule is established, the "reset ratio" must be calculated. This is the ratio between the ranges of the two variables. For instance, the outside air temperature range, from the schedule, is 70 degrees (-10 to 60). And the hot water setpoint range, from the schedule, is 50 degrees (180 to 130). The reset ratio, therefore, is 70 to 50, or 7:5. The two values composing the reset ratio are normally what have to be "plugged in" to the reset controller.

Reset control can be limited outside the range of the reset schedule. If the temperature outside dropped below –10 degrees, we can request that the controller not reset the hot water temperature setpoint upward

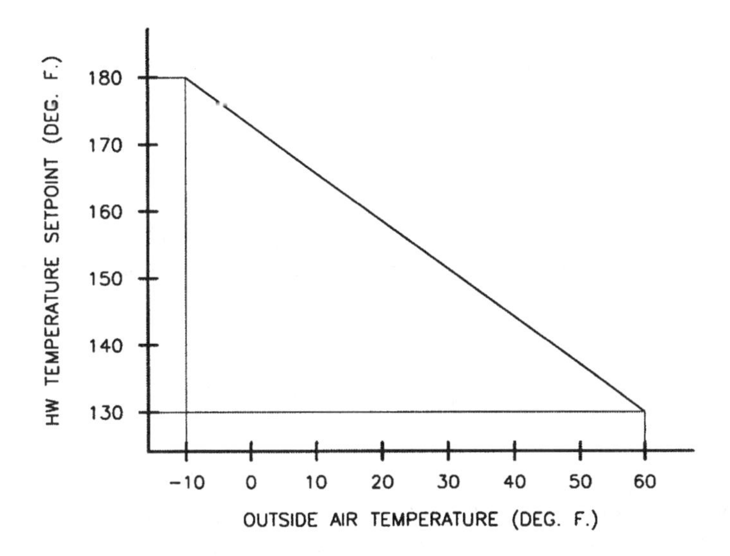

Figure 6-1. Graph depicting a reset schedule.

any further. Likewise with the upper range of outside air temperatures.

Upon first glance, reset control appears to be a substantial "energy saver." Upon further observation, one may conclude that this isn't necessarily the case. For our example, if we did not employ reset control to the hot water system, the setpoint would be 180 degrees, throughout the whole range of outside air temperatures. At outside air temperatures of 60 degrees, the hot water pumped to the system's heating equipment would therefore be 180 degrees. The heating equipment would have very little need for 180-degree water for any period of time. Thus, back at the heat exchanger, very little steam is utilized to maintain hot water setpoint.

So why use reset control in the first place? It becomes more of a control issue. Whereas with straight temperature control, if there were a call for heating by any particular heating appliance, the appliance would get 180-degree water on a day when its 60 degrees outside. That's pretty hot, especially if the heating appliance is a radiator. Needless to say, the zone served by the heater would quickly satisfy, and its control valve would close. Now consider the same scenario, only employing reset control. Take that same 60-degree day, and now upon a call for heating, the appliance would receive 130-degree water. Seems like a much better "match" to the heating load on that particular day, and it is.

Reset control is also commonly used in discharge air applications, where the air handling system tries to maintain a constant "cool" discharge air temperature setpoint, for use by zoned VAV terminal units and/or reheat coils. Upon discovery of problems in colder times (too cold in the building!), reset control may be chosen to be employed as a possible solution to the problem. In this scenario, as the outside air temperature falls, the discharge air temperature is "kicked up." Reset control can be also be done via space temperature, in lieu of outside air temperature.

Night Setback and Morning Warm-up

These two terms apply to air handling system "unoccupied mode" cycles. Thus, both are contingent upon scheduling; if the system in question runs continuously, 24/7, then these terms need not apply. For our example, we will consider a reheat system, where the main air handler operates in the occupied modes to maintain a constant 55-degree discharge air temperature out to all of the reheat coils.

When the system shifts into the unoccupied mode, a few things happen. First, the air handler goes to sleep. With this, so do all of the reheat coils. A space temperature sensor or controller sits in the space, in a general location. A "night setback" space temperature setpoint is established via the control system, typically 60 degrees. The entire system is shut down, and stays that way, unless the space temperature drops below the night setback setpoint. Upon this occurrence, the air handler's supply fan "comes to life." Cooling is locked out, and the air handler's outside air damper remains shut. If the air handler is equipped with heating, then it will be engaged, normally at full thrust (though may be limited by discharge air control). The reheat coils come to life as well, and likely all go to full heating (their setpoints are around 72, and the general space is 60; you do the math!). The entire space served by the air handling system heats up quickly, setpoint is reached, and the system goes back to sleep. This is a typical "night setback cycle."

"Morning warm-up" occurs when the system transitions from the unoccupied mode to the occupied mode. It's kind of like a night setback cycle; the supply fan turns on, cooling is locked out, the outside air damper remains shut, air handling unit heating is engaged (if equipped), and the reheat coils are enabled for operation. Yet all of this only happens if the space temperature, or perhaps the return air temperature, is below the morning warm-up setpoint, which is typically 70-72 degrees. If above the morning warm-up setpoint, the system goes directly into the normal occupied mode operation of discharge air cooling. If not, then the system enters this morning warm-up cycle, and stays in this mode until the morning warm-up setpoint is reached. The purpose of a morning warm-up cycle is to bring the temperature of the general space served by the air handling unit up to "room temperature" as quickly as possible, before transferring to its normal mode of operation (discharge air cooling).

In a typical reheat system, the air handling unit will probably not be equipped with any form of heat, and if it is, chances are that it's not there specifically for night setback and/or morning warm-up. However, at least for our hypothetical example, we can use the air handler's heating coil for these cycles. Care must be exercised though, especially if the air handling unit has plenty of heating capacity in its heating section, and the reheat coils are electric. The concern is that the electric reheat coils will typically have "manual reset" thermal cutout switches. You can probably see where I'm going with this. If the air handler is delivering hot air to the coils, and the coils are also operating at full throttle, then the potential exists for these coils to trip out on their cutouts, which is not a good thing.

The terms "night setback" and "morning warm-up" are normally used to refer to unoccupied mode operation of an air handling unit serving multiple zones. Night setback stems from the notion of "setting back" the space "heating" temperature setpoint during unoccupied modes. Often, an unoccupied mode "cooling" setpoint is set as well, typically around 80 degrees. So that if the general space temperature rises to this setpoint, the air handler will start up and enter a cooling mode of operation, to bring the space temperature back down below setpoint. The appropriate term for establishing an unoccupied mode cooling setpoint is "night setup," and the implementation of unoccupied mode cooling is referred to as a "night setup cycle." Yet, the term "night setback" has evolved to cover both cycles, and has become common terminology to refer to the establishment of unoccupied mode heating and cooling setpoints, and the implementation thereof. This book makes little distinction between night setback and unoccupied mode heating and cooling operation. For here on, it is assumed that the term "night setback" refers to air handler unoccupied mode of control, single zone systems and multizone systems alike.

While the term "night setback" can be used to described the unoccupied mode of operation for both single zone air handlers and air handlers that serve multiple zones, the term "morning warm-up" is mildly out of context, if used in the same sentence as "single zone system." Morning warm-up refers to a heating cycle that occurs before an air handling system, one serving multiple zones and designed to operate to maintain discharge air temperature, begins its normal "occupied" mode of operation (discharge air cooling). With a single zone system, such as a constant volume rooftop unit operating via a programmable thermostat, the system goes straight to its normal "occupied" mode, instantly upon transition from the unoccupied mode. Sure, the unit may go to full heat, if the space temperature is below the occupied heating setpoint of the thermostat, but that's its normal operation, and not some special "cycle." The real difference here is that we are not providing any type of additional control scheme to perform "first thing heating" in a single zone system, whereas we are with a system designed to maintain discharge air temperature during the occupied modes.

Economizer Operation

The upcoming chapter on rooftop units does a pretty good job of describing economizer operation, specifically as it applies to rooftops. Instead of taking up time right here to engage in a full blown discussion on economizer operation, we will simply talk in general terms about what is meant by economizer operation, and the general principles behind it. Figure 6-2 gives a diagrammatic view of an economizer.

To economize means to utilize outside air for "free" cooling. This is done by modulating the outside and return air dampers of an air handling unit, if outdoor conditions permit. The colder (and less humid) it is outside, the more of an opportunity there is to economize.

The decision on whether to economize or not is based on the "condition" of the outside air. If it's 80 degrees outside, there's no opportunity for economizer operation. So the temperature of the outside air is a determining factor; a two-position temperature controller can make this decision, and relay its decision to the economizer control system.

Humidity is another factor to consider in making the decision to economize. Obviously if it's cool yet humid outside, we don't we want to use the outside air for free cooling, if at the same time we would be

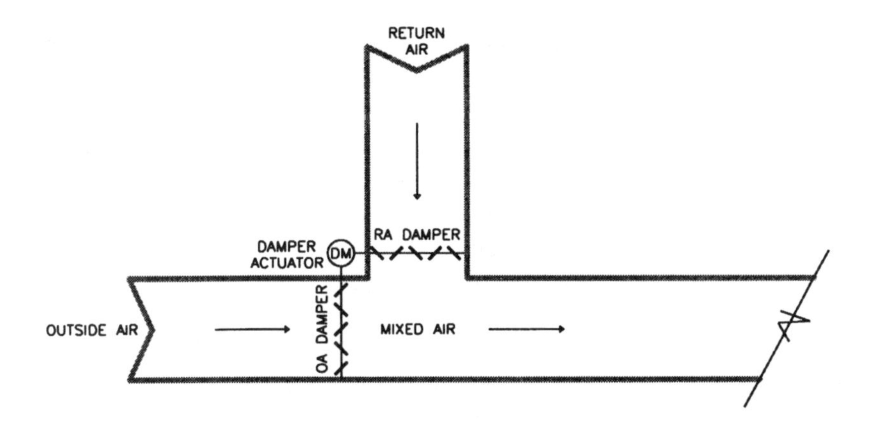

Figure 6-2. Economizer section of a typical air handling unit.

contributing to an indoor humidity problem. On the other hand, if its warm and dry outside, the use of outside air may be at least partially beneficial to our cooling needs. An enthalpy controller can make this decision for us, and is a much better decision maker than a straight "dry bulb" temperature controller. We might even be inclined to stick an enthalpy controller in the return air duct. Comparative enthalpy looks at the return air, compares it with the outside air, and determines whether the outside air has any "advantage" over the return air. If so, economizer operation is permitted, and if not, economizer operation is prohibited.

With a microprocessor-based control system, instead of using an enthalpy controller, we might directly sense both the temperature and the humidity of the outside air, and then make an enthalpy calculation. The result of this calculation would provide the basis for our decision to economize.

Once the decision is made to economize, and a call for cooling has been requested, then economizer operation is implemented. The other means of air handler cooling (chilled water, DX) will not necessarily be entirely disabled, for there may be instances where the outside air, although beneficial to the cooling process, may not be sufficient to completely satisfy the cooling needs. This is especially true for economizer decisions based on enthalpy. For example, if the outside air is 60 degrees and dry, the enthalpy controller may permit the use of the economizer, and the 60-degree air definitely helps the cause. But if we are trying to maintain a discharge air temperature setpoint of 55 degrees, economizer operation alone just doesn't cut it; the outside air damper will modulate fully open without reaching setpoint. In this scenario, additional cooling is required, to bring that discharge air temperature down the remaining 5 degrees.

Economizer operation, as it applies to air handling systems in general, is one of the more misunderstood sequences of control in HVAC. Hopefully this section, and the upcoming chapter's discussion on rooftop unit economizer operation, helps to clear up many of the misconceptions of economizer operation and control.

1/3, 2/3 Control of Steam

Steam is often a difficult medium to control. This is especially true with larger steam capacities. For our example, consider again a steam-to-hot water heat exchanger (Figure 6-3). A steam valve controls the rate of steam entering the heat exchanger, in an effort to maintain hot water temperature setpoint. Sounds simple enough. A single valve modulating to maintain hot water setpoint. The problem is, with larger steam capacities, and consequently with larger steam control valves, control of the steam rate is much more difficult to "dial in," resulting in a control valve that constantly "hunts," and a hot water temperature that swings widely

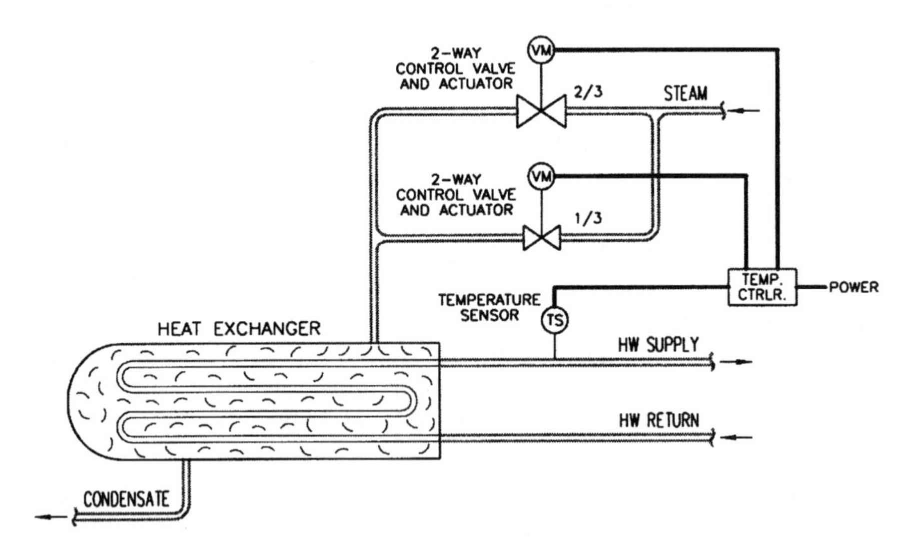

Figure 6-3. 1/3, 2/3 control of a steam-to-hot water heat exchanger, performed by an electronic temperature controller.

above and below setpoint.

A method of control that has been around for a long time, that has been successfully employed as an "alleviation" to this predicament, is that of 1/3, 2/3 valve control. The premise is that we can better control the steam rate with two smaller valves, sized accordingly. One valve is sized for 1/3 the steam capacity, and the other is sized for 2/3 the capacity. The valves are "sequenced"; the smaller of the two valves modulates open first, in an effort to achieve and maintain setpoint. As demand increases, the smaller valve reaches its fully open position, and the larger valve begins to modulate open. This translates into tighter setpoint control, especially during periods of light demand, when the smaller valve is the only one modulating.

The rule of thumb, for choosing this type of control, is that if you require a steam control (globe) valve of larger than 2", you should go with 1/3, 2/3 control. So for a given application, you first need to select a control valve sized for the full capacity, and if your selection turns out to be 2-1/2" or larger, better go with two valves. Selection of the two valves, when sizing them for 1/3 and 2/3 the total steam capacity, will not be an exact science. You won't end up with control valves sized exactly for these capacities. Regardless, whatever you come up with will undoubtedly be better than going with a single control valve. Some engineers and designers are more conservative, and specify this type of control for all critical proportional steam control applications.

Face/Bypass Damper Control

Face and bypass dampers are another method of "Taming the Beast" that is steam. Primarily an airside scheme of control (dampers!), this method is used mostly with make-up air units utilizing steam as the source of heating. So for our example, we consider a make-up air unit, one that handles 100 percent outside air, with a steam heating coil, and face/bypass dampers (refer to Figure 6-4).

The make-up air unit operates to "temper" the delivered air, and thus heats the air when it's below "room temperature." So for all outside air temperatures below say, 70 degrees, the unit must heat the air up to discharge air setpoint (70 degrees), before delivering it to its destination. The steam coil is equipped with a two-position control valve, to be open whenever the outside air temperature is below 70 degrees.

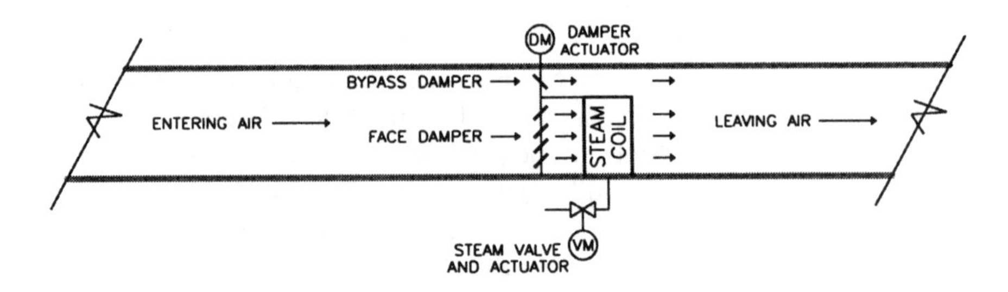

Figure 6-4. Steam coil equipped with a face/bypass damper assembly.

The steam coil does not occupy the entire cross-section of the make-up air unit. There is space up above the coil, for untempered air to pass through. The area of the space above the coil is usually less than the face area of the coil. Each of these two sections (the coil and the open area above) is fitted with a damper. The dampers are linked together and work in unison, in opposite directions, so that when the face damper (coil) is open, the bypass damper (open area) is closed, and vice versa. Of course, the two dampers can assume any position within their entire range of motion, when fitted with a proportional damper actuator. Which is exactly what we'll do.

As we mentioned earlier, the steam valve is fully open when the outside air temperature is below 70, so steam flow through the coil is available for all outside air temperatures below 70 degrees. To maintain discharge air temperature setpoint, we must pass varying amounts of air through the steam coil, and mix it with air bypassed around the coil. We control the face/bypass dampers to meet this end. As the outside air temperature drops, more and more air must pass through the coil, with less and less being bypassed. This is a fairly precise method of control. At least when you compare it with trying to modulate the steam valve to maintain setpoint. With this method, we are directly controlling the medium that we desire to maintain the temperature of.

Face/bypass damper control is sometimes specified in conjunction with proportional steam control. While this may be "overkill" for certain applications, it is not completely without merit, especially when implemented with microprocessor-based control. This will be explored in greater detail in the chapter on make-up air units.

DEHUMIDIFICATION

We know how to provide humidification in a typical HVAC application. We generate steam and inject it into the supply air duct of an air handling system. But how do we dehumidify? How do we take moisture *out* of the air? Seems that it would be a bit trickier than putting moisture into the air. And it is...

Specialized air handling units are available that dehumidify by means of a "desiccant." This is a substance that pulls moisture out of the air. Kind of like a "humidity sponge." The substance is in the form of a wheel that spins continuously. At any point in time, half the wheel is in the air stream of the unit, sucking up moisture out of the air, while the other half is "rejuvenated" or "wrung out." These are rather complex systems, and costly too. There is another way of providing dehumidification, which is what we are going to take a look at here.

The "classic" method of dehumidifying is by cooling the air and heating it back up. As you may know, the ability of air to hold moisture decreases as the temperature of the air decreases. In other words, the colder the air, the less water it can hold. So by cooling down air that is holding a lot of moisture, we force the moisture out of the air, "wringing it out." Of course now the air is too cool for our needs, so we need to heat it back up, or "reheat it."

For our example, we consider a single zone, constant volume air handling unit with refrigeration type cooling and electric heating (refer to Figure 6-5). It is imperative that the refrigeration coil (direct expansion, or DX coil) is "first in line" in the air handler's air stream, with the electric heating coil downstream of the DX coil. A thermostat in the space controls the cooling and heating operation of the air handler. In addition,

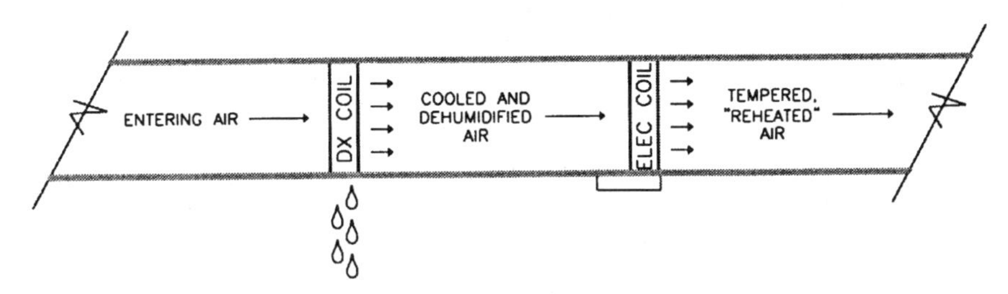

Figure 6-5. Direct expansion and electric heating coils employed in a dehumidification cycle.

there is a dehumidistat in the space, adjacent to the thermostat.

The dehumidistat simply overrides the air handler into a cooling mode upon a rise in space humidity level above setpoint. As the supply air is cooled down and dehumidified, the space temperature tends to drop, the space thermostat calls for heating, and the electric heater is staged as necessary.

In theory, we can implement this cycle in a method "opposite" to what we've done in the last paragraph. In other words, we can have the dehumidistat call for heat, resulting in the space temperature rising above the cooling setpoint of the thermostat. Cooling is engaged, and dehumidification takes place. This method, though seemingly backward, is at least acceptable in theory, and could have its place in some unorthodox application.

What we see with dehumidification is a process that is not very energy conscious. We are using energy to cool and dehumidify the air, and we are using energy to heat the air back up. The legality of cooling down air and "canceling out" the energy used in the process by applying reheat may be contested, depending upon where you're at and what codes apply (for what it's worth). Regardless, this is traditionally how it's done, and unless the customer is willing to spend the money up front and invest in a "specialized" dehumidification system, this is basically the next best thing.

Primary/Backup Operation

The term "primary/backup" implies system redundancy. In simpler terms, it implies that two pieces of equipment are each sized to handle an application, with one operating at a time, and the other serving as a "backup" in case of failure of the "primary." The term can be extended to cover systems with more than two pieces of equipment. This will be touched upon, as it applies to pumping systems, in the upcoming chapter of the same name.

Our example will address redundancy as it applies to the cooling needs of a computer room. We consider two computer room air conditioning units, each sized to handle the maximum possible cooling load in the space. Only one unit is in operation at a given time, with the other sitting quietly, just waiting for the operating unit to fail!

Upon failure of the primary unit, the backup is pressed into operation. This might be a manual procedure; somebody would have to recognize or be alerted to the failure, and turn on the backup unit. More likely, though, in this type of application, operation of the backup unit upon failure of the primary unit is an automatic function, performed by some supervisory control system. Now that the primary unit has failed and the backup unit is running, time to get that primary unit checked out and repaired!

We've just looked at what is meant by the terms "manual backup" and "automatic backup," as they apply to primary/backup systems. Now let's take a look at few more terms relating to this type of system. "Alternation" or "rotation" is a term describing the procedure of taking the primary unit out of service and putting the backup unit into service, as a means of giving both units "equal playing time." Upon alternation, the backup unit becomes the primary unit, and vice versa. "Manual alternation" is done manually, either by physically turning off one unit and turning on the other, or by means of an alternating switch. "Automatic alternation" is done automatically by the supervisory control system, and is normally time-based. In other words, unit 1 runs this week, unit 2 runs next week, and so on. Alternation is basically performed as a means of exercising both units equally. If alternation isn't performed, it is possible that the primary unit will be "overworked," while the backup unit sits dormant and "lives the good life." Upon failure of the primary unit, the backup unit may be too accustomed to its lifestyle, and may very well choose to just not run. Hey, equipment is built to run, right? And if equipment sits too long without operating, can you count on it to come to life when you really need it? Another argument for regular alternation is that if you don't alternate the equipment, you end up "wearing out" the primary unit in a much shorter time frame, thus having to replace it "sooner than expected."

You can take all of the above with a grain of salt, formulate your own opinions, and draw your own conclusions. The preceding paragraph is presented as such simply to drive home the issue of what is meant by these terms, and why we do things like this to begin with.

Lead/Lag Operation

The term "lead/lag" implies staging. It implies that two or more pieces of equipment are sized so that the sum of their capacities equals the maximum system requirements. The term is widely confused with the term "primary/backup," but is distinctly different, as we will see here.

For our example of lead/lag, we will consider two chillers, producing chilled water for a system of fan coil units having chilled water coils and control valves. Each chiller is sized for half the maximum required system capacity. A "lead" chiller is chosen, and is the first to operate. During light load conditions, the lead chiller does all the work, while the lag chiller sits idle. If cooling demand increases to a point where the lead chiller can no longer handle the load, then the lag chiller is brought on to operate as well. As demand falls, the lag chiller will drop off line, provided that the lead chiller can maintain the load.

"Alternation," as it applies to lead/lag systems, describes the procedure of promoting the lag unit to "lead unit," and demoting the lead unit. As with primary/backup systems, alternation is done to give both units equal run time, for pretty much the same reasons as were discussed in the previous section. Whether done manually or automatically, alternation should be implemented. If not, then the unit selected as the "lead unit" will always be in operation, for all levels of demand. The "lag unit," on the other hand, will only be pressed into operation when demand is heavy. Without rehashing the points made in the last section, suffice it to say that alternation of the lead unit is "just good practice."

Lead/lag is commonly utilized "internally" with packaged refrigeration equipment, such as chillers and condensing units. The compressors making up these types of equipment are normally capable of being manually or automatically alternated.

Chapter 7

INTERMISSION

Okay, let's take a breather here, and think about what we have talked about thus far, before we go any further. It is important at this point that an understanding of what has been covered has been established, and a familiarity with the various terms and terminologies has been achieved. The subsequent chapters will make reference to many of these terms without going into any great detail as to the concepts behind them. In other words, these next chapters "assume" that the reader has read and understands the content making up the previous chapters. Okay, so what have we covered?

Chapter 1 provided an introduction to mechanical systems and controls, and defined an HVAC system as a mechanical system plus the associated controls and control system required to operate it. To that end, the chapter attempted to define, in simple and concise terms and concepts, what a mechanical system is, and what is meant to control such a system.

Chapter 2 gave an overview of mechanical systems and of the various equipment making up these systems. The systems and equipment were broken into three categories. Airside Systems and Equipment covered those of which primarily deal with the movement of air, and the conditioning of it. Waterside Systems and Equipment covered those of which mainly deal with the movement and conditioning of water in HVAC applications. And Miscellaneous Systems and Equipment dealt with those types that were categorized as neither airside nor waterside. Hence a separate category for these systems and equipment.

Chapter 3 formally introduced us to controls, by providing the fundamentals of the various control methods in use today. Each section of this chapter, to an extent, built upon the previous section. Two-position control was discussed first, and then staged control. After that, we moved into proportional and floating control methods. This chapter laid the groundwork for the following chapters on actual control components.

Chapters 4 and 5 covered the "hardware" that is used to implement control schemes and build control systems. Chapter 4 discussed sensors and controllers, and took on a similar format as the previous chapter, at least in terms of defining and categorizing controller types. Chapter 5 discussed the various end devices that these controllers ultimately take control of. These two chapters round out the whole topic of controls; by now the reader should have a pretty good base knowledge of the topic, and thus should be well prepared to take on the subsequent chapters on mechanical systems, equipment, and control.

Chapter 6 covered some common control schemes that are used with popular mechanical systems. Each scheme of control was defined, so that we need only refer to its defining term when we discuss these schemes in subsequent chapters.

You with me so far? All right! Now let's take a quick look at where we're going! The rest of this book will follow the format outlined in Chapter 2; each of the following chapters will cover the mechanical systems and equipment originally defined back in that chapter, in the precise order that they were listed there. The first few chapters will deal with the airside, and will cover air handling systems and zoning equipment. The following chapters will handle the waterside, thus covering pumps, boilers, and chillers. The remaining chapters will tackle that "miscellaneous" category.

It is important to note that all discussions up to this point have been necessary, in order to comprehend the material in the upcoming chapters. A "prerequisite," if you will, to the rest of the book. It is these chapters to come that are the prime reason for writing this book, yet in order to understand the content, there needed to have been some groundwork laid. Furthermore, understand that the upcoming chapters are not necessarily meant to provide acute insight into the design concepts behind mechanical systems. Nor are they meant to provide any great amount of detail into the mechanical "inner-workings" of packaged equipment. These chapters are meant to provide a solid understanding of how these systems and equipment should operate, and how they can be controlled to meet that end, with methods that are "practical" and controls that are "available."

One last item needs to be mentioned before we venture ahead. It is important to note that the author of this book is a Chicago suburbanite. Why is that important? It is because this book is written based on HVAC system design that typically uses Chicago "design days" of negative ten and ninety degrees Fahrenheit, as the basis of "worst case conditions." Obviously not everything discussed in these upcoming chapters applies to all climates. For instance, safeguarding steam heating systems from coil freeze-up is likely not an issue in Southern California. Likewise, humidity concerns may be much different in Arizona than they are in the Midwest. In addition, how different regions define certain schemes of control may differ as well. Nevada may base the decision to economize strictly on outside air dry bulb temperature, and simply open the outside air damper fully. That may be their definition of economizer. Midwestern design however usually takes into account the humidity of the outside air, and thus bases the decision to economize on enthalpy (temperature and humidity). When economizer operation is initiated, the outside air damper must be limited in its function, so as not to bring in large amounts of outside air that is below the freezing point. The point is everything covered in the following chapters does not necessarily apply to all parts of the country. Yet, based upon the climate of the Midwest, which ranges from extreme cold to high heat and humidity, these chapters will provide an all-encompassing approach to HVAC systems and control in the USA.

Okay. Intermission is over. Time to grab your popcorn and your soft drink and "get back to the show!"

Chapter 8

ROOFTOP UNITS

Rooftop units are packaged air handling units. They are most often installed, you guessed it, on the roof! Many components make up the package. The three main operational functions, or sections, of a rooftop unit are its supply fan, its method of cooling, and its method of heating. All of these functions are integrated into a single piece of equipment.

The "heart" of the rooftop unit is the supply fan. The fan's function is to draw air through the rooftop unit's heating and cooling sections, and deliver it to its "final destination."

The heating section of the rooftop unit is typically a gas-fired furnace. Rooftop units with electric heating sections are available, for applications in which natural gas is either not available or not the most feasible option. Electric heat is in the form of electric resistance heating elements, which heat up and glow when energized.

REFRIGERATION CYCLE

Cooling is mainly accomplished by refrigeration (mechanical or direct expansion "DX" cooling), and the entire refrigeration cycle is within the unit. For those unfamiliar with what is meant by the term "**refrigeration cycle**," please refer to Figure 8-1, and consider this the "short version":

The refrigeration cycle consists minimally of a compressor, a condenser, an expansion device, and an evaporator. The compressor is the heart of the system, and is used to pump low-pressure refrigerant vapor from the evaporator and compress it to a higher pressure. This hot, high-pressure vapor is delivered to the condenser, where it rejects its heat to the ambient (outside) air. As the heat is removed from the refrigerant, it condenses, and turns to a high-temperature liquid. The refrigerant, now in its liquid

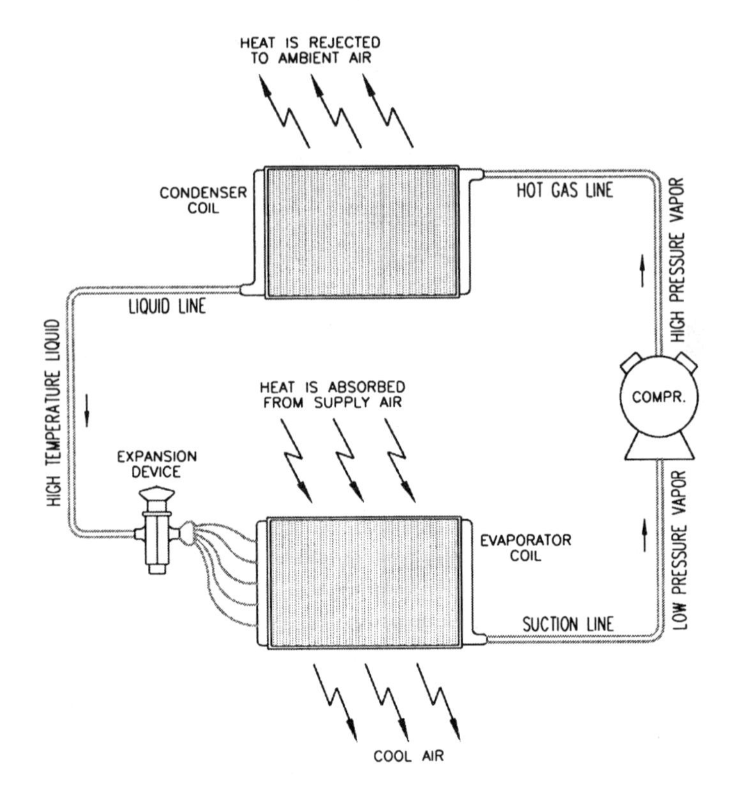

Figure 8-1. Rooftop unit refrigeration cycle.

phase, passes through an expansion device that substantially reduces the pressure and temperature of the liquid refrigerant. The cool, low-pressure liquid refrigerant is then delivered to the evaporator. In this application, the evaporator is a finned coil of which the rooftop unit's supply air passes over. As the air passes over the evaporator, it's cooled down. The refrigerant absorbs the heat from the air, and boils. The refrigerant, now a vapor again, is drawn to the compressor, and the cycle continues...

The smaller rooftop units, those of which are in the range of a few tons or so (one ton equals 12,000 Btuh), are normally single-stage units, meaning they have only one stage of heating and one compressor. The larger units most always have two (or more) stages of heating and cooling. Two-stage heating is accomplished by a two-stage gas valve, which assumes a partially open position upon a call for one stage, and opens fully if there is a call for both stages. Two-stage DX cooling is accomplished by staging compressors, meaning that a rooftop unit with two stages of DX cooling has two compressors.

ECONOMIZER

One of the most clever features of a typical, standard rooftop unit, and perhaps also one of the most misunderstood, is the economizer section (Figure 8-2). The **economizer section** of a rooftop unit, or for any air handler for that matter, is designed to utilize outside air as a source of "free cooling" if outside air conditions permit. The economizer (minimally) consists of the following components:

- Outside and return air dampers
- Damper actuator(s)—only one if the outside and return air dampers are mechanically linked

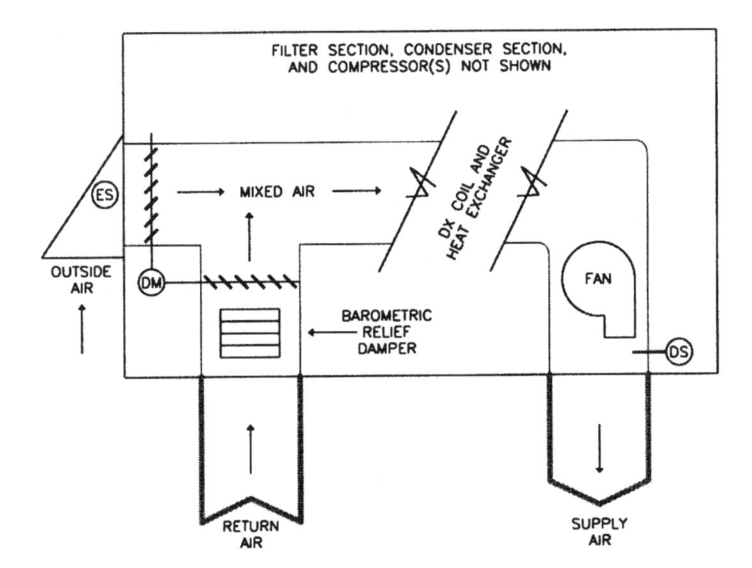

Figure 8-2. Diagram illustrating the economizer of a small packaged rooftop unit. Outside and return air dampers are operated by a single damper actuator. Enthalpy sensor and discharge air sensor are factory mounted and wired into the unit's integrated economizer controller. Relief damper allows air to escape during economizer cycles, when the return air damper can be partially to substantially closed.

- Outside air enthalpy (temperature & humidity) sensor or dry bulb (temperature only) sensor.
- Discharge air temperature sensor.
- Economizer controller

What follows is a sequence of operation for the economizer section of a typical commercial rooftop unit. This sequence can hold true for single zone systems, where temperature control is done via space temperature sensing, and for VAV and reheat systems as well, where temperature control is done via discharge air temperature sensing.

The outside air enthalpy or dry bulb sensor monitors outside air conditions. If outside air conditions are suitable for "free cooling" as determined by the sensor and the economizer controller "enable" setting, then economizer operation is allowed.

Upon a call for "first stage cooling," economizer operation is initiated. Outside and return air dampers begin to modulate (from minimum position) in order to maintain a suitable discharge air temperature (typically between 50 and 56 degrees). If necessary, the outside air damper will modulate fully open in an effort to achieve this temperature, though depending upon the outside air temperature (and the economizer controller "enable" setting) may not be able to. In other words, if the economizer controller is set to allow economizer operation at a "higher" outside air enthalpy or dry bulb temperature, then the temperature of the outside air may be higher than what is trying to be maintained at the discharge, and therefore the outside air damper will modulate fully open without succeeding to achieve this discharge air temperature setpoint. Depending upon the cooling requirements, this may or may not be sufficient to satisfy the call for "first stage cooling."

If economizer operation alone is not sufficient to satisfy the requirement for cooling, resulting in a call for "second stage cooling," then "first stage DX cooling" is engaged. The DX coil becomes active, and the outside and return air dampers continue to modulate to maintain a suitable discharge air temperature as sensed by the discharge air sensor, which is located downstream of the coil. With the DX coil active, the resulting action of the economizer dampers is that they modulate back down toward minimum position in order to continue to maintain the discharge air temperature setpoint (as sensed downstream of the DX coil).

As the requirement for cooling is satisfied, then DX cooling and economizer operation are disengaged in the reverse order that they were engaged.

Unless equipped with low ambient controls, the first stage of DX cooling of the rooftop unit, depending upon the climate, may need to be equipped with an optional outside air temperature "lockout" controller. This controller prohibits compressor operation if the outside air temperature is below the setpoint of the controller (typically 50 degrees). This means that, **if** outside air conditions are suitable for free cooling, **and** the outside air temperature is below 50 degrees, **then** economizer operation alone is permitted (DX cooling is disabled). This is acceptable, because if the outside air temperature is below 50 degrees, then the economizer should be able to maintain discharge air temperature setpoint, and therefore economizer operation should be sufficient to satisfy the cooling requirements.

If a rooftop unit does not have an economizer, then (depending upon the climate,) it should be equipped with low ambient controls, which would allow the compressor(s) to operate at low outside air temperatures. Otherwise, without an economizer section, and without low ambient controls, the rooftop unit has no means of providing cooling when the outside air temperature is below 50 degrees or so.

Outside Air Damper Minimum Position

While we are on the topic of economizer, we should take some time out to discuss "fresh air requirements." We have already made reference to the term "**minimum position**" as it relates to the outside air damper. But what is actually meant by it, and what dictates it? Minimum position, whether code-mandated or "just good sense," is the manner in which we allow a continuous, minimal amount of fresh outside air to be delivered into the spaces served by our rooftop unit (or any air handler). With today's code and IAQ (Indoor Air Quality) driven requirements for increasing amounts of outside air, the concept of minimum position has taken on substantial merit. The handling of any considerable amount of outside air via a packaged rooftop unit impacts the operational requirements of other components within the unit; the more outside air required, the more of an impact on overall system operation, the more of a need for complex control strategies, and the more of a potential for problems.

If a rooftop unit is to provide a 20 percent minimum of outside air, and the return air temperature is 72 degrees, then if the outside air temperature is –10 degrees, the temperature of the "mixed air" (mixture of outside and return air) is 55 degrees. This is fine for a VAV rooftop unit designed to deliver 55-degree air at all times, and is "borderline acceptable" for a single zone unit. However, with outside air requirements any larger than this, we will need to be able to heat or temper the air prior to delivering it. Before we skin this cat, let's first talk about a couple of methods of establishing and setting the outside air damper minimum position.

The first method of setting the minimum position is simply by manual means. The rooftop unit's control system will have provision for setting the value, either by a manually adjustable dial or "potentiometer," or, if the unit is equipped with microprocessor-based controls, through a user-definable parameter. The maximum that this value can be set is typically 50 percent, for anything over that approaches (or surpasses) the unit's practical operating limits.

The second method of setting the minimum position of the outside air damper is automatically, as a function of the "quality" of the air within the space served. How's that done? A carbon dioxide (CO_2) sensor/transmitter can be installed in the space and wired back to the rooftop unit (if the unit's control system has provision for it). The device will measure, in parts per million (ppm), the level of CO_2 in the space. The CO_2 level correlates directly with the number of people in the space, and can therefore be used to calculate the percentage of outside air that's actually required. The rooftop unit's control system can then modify the minimum position of the outside air damper in accordance with the CO_2 level. If the CO_2 level is low, then the damper will be at its absolute minimum position. As the CO_2 level in the space increases, the minimum position of the outside air damper is "reset" upwards. A maximum can be set, so that the damper never opens more than, say, 50 percent (except of course during economizer operation).

Well that makes a lot of sense now, doesn't it? Used to be where, in order to satisfy code and indoor air quality requirements, we would have to manually set the minimum position of the outside air damper for the worst case scenario of "maximum occupancy." At any other levels of occupancy, we would actually be bringing in more than the required amount of outside air, and would be expending energy on heating it (or cooling it, for that matter). Now, with a means of correlating the level of occupancy in a space to a measurable variable, we need only bring in the amount of outside air required to fulfill the needs of the occupants in the space at any given time. This is good news for designers and engineers, and is actually recognized by code and indoor air quality proponents as an acceptable means of satisfying fresh air requirements (conditions permitting).

With all this need for increasing amounts of fresh air comes the need to be able to heat the outside air. As we stated earlier, a 20 percent requirement of outside air on a (-10 degree) design day yields a mixed air temperature of around 55 degrees. For any value greater than 20 percent, the temperature of the mixed air will be less than 55 degrees, and will most definitely need to be tempered. With a rooftop unit, this isn't always an easy thing to do. Since rooftop units designed to fulfill single zone applications will normally have, at most, two stages of "space initiated" heating, if the space isn't "calling" for heating at any given time, the mixed air "is what it is." Unheated, untempered, and delivered to the space, with no regard for the temperature of it. Suffice it to say that there is a definite limit as to the amount of outside air a single zone rooftop unit can feasibly handle, at least on design days.

But wait a minute. What about all that talk about resetting the outside air damper minimum position as a function of the CO_2 level in the space? How can a rooftop unit operate to counter that? Well, it can't, unless the rooftop unit is equipped with the means of tempering the air, or in other words heating the air to an acceptable discharge air temperature level. For smaller, single zone packaged units with staged heating, this is simply not (very) feasible. And for larger, VAV style units, it cannot be done without the proper method of heating (preferably proportional) and the required accompanying control system. We will talk more about this issue in the section on VAV systems. Simply be it known for now that a packaged VAV rooftop unit, one that is designed to maintain discharge air temperature at 55 degrees or so, would have to "heat" to setpoint when mixed air temperatures are below this point. For a "packaged" piece of equipment, this isn't necessarily automatically "do-able," as will become evident upon reaching that section of this chapter.

Exhaust

One last component that a rooftop unit may be equipped with, that's worth discussing in this section, is a **power exhauster**. For small rooftop units, the amount of outside air that's ever introduced by the unit, even during the economizer cycle, may be insignificant in regard to space pressure. The unit will minimally be equipped with a barometric relief damper. If the unit is bringing in a fair amount of outside air, as during economizer operation or because of a large minimum fresh air requirement, the pressure in the space, and hence the pressure in the return air duct leading back to the unit, will tend to increase. The barometric relief damper, installed in the unit's return air section prior to the return air damper, simply provides a means for the excess air to escape the system.

With larger rooftop units, it may be a requirement that this excess air be actively removed from the system, via an optional power exhauster. This is primarily an exhaust fan that energizes when the amount of outside air being brought in by the rooftop unit exceeds some value. This is simply determined by the position of the outside air damper, and not by physically sensing the amount of outside air passing through the damper. For more precise control, a few different options exist, each taking into consideration space static pressure. For each of these methods, a static pressure controller monitors the pressure in the space. A setpoint is established via the controller, typically in the area of .1" W.C. (inches of water column). With each method, steps are taken to counteract increases in space static pressure above setpoint. Space static pressure control of a rooftop unit exhaust fan (in some form or other) is typically standard with VAV rooftop units.

The first method is to have not one exhaust fan, but several fans. The strategy is straightforward staged control; as the space pressure increases, turn more fans on.

The second method entails a single exhaust fan, and a motorized exhaust air damper. The strategy consists of monitoring the outside air damper's position, turning on the fan if a certain point has been exceeded, and then proportionally controlling the exhaust air damper in order to maintain space static pressure setpoint. While this is a fairly precise means of controlling a power exhauster to maintain space static pressure, it may not necessarily be the most "energy conscious," especially when you think of a large exhaust fan blowing against a partially opened damper.

The third method consists of using a variable frequency drive (VFD) to control the speed of the exhaust fan, as a function of space static pressure. This is generally the most precise method of space static pressure control, and the most energy conscious, yet the first cost associated with the VFD may be prohibitive in certain applications.

Now that we've established a basic understanding of the packaged rooftop unit and its components, let us venture ahead and take a look at some of the more typical rooftop unit applications. Systems utilizing rooftop units can be divided into four categories: **Single Zone Systems, Reheat Systems, RTU Zoning Systems**, and **VAV Systems**. How the rooftop unit is controlled, and how the unit should be configured from the factory, is discussed for each type of system. Bear in mind that for each of these four types of systems, the packaged commercial rooftop unit is utilized differently! This is an important point to remember when selecting and specifying a packaged rooftop unit for a given application.

Single Zone Systems

We start this section by formally defining a term that up till now we've touched upon quite a bit yet have not fully addressed. The term **zone**, as it applies in the HVAC industry, is an area of temperature control. A temperature controller or thermostat within a zone can command the heating and cooling equipment serving that zone to be in a heating or cooling mode (or neither). There is only one temperature controller or thermostat per zone, this being perhaps the most telltale evidence of a zone and its existence. A space may be controlled as a single zone, by a single piece of heating/cooling equipment. Or it may be broken down into multiple zones, with each zone having its own rooftop unit.

In its simplest form, if a rooftop unit serves one zone and is controlled by a thermostat within that zone, then it is a single zone system (Figure 8-3). Single zone systems usually consist of rooftop units 30 tons and under, for loads any greater than that are usually broken down into smaller zones.

Although packaged rooftop units are available for multiple zone applications, they were originally developed strictly for single zone applications. A single or multistage heating-cooling thermostat is wired back to the rooftop unit, to control heating, cooling, and fan operation. In this day and age, some rooftop unit manufacturers are opting to go with electronic space temperature controllers that go hand-in-hand with their microprocessor-based rooftop unit control systems. Instead of the traditional thermostat that utilizes temperature switches and contact closures to control heating, cooling, and fan operation, these controllers

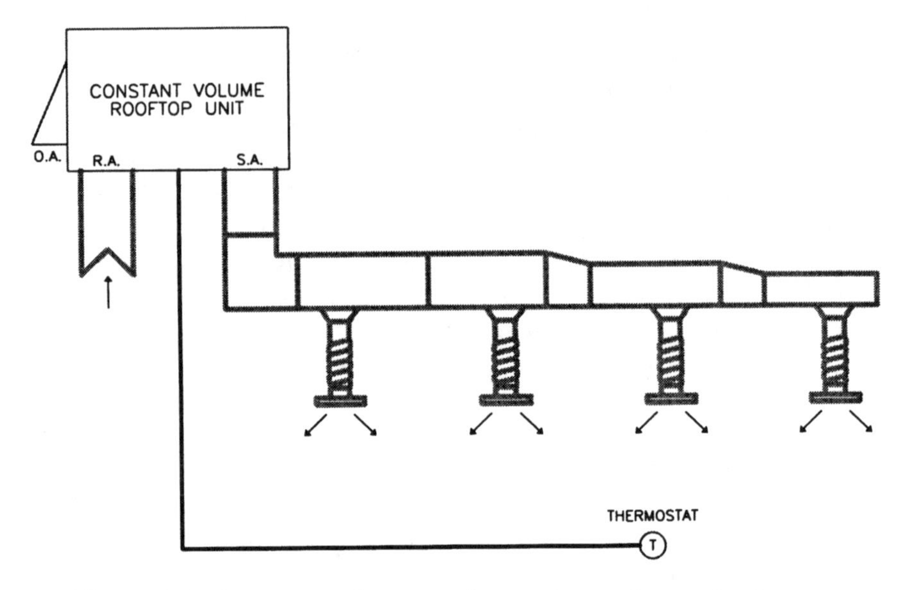

Figure 8-3. Constant volume rooftop unit serving a single zone.

send electronic signals back to the rooftop unit's microprocessor. The signals are processed, and the unit is placed into the appropriate mode of operation. The means of achieving unit operation is different from the conventional thermostat method of control, however, the result is the same. Furthermore, this is not the trend with all manufacturers, nor is it the case for any particular manufacturer's entire offering of rooftop units. The conventional means of controlling a rooftop unit is still very much in use today, in part because it allows the consumer to choose among many of the various brands, types, and styles of thermostats offered, to control their rooftop equipment.

So a single zone rooftop unit system consists of a single zone, constant volume rooftop unit, and a thermostat in the space served by the unit. What else, if anything, is required for system operation? Nothing! That's it! Oh sure, there may be the requirement for a duct-mounted smoke detector or two, to shut the unit down upon detection of smoke in the duct. But other than that, the whole control system design "scope" of a single zone rooftop system is the selection of the thermostat.

Thermostat selection basically consists of two parts: choosing the proper operational functions, such as number of stages, fan and system switches, programmability, etc., and choosing the style of thermostat.

Choosing the number of stages is relatively straightforward, and can be determined by taking a look at the rooftop unit's specifications. One point worth noting here, has to do with small, single compressor units with economizers. Even though this type of unit technically has only one stage of mechanical cooling, a thermostat with two stages of cooling is required, because of the economizer section. Recall previously the discussion on economizer operation. If the outside air is suitable for free cooling, then a call for "first stage cooling" is economizer, and a call for "second stage cooling" is first stage DX cooling. For a single zone system, this means that the thermostat must be able to call for two stages of cooling.

Consider a system that serves one large zone and perhaps several smaller zones, with the large zone containing the thermostat controlling the rooftop unit, and the smaller zones having zone dampers or some form of reheat control. Although technically there are several zones being served by the rooftop unit, this system is still classified as a single zone system, at least from a controls standpoint. The rooftop unit operates via its thermostat, with no regard as to what the "subzones" require, in terms of heating and cooling. The subzones are, in essence, at the mercy of the rooftop unit. The sections on zone dampers and reheat coils will explore further the concept of "subzoning."

Sequence of Operation for Single Zone Rooftop Unit

The following is a sequence of operation of a typical small commercial packaged rooftop unit, controlled by a programmable thermostat:

A programmable thermostat controls the operation of the rooftop unit. Desired occupied and unoccupied heating and cooling setpoints are programmed via the thermostat. Fan operation is determined by the position of the fan "ON-AUTO" switch, as well as by the mode of occupancy. Heating and cooling operation as described herein assumes that the system "HEAT-AUTO-COOL-OFF" switch is in the "AUTO" position.

Occupied Mode

Programmable thermostat signifies occupied mode. Supply fan either runs continuously, or cycles upon a call for heating or cooling, as determined by the position of the fan switch. Outside air damper opens to its minimum position (20 percent adj.) when the supply fan is in operation.

Upon a rise in space temperature above the occupied cooling setpoint, rooftop unit cooling is initiated:
If the outside air is not suitable for free cooling as determined by the outside air dry bulb temperature (or enthalpy) controller, then outside air damper remains at minimum position, and first stage DX cooling is engaged. Upon a further rise in temperature above the occupied cooling setpoint, then second stage DX cooling is engaged (if the rooftop unit is equipped with two compressors). As the space temperature falls back toward setpoint, then the cooling stages are disengaged in the reverse order that they were engaged.

If the outside air is suitable for free cooling as determined by the outside air dry bulb temperature (or enthalpy) controller, then outside and return air dampers modulate to maintain a suitable discharge air temperature (typically 55 degrees). If the outside air damper modulates fully open and economizer operation cannot satisfy the cooling needs of the space, resulting in a call for second stage cooling by the programmable thermostat, then first stage DX cooling is engaged as well. As the space temperature falls back toward setpoint, then DX cooling and economizer operation are disengaged in the reverse order that they were engaged.

Upon a drop in space temperature below the occupied heating setpoint, rooftop unit heating is initiated:
Outside air damper remains at minimum position, and first stage gas-fired heating is engaged. Upon a further drop in temperature below the occupied heating setpoint, then second stage gas-fired heating is engaged (if the rooftop unit is equipped with a two-stage gas valve). As the space temperature rises back toward setpoint, then the heating stages are disengaged in the reverse order that they were engaged.

Unoccupied Mode
Programmable thermostat signifies unoccupied mode. Supply fan cycles upon a call for heating or cooling. Outside air damper opens to minimum position upon a call for heating or cooling. Unoccupied heating setpoint is lower than occupied heating setpoint (60 degrees adj.). Unoccupied cooling setpoint is higher than occupied setpoint (80 degrees adj.). Rooftop

unit heating and cooling operate as described above in Occupied Mode, except as to maintain unoccupied heating and cooling setpoints.

Upon reading this squence of operation, a few things should be pointed out. The first has to do with fan operation. For most commercial applications, the supply fan is supposed to run continuously in all occupied modes. This is an indoor air quality issue and is generally code-mandated, the issue being that the space served by the rooftop unit needs to be continuously supplied with fresh air, in the form of outside air, as introduced via the outside air damper.

The term "**cycling**," as it applies to the supply fan, simply means "turning on and off." So that when the operation of the supply fan is described as "cycling upon calls for heating or cooling," it means that the fan turns on upon a call for heating or cooling, and turns off when the call for heating or cooling has been satisfied. This is also referred to as "**intermittent**" fan operation.

A point that is interesting to note, that may not be common knowledge to the average HVAC "layman," has to do with rooftop unit fan cycling. The thermostat, in its traditional form, is designed for use with packaged HVAC equipment with gas-fired heating. How the thermostat traditionally controls the supply fan, during intermittent operation, requires some explanation. If a thermostat is set for intermittent fan operation, or if the thermostat (programmable) is simply in the unoccupied mode, the supply fan cycles upon calls for heating and cooling. If there is a call for cooling, the thermostat will turn on the fan and engage the first stage of cooling at the same time. When the call for cooling is satisfied, the supply fan and the cooling are then disengaged by the thermostat. If there is a call for heating, the thermostat will engage the first stage of rooftop unit heating, *but will not turn on the fan*. The rooftop unit's gas-fired heat is activated, and its heat exchanger begins to heat up. A "fan cycling" temperature switch, integral to the gas-fired unit, makes when the heat exchanger heats up to a certain temperature, and then the supply fan is activated. When the call for heating is satisfied, heating is disengaged by the thermostat, yet the supply continues to run. The fan remains in operation, until the residual heat is dissipated from the heat exchanger, and the fan cycling temperature switch breaks.

This is fairly typical with most thermostats and most packaged HVAC equipment with gas-fired heat. When dealing with rooftop units with electric heat, or perhaps when trying to apply a thermostat to a non-conventional system, this issue must be addressed at the thermostat, either with the thermostat itself, or with some supplemental controls components. Many of the commercial programmable thermostats on the market today allow for the installer to set the thermostat for "automatic fan operation on calls for heating."

The last point to note here has to do with the minimum position of the outside air damper. The position of the outside air damper is dictated by fresh air requirements, whether it be code-mandated, or simply "good practice." The adjustment of this parameter is via the rooftop unit's control system, whether it's done at the damper motor itself, or elsewhere in the unit's control system. The purpose of establishing an outside air damper minimum position, of course, is to continuously provide some fresh air to the occupants of the space being served by the rooftop unit, during occupied modes.

For a *typical* commercial rooftop unit, the outside air damper opens to minimum position when the supply fan is commanded to run, and closes when the supply fan is commanded to turn off. It has been explained that, if the unit is controlled by a programmable thermostat and is in the unoccupied mode, as dictated by the thermostat's time clock, the supply fan operates intermittently, cycling upon calls for unoccupied heating or cooling. If there is a call for, say, heating, the supply turns on, and unoccupied heating is engaged. At the same time the supply fan starts, the outside air damper opens to its minimum position. Yet this is the "unoccupied" mode. Theoretically, there are no occupants in the space; the unit is coming on just to prevent the space from becoming too cold. Why then, would the outside air damper need to open? Why use the energy to heat up that outside air component of the supply air, if there is no requirement for outside air? Especially if our outside air damper minimum position is 30 percent, and it's negative ten degrees out there! Sounds like a waste of energy. Why not keep the outside air damper closed during the unoccupied mode? That doesn't sound like too much to ask for, and it sure makes a lot of sense.

While with fan coil systems and other types of built up systems, this design criteria is often implemented, with conventional packaged rooftop units (those without advanced microprocessor-based control systems), it just isn't all that feasible. When using a programmable thermostat to control a packaged rooftop unit, all we are really doing is modifying setpoints and fan mode operation when we transition from occupied mode to unoccupied mode. The rooftop unit's control system does not know what "mode" the thermostat is in. All it knows of is what its heating and cooling setpoints are, and how its fan should operate. In other words, the programmable thermostat is telling the unit this information, without actu ally letting the unit know explicitly what mode the thermostat is in (occupied or unoccupied). How then, can we have the outside air damper remain closed in the unoccupied mode, if the rooftop unit's control system knows nothing about which mode the thermostat is in?

There is a way around this, yet it's probably more trouble than anyone ever cares to go through. By conventional rooftop unit/thermostat connections, it's true that the rooftop unit knows nothing about the thermostat's mode. Many programmable thermostats, however, offer a time-of-day output. This normally takes the form of a relay output, which changes state upon thermostat time clock transition from mode to mode. This relay output, hence, provides us with information as to which mode the thermostat is in; occupied or unoccupied. We can convey this information to the rooftop unit, via an extra pair of wires, and feasibly break into the outside air damper control circuit with this thermostat relay output. In other words, during occupied modes, the time-of-day relay contact that we would utilize from the thermostat would be closed, and the damper actuator control circuit would be intact. Upon transition to the unoccupied mode, this contact would open, and thus disable the damper actuator control circuit. The damper would spring shut, and would remain closed for the entire unoccupied period, regardless of any calls for unoccupied mode heating or cooling.

With all that has been covered here, suffice it to say that there are quite a bit of application considerations to be aware of, even with such a seemingly simple topic as this. Read on, for it only gets better (or worse?)...

Reheat Systems

Figure 8-4 shows a typical rooftop unit reheat system. Reheat systems are constant volume, multiple zone systems. In a reheat system, the air handler normally operates to maintain a constant discharge air temperature setpoint of around 55 degrees (there is no space thermostat controlling the unit). The air handler serves not one zone but many zones, with each zone having some form of "reheat" in its distribution ductwork. Each zone has a space temperature controller that controls its respective reheat device. It is in this manner that any zone can receive either cooling or heating as required by the zone. If a zone's space temperature is above setpoint, then no amount of reheat is activated, and cool air is delivered into the zone. As the zone's space temperature drops, the cool air is then reheated as required to raise the space temperature to setpoint. Although this type of system isn't very energy efficient (or in some cases even legal!), it does provide for good multizone comfort control. Regarding its legality, you would have to talk to an engineer, one familiar with the mechanical codes governing the design and installation of energy-based systems, as to its acceptability in particular applications.

Application Considerations

As we stated in the above paragraph, a reheat system is a constant volume, multiple zone system. When a rooftop unit is selected as the air handler of choice for the given reheat system application, caution must

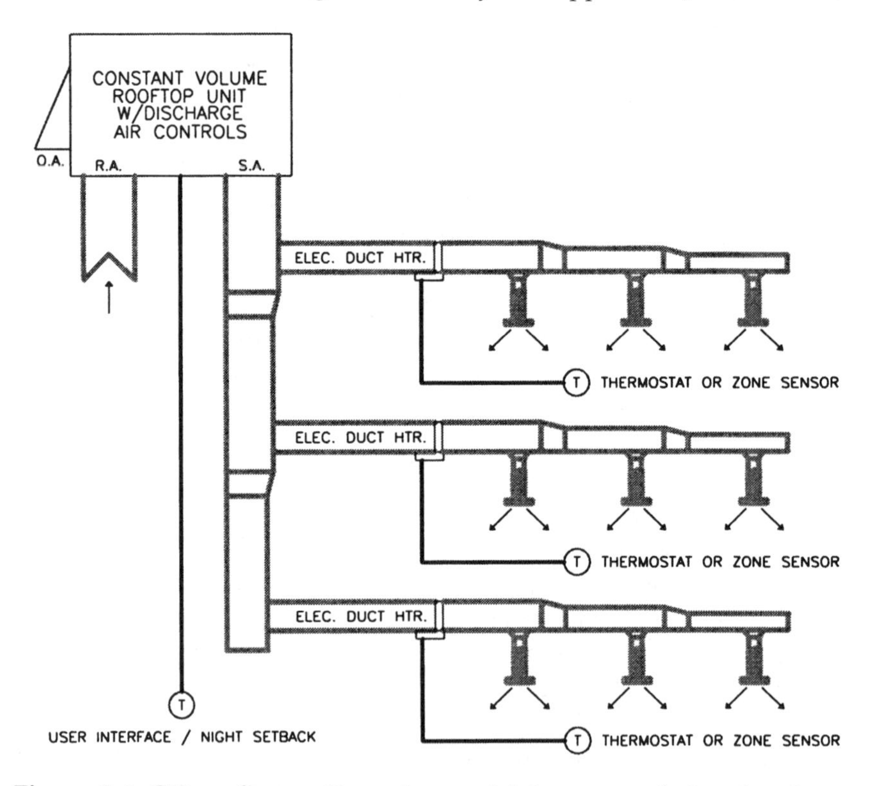

Figure 8-4. CV rooftop unit serving multiple zones of electric reheat.

be exercised in selecting the unit. The rooftop unit selected for this application must be a constant volume unit, and also one that is designed to operate to maintain discharge air temperature setpoint. The term "constant volume," in many manufacturers' utilization of the term, implies a single zone rooftop unit, one that is controlled by a thermostat in the space. Yet our requirement for this type of system is for the unit to operate via discharge air controls. If we simply convey to the rooftop unit manufacturer that our requirement is for a constant volume unit, we may not get what we really need for this type of system. We need also to convey our requirement for control of the unit based on discharge air temperature, and not space temperature. The manufacturer should be able to provide us with the correct rooftop unit configuration. In fact, the simplest way to specify this may be to ask for what the manufacturer calls a VAV unit, yet request that it come with no form of "air quantity control." In manufacturer's lingo, a VAV rooftop unit is one that is applied to VAV systems, one that operates to maintain constant discharge air temperature, and has some form of supply fan air quantity control. Air quantity control can come in the form of inlet guide vanes (IGVs), which vary the amount of air that the supply fan can physically deliver, or a variable frequency drive (VFD), which directly varies fan speed. If the manufacturer can offer his VAV rooftop unit with neither IGVs nor a VFD, then the specified unit will perfectly suit our reheat system application. At the very least, the unit can be ordered with the IGVs, and the IGV actuator can be disabled upon installation.

"Why can't I just buy the manufacturer's single zone, constant volume rooftop unit, and throw the thermostat in the supply air duct?" The question is not an uncommon one, and isn't really all that out of line. Can't we control a unit that is designed to be operated by a space thermostat by putting the thermostat in the discharge? And if not, why? Well, the answer to the first question is simply "no." A rooftop unit that is designed to operate via space temperature sensing is normally equipped with a maximum of two stages of cooling. The unit is designed to operate to take advantage of the "thermodynamic lag" inherent in space temperature control. It takes a finite amount of time for the space thermostat to sense the result of its call for cooling, let's say. The thermostat calls for cooling, and the rooftop unit responds by activating first stage cooling. The thermostat does not immediately feels the effects of the rooftop unit's cooling mode. In time, the space will either drop in temperature, as a result of the call for first stage cooling, or will rise in temperature, the result of an increased load in the space. In either case, the space temperature is not changing rapidly, and hence the thermostat does not instantaneously register the changes. The result is that control from the thermostat is slow-acting, with the thermostat engaging and disengaging stages of cooling over a lengthy period of time.

Now consider if we took the space thermostat, set it for 55 degrees, and threw it in the discharge air duct (of a single-stage unit). As soon as we placed the unit into operation, the thermostat would call for cooling. The unit would enter the cooling mode, and the discharge air temperature would begin to decrease toward setpoint. Once it dropped below 55, the thermostat would disengage cooling. As soon as cooling is disengaged, the discharge air temperature begins to rise. It only takes moments for the discharge air temperature to rise back above the cooling setpoint, at which point cooling would be re-engaged by the thermostat. This cycle theoretically repeats itself indefinitely, with the compressor turning on and off and on and off, as the means of maintaining setpoint. As compressors don't like to be cycled rapidly, there are most likely compressor anti short-cycling controls built into the rooftop unit. So that what realistically happens is, the thermostat calls for cooling, the compressor starts, the discharge air temperature drops below thermostat setpoint, the compressor turns off, and is not allowed to come back on for a minimum amount of time as dictated by the compressor's anti short-cycle timer. The discharge air temperature drifts upward, until the timer times out and the compressor is allowed to operate again. The compressor starts, and the temperature is pulled back down to setpoint, at which point the compressor turns off again. The discharge air temperature resulting from this type of cycling basically swings from 55 degrees up to 70 or so degrees, back down to 55 and back up to 70, and so on, and so on, and so on...

The above scenario plays out similarly with a two-stage unit and thermostat, though the results are not as exaggerated. Still, this is not the intended method of operation for a single zone, constant volume unit. On the other hand, a rooftop unit specifically designed to operate via discharge air temperature sensing, for one thing, will almost always have more than two stages of cooling. In addition, the compressors themselves may be equipped with unloaders, which allow the compressors to operate at partial capacities. A specialized control system, based on discharge air temperature control, is also built into the unit, and utilizes special control algorithms, interstage differentials, and discharge air sensing and anticipation (i.e., how fast am I approaching setpoint, and what course of action should be taken?). These features, as well as other control schemes, ensure that a proper balance is maintained between smooth operation and tight temperature control.

Is there a way to feasibly operate and control a rooftop unit, one that is designed as a single zone constant volume unit, in a reheat application? Well, there is an alternative, albeit not a very appealing one. We could take our single zone unit, and control it from a return air temperature controller. The return air temperature, like the space temperature, is subject to thermodynamic lag; it takes a finite amount of time to register a change in the return air temperature, upon commanding a change in the mode of operation of the rooftop unit. In addition, the return air is the sum total (minus any exhaust) of the air being supplied to all of the reheat zones, and is *theoretically* a good indicator of the average temperature needs of the zones. The return air temperature controller can be set at 74 degrees. If the air coming back from the zones rises above 74, the controller will call for first stage cooling. This of course causes the supply air temperature to drop, which is the result that we are looking for. If the return air temperature continues to rise, the controller will call for second stage cooling, thus dropping the supply air temperature even lower. As zone cooling needs are satisfied, the return air temperature will drop, and the controller will disengage the stages of cooling in the reverse order that they were engaged.

While this seems to be a pretty reasonable method of control, in taking a unit that is designed to be controlled from a space thermostat, and controlling it with some other method, potential problems exist. Excessive compressor "on" times, coil frost-up, varying supply air temperatures, excessive energy usage and equipment wear, and unstable operation, are just a few of the problems that can arise. Extra design time may also be necessary in order to "redesign" the system controls in order to minimize any possibility of potential problems. Again, if at all possible, when considering a rooftop unit reheat application, choose a rooftop unit that is designed to operate to maintain discharge air temperature. Oh, yeah, and no air quantity control!

Time-of-Day Control

What else is there to know about or be aware of, in regard to the operation and control of a rooftop unit in a reheat application? What about time-of-day operation and night setback/morning warm-up control? Without a programmable thermostat, how do we establish occupied and unoccupied modes of operation for the system?

First, if the system is not to be in continuous operation, there needs to be a time clock function for scheduling the unit. For a rooftop unit designed to be applied to this type of system, there will not necessarily be a programmable thermostat associated with it, at least not in terms that we've become familiar with. However, most manufacturers will offer, as part of the package, a space mounted "human interface module." This device is really not much more than a time clock and a built-in temperature sensor. Time-of-day operation is set via this controller, as well as occupied discharge air temperature and unoccupied space temperature setpoints. The sensor is utilized as the means of employing night setback and morning warm-up cycles. Space temperature reset control of discharge air setpoint may also be employed, if available as a configurable parameter through the human interface.

If the manufacturer does not offer such a device, at the very least there will be a means of starting and stopping the rooftop unit, perhaps by an input specifically labeled as "time clock input." This being the case, a field supplied time clock must be installed and wired back to the rooftop unit. A "night setback thermostat" must also be field supplied and installed, if night setback (and/or morning warm-up) operation is required.

Freezestat Requirement

For hot water reheat systems, a freezestat should be installed at the discharge of the rooftop unit, and wired back to the unit. The freezestat shuts down the unit, in order to protect the downstream hot water coils, if the discharge air temperature drops below its setpoint.

RTU Zoning Systems

It is not the purpose of this section to thoroughly explore the application and operation of rooftop unit zoning systems (the chapter on this will serve that purpose). Rather, the focus of this section is on the rooftop unit itself, and its requirements in regard to controls.

As we stated in the brief overview on RTU zoning systems back in Chapter 2, an RTU zoning system is a system of components designed for operation with a single zone, constant volume rooftop unit. The rooftop unit need not be anything special. A rooftop unit designed to operate from a space thermostat is the type of rooftop unit that is to be utilized in this type of application. The zoning system control panel replaces the thermostat, and is wired to the rooftop unit terminals. The system control panel, therefore, has complete control of the rooftop unit, just as a thermostat would.

That's it! For a rooftop unit zoning system, simply specify a single zone, constant volume rooftop unit, and you're in business! RTU zoning system requirements, guidelines, and rules of thumb will be thoroughly discussed in the chapter devoted to this topic.

VAV Systems

VAV (Variable Air Volume) systems are variable volume, multiple zone systems. As with a reheat system, the VAV air handler normally operates to maintain a constant discharge air temperature setpoint of around 55 degrees. Unlike with a reheat system, the system supply fan does not operate at constant volume, but rather at a variable volume, as dictated (indirectly) by the cooling requirements of the system. A rooftop unit serving a VAV system is illustrated in Figure 8-5.

As with the previous section, the purpose here is not to discuss VAV systems as a whole, but to focus on the rooftop unit and its requirements. Yet, in order to better understand the role of the rooftop unit in a VAV system, we must at least briefly discuss "what's going on" out at the "terminal units" (VAV and/or fan powered boxes).

Common to all VAV and fan powered boxes is a damper that can regulate the flow of cool air from the "main trunk" into the respective zones served by the boxes. A space temperature sensor (one per zone/box) is wired up to the VAV box controller; as the zone temperature increases above setpoint, the controller modulates the damper in the box open, and as the zone temperature decreases toward setpoint, the controller modulates the damper closed, toward its minimum position. Now, as the cooling needs of the zones are satisfied and dampers close off, the pressure in the main trunk will tend to increase. There must be a way to alleviate the pressure in the main trunk, in order to maintain the system at a safe and efficient level of operation.

A rooftop unit designed as a VAV unit will have the means of alleviating the pressure in the main trunk, as terminal unit dampers close

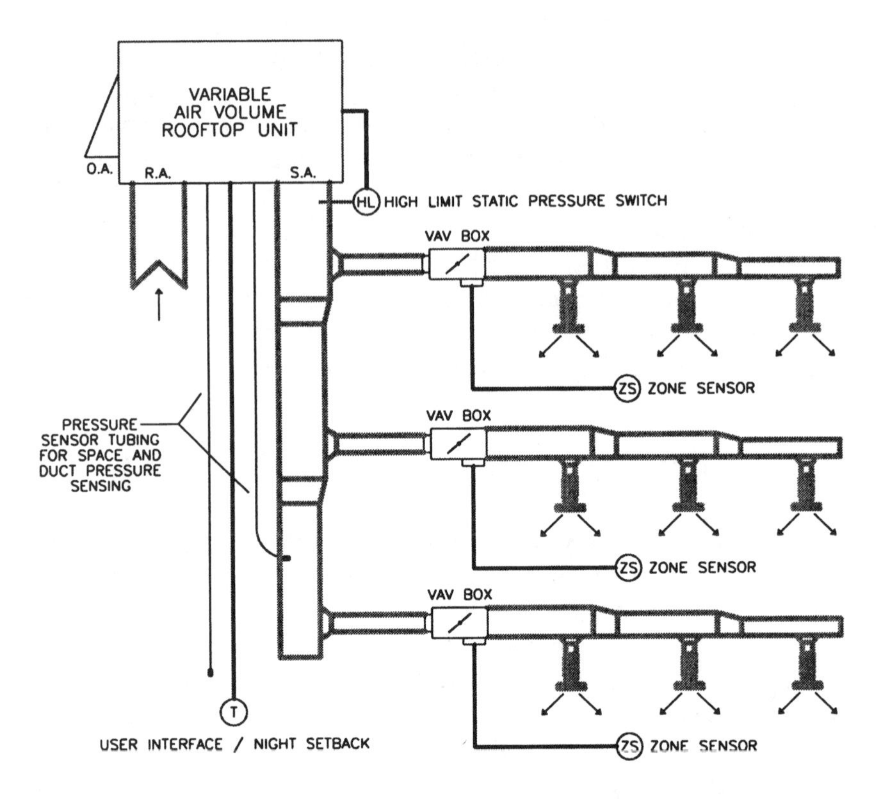

Figure 8-5. Variable Air Volume rooftop unit serving a VAV system.

off. This is most always in the form of inlet guide vanes (IGVs) or a variable frequency drive (VFD). Whereas IGVs are the traditional method of air quantity control, VFDs have come into their own as a viable alternative to IGVs. With either method, static pressure in the main trunk is measured, a setpoint is established via the rooftop unit's integral duct static pressure controller, and the method is employed in order to achieve and maintain setpoint (1.5" W.C. typical). Note that the rooftop unit will have an "integral" factory wired pressure controller, yet tubing must be run from the unit's controller to a point in the duct system, typically halfway downstream (or further) from the unit, in the main trunkline.

IGVs are simply dampers on the inlet side of the supply fan, operated by a damper actuator. As pressure builds in the main duct, the actuator modulates the dampers closed, thus reducing the quantity of air that the supply fan can deliver. The VFD is a more energy conscious alternative, albeit a more expensive one, in terms of first cost. The VFD directly controls the speed of the supply fan motor, in response to the signal received by the pressure controller. Either method of duct static pressure control is acceptable by today's standards, yet with the falling cost and rising reliability of VFDs, this is tending to become the more popular option, especially when energy savings and "payback" are driving forces behind the decision making process.

So we've defined the basic requirements of the VAV rooftop unit. The unit must be designed to operate via discharge air temperature control, and also must have some means of duct static pressure control. These requirements come automatically, as part of the package, when you specify a rooftop unit for a VAV application. In addition, a couple of field installed devices may be required as well, for acceptable operation of the VAV system. Maybe "protection" is a better term (in lieu of "acceptable operation"), for the two devices that come to mind are safety devices.

Like with the reheat system, if the VAV system consists of terminal units equipped with reheat coils, and these coils are hot water coils, then a freezestat should be installed at the discharge of the rooftop unit, and wired back to the unit. The freezestat is necessary to protect the downstream hot water coils, and is definitely worth the cost and the labor required to install it.

A high limit static pressure switch must be installed in the main trunk and wired back to the rooftop unit. The purpose of this switch is to shut down the unit if the pressure in the main duct becomes excessive. Excessive pressure can be caused by a failure of the duct pressure control system, as illustrated by the IGVs or VFD allowing the supply fan to "run wild," with no regard to duct static pressure.

In the introduction to this chapter, we talked a little bit about fresh air requirements and the accompanying requirements for heating. We "cracked the egg," so to speak, and now it's time to "break it open."

Fresh Air Requirements and Considerations

In addition to the CO_2 sensing method, another method of establishing the "minimum position" of the outside air damper, that we did not discuss back then, is unique to VAV systems. It is the process of maintaining the actual quantity of outside air introduced via the outside air intake "at a constant value," regardless of what the supply fan is delivering. For example, if the rooftop unit maximum deliverable cfm is

10,000, and the "code-mandated" outside air requirement is 20 percent of this, then the outside air brought into the system will be 2,000 cfm, always. If the unit is operating at full capacity, then the percentage of outside air is 20 percent of what the rooftop unit is delivering to the terminal units. If the unit is operating at half capacity, however, the percentage of outside air is 40 percent of what the system is delivering to the units. On a design day, this mixture of 40 percent outside air and 60 percent return air, untempered, can be less than 40 degrees!

The above method of meeting fresh air mandates is something that packaged rooftop unit manufacturers can typically offer as an option. This is accomplished, via the factory furnished control system, by monitoring the outside air cfm value, and controlling the outside air damper to maintain the prescribed value. As far as meeting code, this method does it, no doubt about that! Yet maintaining a constant amount of outside air with no regard to actual occupancy requirements likely results in periods of "too much fresh air." In other words, providing a predetermined fixed amount of outside air when building occupancy is less than maximum not only meets, but exceeds code requirements, and excessive energy will be used in conditioning this component of the delivered air.

Anyway, with that said and done, we now can talk more in depth about the need for heating when outside air requirements are high (more than 20 percent of the delivered air). Whatever method is employed of establishing the minimum amount of fresh air, we must be able to heat the mixed air up to setpoint, in addition to being able to cool it down to setpoint. Traditionally, the heating section of a VAV air handling unit is there for night setback and morning warm-up cycles only, and is typically "all or nothing." Only with the advent of these increasing requirements of outside air have we begun to look upon the heating section of a VAV air handler to provide mixed air "tempering." It is not wise to blindly assume that a given rooftop unit's factory installed heating section can perform this function. The heat would, first and foremost, have to be in the form of proportional heat or at the very least, staged heat with a good amount of staging. Second, the rooftop unit's control system must be able to recognize the need for mixed air tempering, in order to provide. This would need to be in the form of a mixed air sensor or "changeover" temperature controller. If these two requirements are not met, then the unit's heating section cannot and will not provide mixed air tempering.

The rooftop unit manufacturer can offer, theoretically, proportional gas-fired heat or staged electric heat, for supply air tempering purposes. Proportional gas-fired heat comes in the form of a modulating gas valve. The valve, however, is not allowed to modulate fully closed, so you do not get "full modulation." "Turndown ratio" is a term describing just how much modulation you get from a gas-fired heating appliance, and is worth defining at this point, by way of example. A turndown ratio of 3:1 means that the control valve can modulate down to provide "1 over 3", or 1/3 (33 percent) the maximum capacity. Likewise, a turndown ratio of 8:1 means that the valve can modulate down to "1 over 8", or 1/8 (12.5 percent) the maximum capacity. The greater the turndown ratio, the better the chances that the heating section can effectively temper the mixed air under all circumstances. Likewise with electric heat. In other words, the more stages of electric heat, the better suited the heater is for mixed air tempering.

If the rooftop unit selected for a given application "falls short" in terms of the above requirements, other methods of heating the outside air component of the mixed air must be explored. The outside air can be tempered directly, before it reaches the unit, or the supply air can be tempered, after it leaves the unit. Each method has its issues, and either method must be suited to the application, in terms of sizing, operation, and integration with the rooftop unit's control system.

Contrary to popular belief, packaged rooftop units do have their limitations, as far as applications go. And while you get a lot of nice features with the package, it is imperative that the equipment specifier know the "ins and outs" of the application at hand, so that the package is not misapplied to a mechanical system that doesn't quite know what to do with it...

Chapter 9

Make-up Air Units

Make-up air units replace, or "make up" the air that's removed from a space by some exhaust system. Typically designed and built as heating units, they can also incorporate cooling capabilities, if the application requires it. These are systems primarily designed to operate via discharge air temperature control; as the air is removed from the space by the exhaust system(s), tempered "make-up air" is introduced into the space, typically at space temperature. For example, if the make-up air unit is serving a warehouse, and the desired temperature within the warehouse is to be no less than 60 degrees, then the make-up air unit operates, whenever the exhaust system is in operation, to provide 60-degree air (of the same quantity that's being exhausted) into the space.

In addition to discharge air temperature control, space temperature "reset" or "override" of discharge air temperature setpoint is typically implemented as well. For these control schemes, a temperature sensor/controller must reside in the space being served and be wired back to the make-up air unit. With either scheme of control, the means of discharge air temperature control is affected by a drop in space temperature from space temperature setpoint. With reset control, the discharge air temperature setpoint is adjusted, or "reset" upward as the space temperature falls below setpoint. With override control, the make-up air unit simply goes to "full fire" (or to a pre-established high fire setpoint) upon a drop in space temperature below setpoint. Both methods allow for some degree of space temperature control, and either method is normally required for most applications.

Make up air units can be purchased as packaged units, or as built up systems. Packaged make-up air units are normally custom-built equipment. There may be a standard "footprint" of which a manufacturer starts with, yet from there the unit is custom designed and built to fit the given application, by adding options and features which tailor the unit for the application. As such, it is important to convey the application at hand to the make-up air unit provider, prior to placing the order, so that what is required and what is received are one in the same.

A make-up air application, in its simplest form, consists of a single exhaust fan and a make-up air unit matched to the fan. The fan is controlled in some arbitrary manner, whether it be manually by a user switch, via a time clock, or by some other means. In this application, the make-up air unit is directly "interlocked" to the operation of the exhaust fan. This simply means that the make-up air unit is to operate whenever the exhaust fan operates. For packaged units, a "point of interlock" is often provided at the make-up air unit terminal strip, to accept the termination of exhaust fan starter auxiliary contacts. "Aux" contacts are normally open or normally closed contacts that are part of the exhaust fan starter, that change state when the starter is energized. By connecting these contacts to the make-up air unit's control system, the unit can know when the exhaust fan is in operation, and will thus operate whenever the exhaust fan operates. Hence the term "interlock."

More complex make-up air applications consist of those where there may be multiple exhaust fans, all of whose operation is arbitrary. A single make-up air unit may be sized to match the total exhaust of all of the fans. Yet with the operation of the fans being arbitrary, there must be some means of making up only the amount of air being exhausted by the fans that are in operation at any given time. This is typically achieved by monitoring space pressure, and then controlling the make-up air unit in some manner, in order to maintain a suitable (neutral) space static pressure.

The following sections will cover many of the various types of make-up air units and applications that abound. We will first take a look at packaged units, with focus on application rather than on the internals of the package. Then we will take a broad look at built up systems. The focus here will be on actual unit controls and the implementation thereof. Finally, we will explore the integration of "cooling" to make-up air units, as it applies to both packaged and built up systems.

Packaged Make Up Air Units—100 Percent Outside Air

Figure 9-1 generically illustrates a "100 percent O.A. packaged make-up air unit," with some of the external appurtenances that might be required, depending upon the application. A packaged make-up air unit designed to handle nothing but outside air will minimally have a supply fan, a gas-fired heating section, and a motorized two-position outside air damper. The unit will come with a discharge air temperature control system, and also typically with some form of space temperature control (reset or override). When pressed into operation, the outside air damper opens fully, the supply fan starts, and unit heating is engaged as required. Most packaged make-up air units will come with a remote mounted control panel. This panel typically has pilot lights to indicate fan and heat status and failures, user switches to allow the unit to operate in a heat mode or strictly in a ventilation mode (labeled SUMMER-WINTER or HEAT-VENT), and a space temperature controller. The space

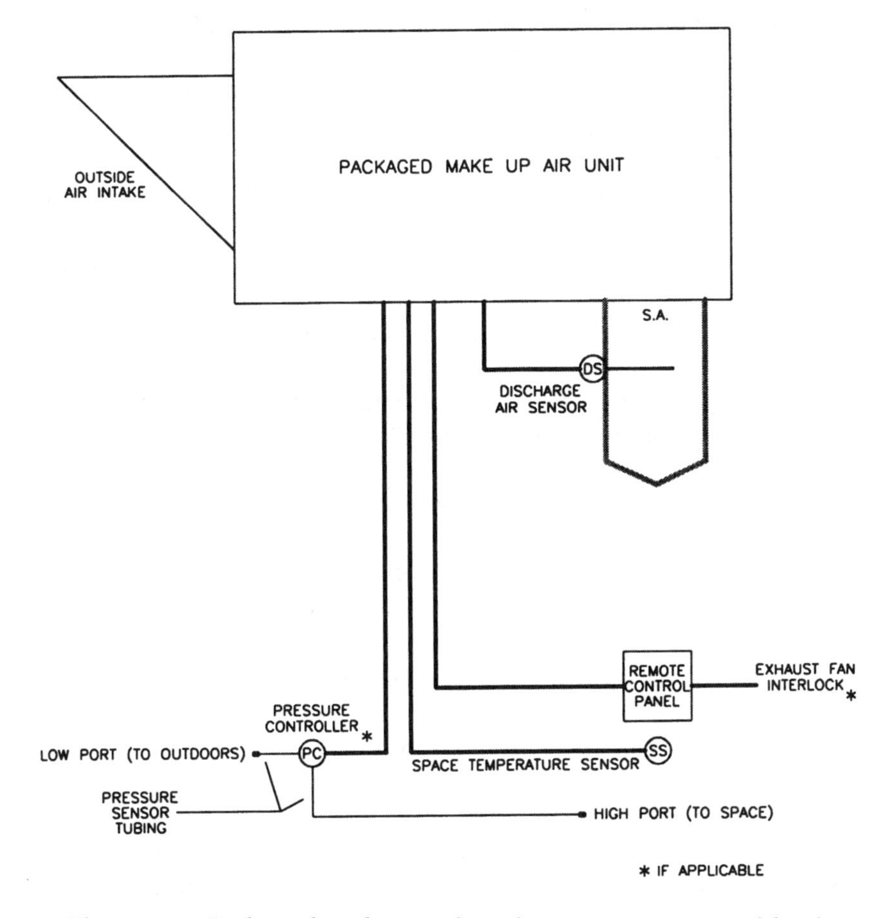

Figure 9-1. Packaged make-up air unit—100 percent outside air.

temperature controller may be an integral part of the control panel, or may be a separate device, requiring field mounting. This being the case, the temperature controller can be mounted in the space, and the control panel can be mounted in perhaps a more convenient location.

If such a unit is to provide make-up air for more than one exhaust fan, and the operation of these fans is arbitrary with respect to each other, then space static pressure control should be implemented. This is something that can be done with the make-up air unit, if ordered as such. The unit would in essence require either one of the two methods of air quantity control that we discussed in the last chapter, IGVs or a VFD. With air quantity control furnished as part of the package, the make-up air unit can (theoretically) be in continuous operation, with no physical interlock to the exhaust fans. The unit will operate, via space static pressure control, to maintain the space at a constant (neutral) static pressure; as exhaust fans are turned on and off, the space static pressure control system tracks the changes in pressure, and imposes air quantity control as required to maintain space pressure setpoint.

Note that we used the word "theoretically" in the last paragraph, regarding the continuous operation of the make-up air unit with no physical interlock to the exhaust fans. The above sequence of control assumes that the unit's air quantity control system can vary down to zero. In other words, if all exhaust fans are turned off, the unit can remain in operation and basically "idle" at an air delivery rate of zero. In the real world, this isn't very feasible; both IGVs and VFDs have some minimum allowable point of operation, in terms of air quantity delivered. For the VFD, this minimum may be 15-25 percent of the maximum deliverable air quantity, and is established through the VFD control parameters. For the IGVs, this minimum is even greater, perhaps more like 30-40 percent of the maximum. If all of the exhaust fans were turned off, the make-up air unit would assume this minimum point of operation, and would tend to pressurize the space. This may be all right for certain applications, yet for other applications may not be acceptable. This must be considered on a job-to-job basis, and if unacceptable, it must be addressed. If exhaust fan operation is completely random, as may be the case if the space served is a laboratory with a dozen fume hoods, each with its own exhaust fan, it may be required to monitor the status of each and every exhaust fan (by means of current sensing switches or starter auxiliary contacts). With this knowledge, the make-up air unit can be pressed into operation only if at least one of the exhaust fans is turned on. This basically describes the task of interlocking the make-up air unit to each and every exhaust fan.

Packaged Make Up Air Units—Return Air Capabilities

A packaged make-up air unit with return air capabilities will have everything that one without it has, with the addition of a return air connection and damper (Figure 9-2). The return air damper, and the outside air damper as well, will be equipped with proportional damper

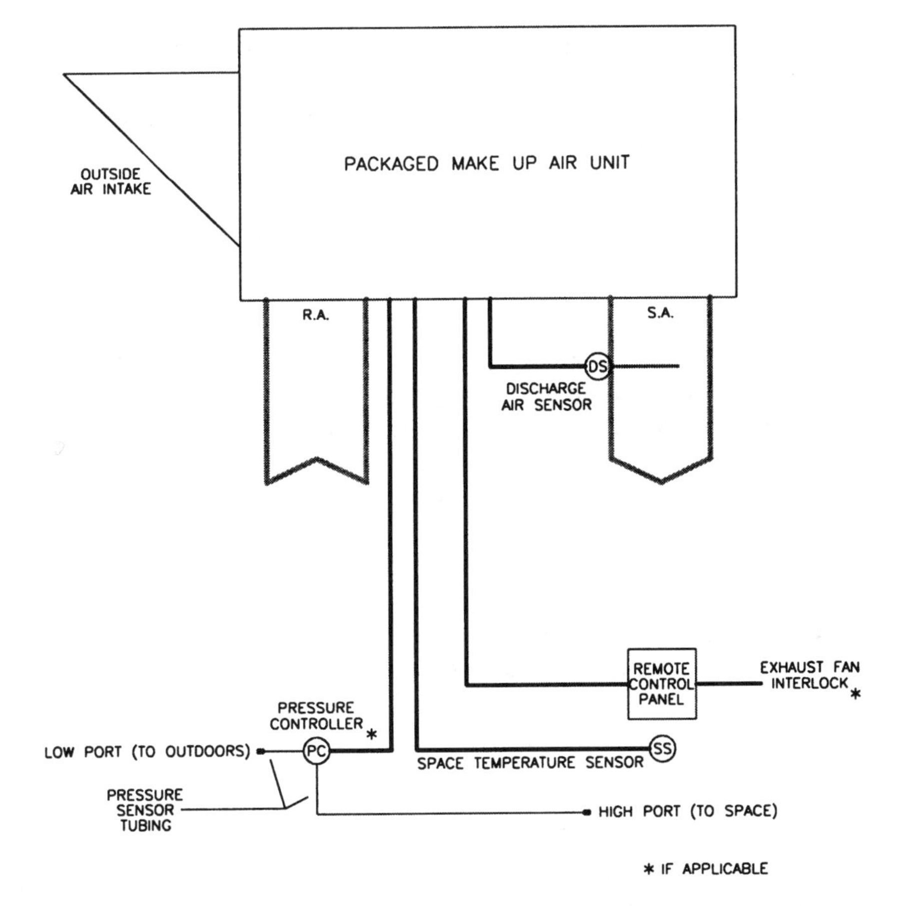

Figure 9-2. Packaged make-up air unit with return air capabilities.

actuators, to be controlled in unison. How the dampers are controlled is the primary topic of this section.

A make-up air unit with outside and return air dampers has space static pressure capabilities built right into it! The dampers themselves can be controlled to maintain space static pressure. Take the example from last section, of the arbitrary operation of exhaust fans within the same space. As exhaust fans are turned on and off, the unit's outside and return air dampers can be modulated to maintain space static pressure; as more fans are turned on, the outside air damper modulates open to allow more outside air, and the return air damper modulates closed. Likewise, as exhaust fans are turned off, the outside air damper modulates closed, and the return air damper modulates open.

What if all of the exhaust fans are turned off? Well, theoretically, we should be able to modulate the outside air damper fully closed, and the return air damper fully open. However, if the packaged make-up air unit is a "direct-fired" unit, we are not allowed to modulate the return air damper open more than 80 percent. Since we have not previously defined the term "direct-fired," now's as good a time as any!

Actually, we need to look at two terms, as they apply to gas-fired heating: "direct-fired" and "indirect-fired." Direct-fired means that the "flame" is directly in the air stream, whereas indirect-fired means that the flame is contained inside a heat exchanger, of which the air passes over. So with direct-fired gas heating, the air is heated up directly by the flame, and with indirect-fired heating, the flame heats up a heat exchanger, which in turn imparts the heat to the air.

Direct-fired make-up air units that have return air capabilities are not allowed to bring in outside air quantities of less than 20 percent. The reason is that the combustion process yields certain "by-products" that are not healthy for human consumption. These by-products must be diluted by outside air to a level that can safely be exposed to human life. A safe dilution rate has been established to be at least 20 percent of outside air. Expressed in other terms, the quantity of air recirculated from the space in a direct-fired heating system must not exceed 80 percent, for safe dilution of combustion by-product. For this reason, direct-fired make-up air units with return air capabilities are commonly referred to as "80/20 units."

With direct-fired make-up air units that employ outside and return air damper control as a means of space static pressure control, we run into the same situation as we did in the last section, with the IGV or VFD method of space static pressure control. The "minimum allowable point of operation" might be of concern, inasmuch as over-pressurizing the space during periods of no exhaust. As with the IGV and VFD methods of control, the application at hand must be evaluated, and the proper solution to the predicament be implemented.

Built Up Make Up Air Units with Electric Heating

Now that we've taken a look at the basic purpose of make-up air units, and of some of the features available with packaged units, we turn our attention to built up systems. Our discussions of these systems will be limited to make-up air units that recirculate no air (100 percent outside air systems), for if we start talking about built up units that are capable of returning air, we begin to "cross that line" between built up make-up air units and general built up air handlers.

With built up make-up air units, we are usually heating by either electricity or by steam. Hot water heating is *typically* not applied in make-up air applications; the potential for coil freeze-up is much greater with hot water than it is with steam, so we utilize steam to minimize this potential, for 100 percent outside air applications. Saving the discussions on steam-powered make-up air units for the next section, we focus our immediate attention on make-up air units with electric heating.

A built up make-up air unit will minimally consist of the following components: motorized outside air intake damper, electric heating section, and supply fan. The conceptual layout of such a unit is illustrated in Figure 9-3. The outside air intake damper is equipped with a two-position, spring return damper actuator; the damper is powered open, and springs closed upon removal of power from the actuator. When called into operation, the make-up air unit's supply fan may not start immediately. The first thing that typically happens is that the outside air damper is powered open. Once substantially open, an end switch on the damper actuator makes, thus allowing the operation of the supply fan and the electric heat. The supply fan starts, and the electric heat is enabled for operation.

The electric heating section is controlled by a discharge air temperature control system, with or without space temperature control (reset or override). In order to understand how the electric heater should be con-

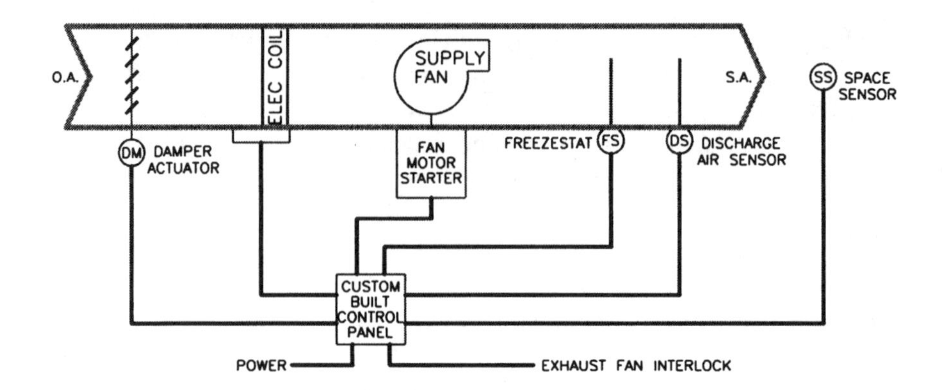

Figure 9-3. Built up make-up air unit with electric heating.

trolled, we need to have some insight as to how it is selected. Electric heaters are generally staged equipment, with each "stage" consisting of a contactor that allows power to an electric resistance heating element. For instance, a three-stage electric heater will have three contactors and three electric heating elements. A call for one stage of electric heat results in one of the three elements being energized. A call for all three stages results in all three elements "glowing."

The heating sections of make-up air units are liberally sized. With outside air temperatures reaching –10 degrees (and colder, for you Midwesterners!), is it not uncommon to find the heating section of a make-up air unit sized for a "**delta T**" of 80 degrees or more. Delta T simply refers to the increase in temperature that the heater can impart to the air passing through it. A delta T of 80 degrees can take outside air at a temperature of –10 degrees, and heat it up to room temperature (70 degrees). Delta Ts in excess of 90 degrees are normal, if the make-up air unit is to have the capability of providing "heating air" at extreme outside air temperatures.

With regard to staging, the delta T plays a significant role in determining an acceptable number of stages. There is a frequently used rule of thumb stating that, for "mediocre" control, the number of degrees per stage should not exceed 10. For our electric heater having a delta T of 80 degrees, this corresponds to "at least" 8 stages of control! That may sound like an awful lot, but believe it or not, this is the typical requirement of the electric heating section of a built up make-up air unit. More critical applications may call for even more stages of control, or for SCR control (Silicon Controlled Rectifier). An SCR is an electronic device that can accept a control signal and proportionally control the power output of the entire electric heating section, as a function of the control signal received. Both staged control and SCR control of electric heat will be discussed more thoroughly in the chapter on reheat coils.

The stages of the electric duct heater should be controlled (via a multistage controller or sequencer) in order to maintain discharge air temperature setpoint. In addition, a space mounted temperature controller can provide the means for space reset or override control. In other words, if the space temperature drops below the setpoint of the space temperature controller, the discharge air temperature setpoint is "reset" upwards, or overridden altogether, thus driving the electric heater to a higher operating point. With electric heat, it is important that the discharge air temperature is not allowed to reach the electric heater's high limit thermal cutout switch. For this reason, "high limit" control is normally implemented, thus limiting the maximum discharge air temperature. This is especially important during milder outside air temperatures, when the temperature of the air entering the electric heating section is relatively warm to begin with, and yet there is still a call from the space for heating.

Speaking of safety devices (i.e., thermal cutout switch), we should talk about other safety devices that are associated with this type of system. An air proving switch, in the form of a pressure switch, will also be a part of the electric heater's control circuit. This device, as well as the thermal cutout, will be factory furnished and installed as part of the electric heater. With either of these devices "not made," the electric heater's control circuit is broken, and heater operation is disabled.

A freezestat may be installed, at the discretion of the system designer. In this type of application, the freezestat does not protect any coils from freeze-up; it primarily serves as a low limit, manual reset "ductstat." If there is a failure with the electric heater and/or the associated temperature control system, the freezestat shuts down the entire system, to prevent cold untempered air from being delivered into the space.

As far as pressure control goes, the supply fan of the built up make-up air unit may be equipped with IGVs or a VFD. Operation is similar to what's done with a packaged unit; the selected method is simply implemented to maintain suitable space static pressure. If there are many exhaust fans whose control is arbitrary with respect to one another, this is something that may want to be implemented. The concern again, as with packaged units, is that "minimum allowable point of operation." An added concern is that the electric heating section must have a minimum amount of air flowing through it to be operational; the heater's air proving switch will not make if airflow through the heater is insufficient. This, however, is a moot point if the IGVs' or VFD's minimum point of operation is greater than the airflow needed to make the heater's air proving switch.

Control System Requirements

With a built up make-up air unit comes the requirement for a built up control system. We have spoken about the various control schemes required for the proper operation of the make-up air unit. However, we have not really talked about pulling all of these schemes together in order for the unit to operate as one complete system.

Actually, with this built up system, the level of complexity isn't that high. If we think about all of the controlled devices, and the required controllers, we begin to understand how the system is to operate as a whole. Upon a call for unit operation (exhaust fan interlock), the damper motor is energized and the damper strokes open. When the damper is substantially open, an end switch on the damper motor makes, thus allowing operation of the supply fan and electric heat. The supply fan starts, and the heater is operated to maintain discharge air temperature setpoint. The safety and limit devices monitor the unit for acceptable operation, and will shut down the necessary components (or the entire unit) upon detection of an unacceptable condition.

The heater itself may come with a factory furnished electronic sequencer, or "step controller," mounted in its controls compartment. If this is the case, then all that is really required (for control of the heater) is to wire a discharge air temperature sensor directly to the step controller, and allow operation of the heater only when the supply fan is running. The logic required to implement the remainder of the various control schemes here can be performed with simple relay logic; the relays performing the logic will be housed in a custom-built control panel, with all controllers and controlled devices wired back to this panel. This is typically the way its done, although certain controllers and controlled devices may be able to "bypass" the control panel altogether. The control panel also typically serves as a human interface; the panel may have indicating lights and switches that allow the user to monitor certain conditions and impose a certain amount of "supervisory control" over the system. The control system designer will normally produce a control panel wiring diagram, along with a "point-to-point" wiring diagram, showing how all of the various components get tied back to the control panel. These diagrams are then turned over to the installing electrician, for panel fabrication and field installation.

If the electric heating section is specified not to come with its own factory furnished and installed sequencer, one would have to be provided. This can be a "store-bought" sequencer, straight off the shelf from the local controls supply house. The sequencer would be incorporated into the control system design, and would take its place inside the control panel (move over, relays!) Another option is to perform the sequencing of the electric heater with a programmable (DDC) controller. By going this route, the relays in the control panel "go away," for all relay logic and control schemes can be performed by the programmable controller. The use of a programmable controller makes even more sense if space static pressure control is required. This being the case, a pressure controller would not be necessary. A space pressure transmitter would send the programmable controller a signal corresponding to the space pressure, the controller would process it, and send out the appropriate control signal to the IGVs or VFD. The programmable controller can virtually tie together and integrate all facets of unit operation, though programming, so that the unit operates as one complete system.

In conclusion, built up make-up air units (and any built up air handling unit for that matter) will normally require the design and fabrication of a custom-built control panel, for integration of the various schemes of control into a single, properly operating system. At the engineer's and/or the control system designer's discretion, the control system can be implemented using electrical, electronic, or even microprocessor-based controls and control methods. The order of system complexity is often a driving factor on which methods to employ. The simpler systems can most always be accomplished entirely with electrical controls (relays). The more complicated systems may require the use of both electrical and electronic devices, in order to accomplish the required control schemes. The more complex the control system, the more justification there is for using microprocessor-based programmable controllers.

Built Up Make-up Air Units with Steam Heating

A built up make-up air unit utilizing steam as its heating source must be equipped with face/bypass dampers in addition to the steam coil and requisite control valve (Figure 9-4). With the exception of its method of heating, the operation of a steam-utilizing make-up air unit is pretty much the same as that of one with electric heat. With all other things being equal, we will focus on control schemes and operating requirements relating to the heating section. Please refer back to the appropriate section of Chapter 6 for a refresher on face/bypass damper control.

Control of the Heating Medium—The Case for Face/Bypass Dampers

The first statement in the last paragraph is a pretty powerful statement! A steam-utilizing make-up air unit, one that handles 100 percent outside air, *must* have face/bypass dampers? Says who? Well, let's just say that that's the "general consensus" among HVAC engineers, at least for the larger systems (in colder climates). To understand the reasoning behind that statement, let us consider the alternative method of control (proportional control of the steam), in an application where the make-up air unit is sized to provide heating in addition to supply air tempering.

For the make-up air unit to be able to provide heating for all outside air temperatures, and assuming that the "worst case" outside air

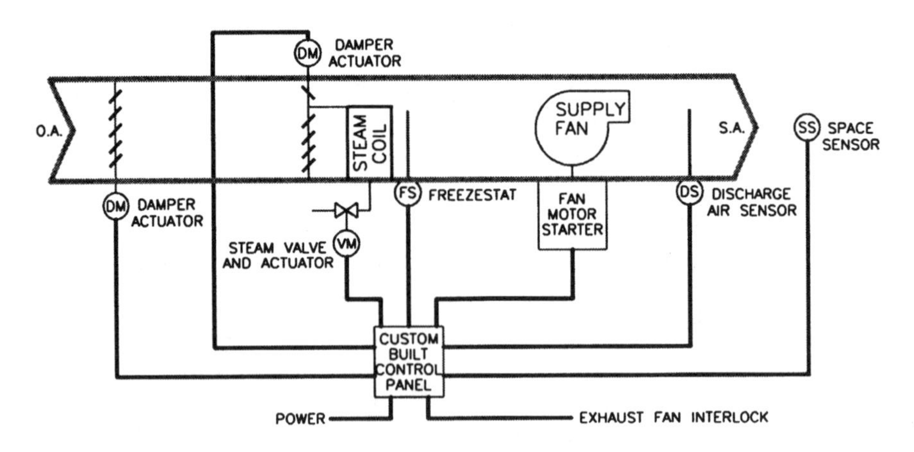

Figure 9-4. Built up make-up air unit with steam heating.

temperature is –10 degrees (Chicago in February!), the capacity of the steam coil must be such that it can impart a delta T of 90 degrees or more to the air passing through it. In other words, the coil must be sized for the worst case scenario, and thus must be able to raise –10 degree air to 80 degrees or more. We had mentioned in an earlier section that steam is a rather difficult medium to control directly, and for larger capacities, we need to "break down" the control into two separate valves (1/3, 2/3 control). So for the sake of this example, we assume that the make-up air unit's steam capacity falls under this requirement.

The first concern is that we are trying to control the temperature of the air passing through the steam coil by regulating the amount of steam passing through the coil. In other words, we are trying to control the temperature of one medium (air) by varying another (steam). Which in itself isn't out of the ordinary, when you think that that's the same thing we do when we try to control the air temperature passing through hot and chilled water coils. However with steam, the shear difference in the temperature of the steam and of the air passing through the steam coil makes it often rather difficult to maintain precise temperature control. What frequently happens in this situation is that the control valves hunt, opening and closing, opening and closing, without ever really dialing in on the setpoint. The result is not a constant discharge air temperature, but one that varies in some broad range about the setpoint.

This in itself is not necessarily a good thing, but when you combine it with the scenario that we are about to unveil, you will be able to see why face/bypass damper control of discharge air temperature in a 100 percent make-up air application makes a whole bunch of good sense!

When the outside air temperature is substantially below freezing, the steam coil is operating at or close to full capacity, with the control valves pretty much fully open (more or less). The steam therefore runs wild through that coil (with no looking back!). At this operating point, there is little or no risk of any kind of a "freeze-up" condition, because of the operating capacity of the coil. Of course, when the outside air temperature is above 32 degrees, there is no risk either.

Oddly enough, when the outside air temperature is at or slightly below the freezing point, there is considerably more risk of a coil freeze-up than when it's zero! At these outside air temperatures, the coil is virtually operating at less than half capacity; the operating control valve(s) are regulating the steam flow through the coil, in an effort to maintain discharge air temperature setpoint. The result is that the steam flow through the coil is relatively "tame," as compared to when it's at full capacity. Without getting too technical here and opening up a whole discussion on steam, we at least need to bring up the topic of steam traps and their function. A steam trap resides on the "leaving side" of the steam coil. The trap's function is to disallow steam from returning to the "return line" until it is in the form of "condensate," or until it turns back into water. Well, when the temperature of the air passing through the coil is at or slightly below 32 degrees, and the steam flow through the coil is rather tame, some of the steam cools down and turns to condensate in the coil, before ever reaching the steam trap. The steam flow through the coil is not violent enough to force the condensate out the coil and through the trap. The condensate, therefore, sits in the bottom of the coil, and begins to freeze. Before you know it, the return opening of the coil is frozen shut, and the problems begin!

Actually there are specialized coils available for steam that minimize the potential for this scenario. Of course, they are more expensive and aren't necessarily specified for every application that runs the risk of coil freeze-up.

Specialized coil aside, this is a very real scenario, and is a pretty good argument for using face/bypass dampers, wouldn't you say? Sometimes the use of both proportional control methods is specified; for outside air temperatures of 35 degrees and above, the face damper is fully open (bypass damper fully closed), and the control valves modulate in sequence to maintain discharge air temperature setpoint. When the outside air drops below 35 degrees, the valves go fully open, and face/bypass damper operation is initiated, to maintain setpoint. This may sound like overkill, but is nonetheless specified often enough. One reason has to do with the "radiant heat" generated by the steam coil, as we will discuss in the next paragraph.

Face/Bypass Dampers *and* Modulating Steam Valve?

If we solely employ the use of face/bypass dampers as a means of controlling discharge air temperature, then we basically use a two-position control valve to either allow or disallow steam to flow through the coil, and base this decision on outside air temperature. In other words, if the outside air temperature is below 70 degrees or so, the two-position valve should be open. The steam allowed into the coil cannot leave until it turns to condensate, as regulated by the steam trap. So when the outside air temperature is not much below 70, what you have is a coil full of steam, very slowly turning to condensate (because of the very small heating load), and radiating a heck of a lot of heat. Now, even though the face damper is substantially or even fully closed, and the air is mostly bypassing the coil, the bypassed air still can pick up some heat off the coil, thus raising the discharge air temperature above setpoint. In short, the tendency to "overheat" is a possibility when the outside air temperature is relatively mild.

To alleviate this predicament, the control valve becomes a proportional control valve (or two; let's stick with one for a moment), and the face damper is positioned to be fully open for outside air temperatures above 35 degrees. Thus, when it's slightly below 70 degrees outside, the proportional control valve (in theory) regulates the flow of steam entering the coil, more or less "matching," or even "undercutting" the rate that the steam is turning to condensate. In other words, the position of the valve is closely matched to the steam trap's passing of condensate, and the likelihood of overheating the supply air is minimized. Now in reality, controllability of course is dependent upon how the control valve is sized, and how well it can function with respect to its load. A single valve may not be cut out for the job, and so now you go to two proportional control valves, as per the trusty old rule of thumb. Face/bypass dampers and two proportional control valves. Overkill? Well...

Here's a better solution. If you're worried about overheating the supply air on warmer days, and you run into the above dilemma, go ahead and use two control valves, selected for 1/3 and 2/3 capacities, as sized for proportional control. Make the smaller valve a proportional control valve, but fit the larger valve with a two-position actuator. Select the outside air changeover setpoint, that point at which you switch control from the valves to the face/bypass dampers, the temperature at which the small valve tops out, capacity-wise. For instance, if the steam coil is sized to raise –10 degree air to 80 degrees, which is a 90-degree delta T, then theoretically the valve sized for 1/3 the total capacity will "max out" when the outside air temperature drops to around 50 degrees. This is the point at which the larger control valve, if fitted with a proportional valve actuator, would begin to modulate. But why bother? At that point, go ahead and change over to face/bypass damper control, and throw that two-position valve wide open (along with the smaller one). Your discharge air temperature setpoint is 70, and the outside air temperature is 50. The chances of overshooting setpoint as the result of bypassed air picking up radiated heat off the coil are nil, and you spare the expense and the hassle of an additional proportional control loop.

Another "alleged" solution to the "radiant heat" scenario is to use (in conjunction with face/bypass dampers) two two-position control valves, at capacities of 1/3 and 2/3, and control them both based on outside air temperature. For instance, at temperatures from 70 down to 40, simply open the smaller valve, and don't open the larger valve until the outside air temperature is below 40. Upon first glance, this seems to be a fairly feasible option, and one that saves the cost and spares the complexity of proportional control. If the outside air temperature is close to 70, only the small valve will be open, and the potential of overheating due to radiation will be minimized.

This is simply untrue. The reason is that, if the steam control valve is at a position to where it can allow a rate of steam into the coil that is greater than the rate of which it can turn to condensate, then it's the steam trap that's regulating the steam flow, and not the valve. This would be the case, for instance, if the outside air temperature is 60 degrees, and the 1/3 valve is fully open. The trap is regulating the steam flow through the coil, not the valve. We could open the 2/3 valve, and this would not change things a bit! So if this type of control scheme is employed as a means of trying to minimize the potential of radiation, the specifier of this control scheme may be a bit disappointed (and befuddled!), when all is said and done, to find that the very problem he attempted to avert has indeed become a reality.

If all of that was a bit difficult to digest, don't worry. The important thing to take away here is that, for steam make-up air unit applications that handle 100 percent outside air, the use of face/bypass dampers is usually not just a good idea, it's often a must! From there, the decision to go with proportional control on the valve side of things is up to the discretion of the designer, on a job-to-job basis. Factors that play in this decision are job budget and first cost, "criticalness" of the application (is that a word?), and of course, recognition of the potential for problems and the accompanying "worry factor."

Make Up Air Units—Addition of Cooling

Cooling may be a requirement in a given make-up air application. An application where a make-up air unit is interlocked to an exhaust fan in an unconditioned warehouse would be typical of one in which cooling is not required. A unit providing make-up air into a corridor or hallway of a wing of hotel rooms, however, would likely be an application requiring cooling as well as heating. Air being exhausted from the bathrooms of each room (by some central exhaust system) is made up through the corridors, by the make-up air system. The air dumping into the corridors, hence, better be conditioned year-round, or guests will be complaining! Can you imagine walking to your room, on a 90-degree day, and the corridor is as bad as it is outside? Needless to say, this is an application requiring cooling in the warmer times.

Packaged MAUs

With packaged make-up air units, the unit manufacturer may be able to offer cooling as an option, as part of the package. If not, a cooling coil would need to be installed in the duct downstream of the unit. The coil could be a chilled water coil (if chilled water is available), or a DX coil (requiring a remote air cooled condensing unit). With either method, additional controls need to be provided, and integrated with the packaged unit's control system (see Figure 9-5). This can tend to become very complicated, and merits some consideration.

Taking either method of cooling on its own, we see that we need to control the cooling in order to maintain desired conditions, when outside air temperatures are in excess of 70 degrees or so. With chilled water cooling, this comes in the form of modulating a chilled water control valve. With DX cooling, the condensing unit associated with the DX coil is staged. With either method, the task of controlling the cooling is relatively simple and straightforward, when nothing else is considered. Given the task of integrating this scheme of control with the operation of the packaged make-up air unit, the control systems designer has several things to consider. For instance, how is "changeover" between heating and cooling modes accomplished? How should the cooling be controlled? To maintain a discharge air temperature setpoint? Or a space temperature setpoint? Or perhaps simply as a function of the temperature outdoors?

As far as changeover goes, several options exist. The make-up air unit remote panel may already have a selector switch, one labeled SUMMER-WINTER or HEAT-VENT. One could feasibly break into the control circuit of this switch, and have the SUMMER or VENT position of the switch enable cooling. Of course, changeover accomplished with this method is a manual procedure, requiring human intervention.

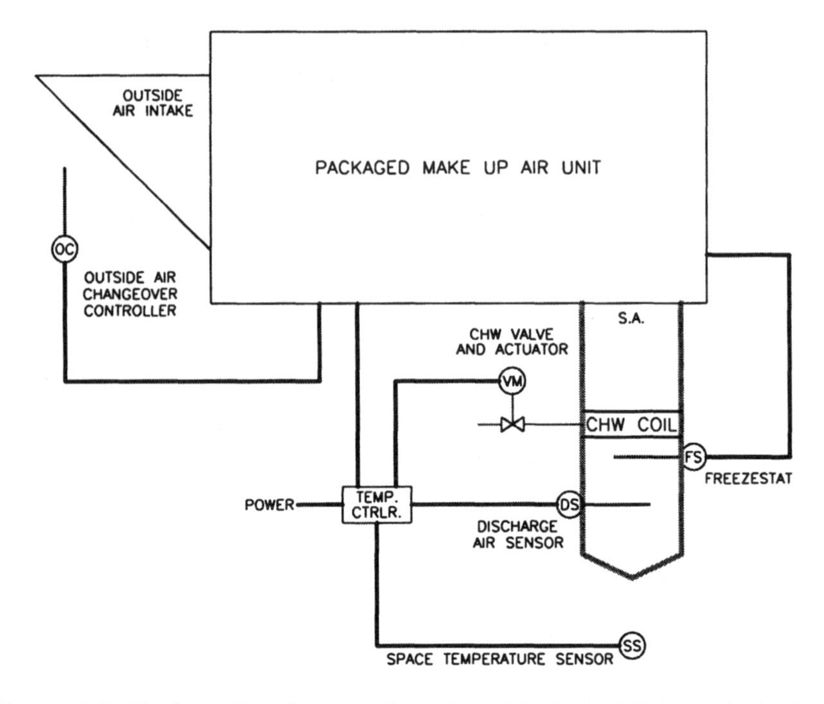

Figure 9-5. Packaged make-up air unit with the addition of chilled water cooling. A separate temperature control system must typically be furnished, as the packaged "heating" unit will have no provision for cooling control. The figure shows what is required, in addition to the heating controls.

Automatic changeover can be accomplished by means of an outside air temperature controller, set for 70 degrees. The mode of operation thus automatically switches over to cooling when the temperature outdoors exceeds 70. A three-position selector switch can be incorporated as well, allowing the operator to select between manual and automatic changeover. This switch would have the labeling of SUMMER-AUTO-WINTER or HEAT-AUTO-COOL.

With either method of changeover, we see the need to at least minimally "break into" the packaged unit's factory wired control system. This is a design issue required to be performed by the control systems designer; he must be able to read and understand the packaged unit's wiring diagram, and determine an appropriate means of "interface."

The second issue that the designer is faced with is determining the method of cooling control appropriate for the given application. For chilled water cooling, this is fairly simple; proportionally control the valve to maintain discharge air temperature setpoint, perhaps with space reset of setpoint. For DX cooling, we are dealing with staged control, and we cannot blindly go with discharge air temperature control without evaluating the application and considering the options.

The number of stages that the remote air cooled condensing unit can be equipped with is of major concern in the decision making process; "the more the merrier," as far as staging goes. Realistically, though, the condensing unit will be limited to six stages max, and more likely to four or even only two. With six stages of control at our disposal, we can control via discharge air and do a pretty good job maintaining a fairly constant temperature that's always close to setpoint. As the number of staging becomes smaller, we have a harder time controlling to maintain discharge air temperature setpoint, and we might want to consider the alternatives.

Space temperature control might be an acceptable alternative. To determine this, we should briefly take a look at the "cooling end" of a make-up air application. For the Midwest, a summer "design day" is 90 degrees. The cooling coil of a make-up air unit should be able to take 90-degree outside air down to a temperature of 70 degrees or lower. If the condensing unit is equipped with four stages of control, this would translate to a delta T per stage of approximately 5 to 8 degrees. With only two stages of control at our disposal, the delta T per stage approaches 12 to 15 degrees or more. To control to maintain a constant discharge air temperature setpoint with only two stages of control is unrealistic, to say the least.

By controlling the cooling via space temperature, then the discharge air temperature will be what the outside air temperature is, if there is no call from the space for cooling. When a call for cooling is initiated, the first stage is activated, and the discharge air temperature is brought down accordingly. Upon a further call for cooling from the space, the next stage is activated, and the discharge air temperature is brought down even further. As the cooling needs of the space are satisfied, the stages are disengaged in the reverse order, and the discharge air temperature rises back up.

While this may be okay if the temperature outdoors is right around changeover (70 degrees), what if it's 80 outside? Sure, when the space is calling for cooling, the discharge air temperature is acceptable, and all is well. But what if the space cooling needs are satisfied? At that point, the make-up air unit will deliver 80-degree air into the space, with no regard to discharge air control. Yeah, eventually the space will warm up to the cooling setpoint, and cooling stage one will be activated, but what about until that happens? Is this acceptable operation, to allow the discharge air temperature to drift like that, with no regard to discharge air temperature control? For certain applications, the answer to that question might very well be "yes." For others, though, we may want to build in some "high limit" scheme of control into our design, to prevent the discharge air temperature from ever becoming too warm.

A simple way to accomplish this is to actually "lock on" stages of cooling as a function of outside air temperature. For instance, with a two-stage air cooled condensing unit, we could lock on the first stage whenever the outside air temperature is above 75 degrees, and lock on the second stage when it's above 85 outside. Of course, now we've taken away control from the space, and given it to the outside air. Well, at this point, we can still give some control to the space, in the form of "override"; if the space becomes too warm, turn on the "next" stage of cooling.

There are different philosophies on all of this, ranging from personal preference, to equipment wear and coil freeze-up concerns, to opinions on energy usage, etc. The bottom line is, with only two stages of control in this type of application, there ain't a whole lot you can do, and maybe you should be looking at the equipment selections a little closer, rather than trying to make something work "better" that it actually can. Remember, the more stages of control at your disposal, the better the chances of being able to maintain stable and desirable temperature control, and the less "callbacks" from the customer!

Needless to say, the incorporation of cooling into a packaged, heating only make-up air unit requires "just a little bit" of consideration and foresight. Also adding to the complexity is the need for not one, but two separate temperature control systems. Sure, the packaged make-up air unit will have all the fixings required to make hot air. But the unit's control system will likely have no provisions for controlling "after the fact" cooling equipment. The unit's remote control panel, for instance, may have a space temperature controller associated with it. But that controller is for heating. We can't necessarily make that controller perform our cooling functions, and most often will be forced to install a separate, cooling, temperature controller, if our application calls for it. Likewise with discharge air temperature control. We can't employ the packaged unit's discharge air temperature control system to perform temperature control of the cooling equipment. We thus are normally forced to provide a separate controller for cooling. So what we end up with is two temperature control systems, one for heating, and one for cooling, integrated into the same overall system. Sounds like a lot of extra stuff, but that's normally the best we can do in these types of applications. And really, when all is said and done, as long as the end user doesn't have a problem seeing "two" space mounted temperature controllers, and as long as the system functions properly and maintains temperature, then everybody is happy and life goes on.

Built Up MAUs

Now for built up make-up air units, addition of cooling. Sounds simple enough. We as designers are creating the control schemes for the built up unit, so we should pretty easily be able to incorporate cooling into our design. Yes, it's true that the designer has added flexibility in this scenario, and can incorporate the cooling scheme(s) into the overall control system. Yet with the addition of another control loop comes an increase in control system complexity. This increase in complexity may altogether prohibit the use of simple control methods, and may warrant the use of electronic or even microprocessor-based controls. The designer may opt to use an electronic temperature controller to handle the heating and cooling of the unit, and may choose to support this controller with relays and relay logic, to handle the other control schemes (fan operation, damper control, safety and limit circuits, etc.). Or he may simply choose to go with a "stand-alone" DDC controller, and perform all temperature control and logic with the controller. For the added cost of the controller, a world of flexibility is gained; design time is reduced, and the relay logic disappears and is replaced with programming. Commissioning time should also decrease, as checkout of the various control schemes should be simplified by the use of the human interface device that's required to connect to the controller. The designer can monitor and verify temperatures, validate control cycles, and troubleshoot bugs, all through the assistance of software.

Chapter 10

FAN COIL UNITS

To begin our discussion on fan coil units, we will start by saying that the term "fan coil unit," as it applies to air handlers, encompasses a broad array of singular components and the combination of them. To define what is meant by the term, or at least how this writing uses the term, it is important to first categorize it as an "air handler," before we qualify it and differentiate it from the general term.

As an air handler, a fan coil unit will have, at the very least, a fan and a means of either heating or cooling. The heating or cooling is "usually" in the form of a coil, however a fan coil unit with gas-fired heat will not consist of a coil, but of a gas-fired furnace. A fan coil can and most often does have the means to both heat and cool. So a fan coil, minimally, consists of a fan, and one or two "coils." The fan coil may be strictly for recirculation, with only a return air connection. Yet many commercial applications call for at least a minimal amount of outside air, thus requiring an outside air intake and a two-position motorized outside air damper, interlocked to the operation of the fan. If economizer operation is desired, the fan coil unit will have a "mixing box," which consists of two dampers (outside and return air), typically linked together so as to be able to be operated by a single damper actuator.

Fan coil units can be purchased as packaged equipment, with the only items requiring field mounting being the thermostat, and perhaps a control valve or two. Our focus is not on packaged fan coil units, but on built up units, those of which need to be fully furnished with controls and a properly operating control system.

Now that we've defined the term, let us attempt to qualify it and at the same time, differentiate it from the term "air handler." A fan coil unit is relatively small in size, normally being able to hide above a ceiling or perhaps behind a wall. It is always a single zone, constant volume unit, meaning that it serves one zone and has no automatic fan speed or air quantity control. The deliverable cfm is on par with that of a small single zone rooftop unit, ranging from a few hundred to a couple thousand. As such, the fan motors are most often fractional horsepower, single-phase motors (though can be three-phase as well). The percentage of outside air that these units will handle is small, typically limited to no more than 20 percent (except during economizer cycle). The complexity of these systems matches their physical size; control schemes applied to fan coil units are kept simple, for more complicated control schemes might merit a control system that is in a sense "overkill" for a unit of such size. Should a unit that meets most of the above requirements for a fan coil unit be used in such a demanding application that it requires more complex (and more costly) control strategies, then it may be time to cross that line and call it a built up air handler. Hence, control system complexity may be the biggest discerning factor of all, with regards to fan coils and air handlers.

The term fan coil unit, as it is used here, is a very broad term, meant to encompass many different configurations. The fan coil unit doesn't even have to physically be a single "piece" of equipment. A fan coil unit with DX cooling and electric heating may take the form of two separate pieces; one piece being the supply fan, DX coil, and mixing box, and the other piece being a downstream electric duct heater. Likewise, a fan coil unit with chilled water cooling and gas-fired heating may consist of one piece being the supply fan and furnace, ducted to a downstream chilled water coil. It's not important how the equipment comes out of the factory. The important thing is how all of the singular components are pieced together, and how they are to operate as one "unit." When all is said and done and the components come together and unite as one, as long as the final form meets all of the above qualifications (last paragraph), then we call it a fan coil unit, and we give it fan coil controls.

The various forms of heating and cooling that a fan coil unit can take on allow for quite a number of different combinations! Rather than spending time on each and every possible combination, we will list the various forms, and then consider, in general, heating and cooling combinations of staged control and proportional control. This narrows our discussion to three categories, as we will see shortly.

But first, a list of heating and cooling methods, as applied to fan coil units:

- Hot water heating (two-position or proportional)
- Electric heating (staged)

- Gas-fired heating (staged)
- Chilled water cooling (two-position or proportional)
- DX cooling (staged)

The first thing that we notice from the list is that steam heating isn't on it. If you see a built up fan coil unit application utilizing steam, let me know, because I have yet to see it. I suppose it's not a completely off-the-wall idea. It apparently just isn't a very popular one.

The second thing that we notice from the list is that there are three forms of heating, and only two forms of cooling. And Last, the only forms of heating and cooling available as proportional are those that utilize water.

Hot water heating and chilled water cooling can be done as two-position control or proportional control. The remainder of these methods of heating and cooling are performed as staged control, either one or two stages, and seldom anything more than that. Electric heating is performed with a one or two-stage "duct heater." Gas-fired heating is done with an indirect-fired furnace and a one or two-stage gas valve. And finally, DX cooling is performed with a remote air cooled condensing unit that has one or two stages of capacity control.

For the sake of simplicity, we will generalize things and categorize two-position control under the more general term of staged control. To repeat a previous sentiment, two-position control is staged control with the number of stages equaling one.

Now we can finally get on with the rest of the chapter and talk about controlling these things! For our purposes, we categorize fan coil units as follows: **staged control of both heating and cooling, staged control of one and proportional control of the other, and proportional control of both heating and cooling**. The following sections assume that the fan coil units under consideration have both heating and cooling capabilities. Hence, it should be simple to visualize a fan coil unit with only one mode, with perhaps proportional heating only, or staged cooling only. Just disregard or ignore the other part of the combination (and anything related to it) for your particular visualization.

Staged Control of Both Heating and Cooling

This is perhaps the most fundamental form of single zone control available, comparable to a small single zone constant volume rooftop unit, in which a simple multistage heating-cooling thermostat can operate the entire system: fan, heating, and cooling. If the means to heat and cool is hot and chilled water, then we're talking a single-stage heating, single-stage cooling thermostat. One stage only, folks. If your application demands something more precise in terms of temperature control, you might not be looking at this as an acceptable alternative. With any other combination of staged heating and cooling, you can in theory get at least a couple of stages of each mode, as with for example, electric heating and DX cooling. Of course, the smaller the fan coil unit, the less of a chance that the selected means of heating and/or cooling can be provided with more than one stage. For instance, a small condensing unit will only have one compressor, and therefore only one possible stage of cooling. Small duct furnaces will not be offered with two-stage gas valves, and therefore will have only one available stage of heating.

All of this may be a moot point, for the smaller the fan coil unit, the less deserving it is of a two-stage heating/cooling thermostat. There does come a point in size, though, where two-stage heating and/or cooling methods become available with the equipment, and some consideration should be taken as to the application and its requirement for precise temperature control. Sure, two stages of control (per mode) isn't going to give you the best of all worlds, but it's better than one, and you still are in the realm of simple fan coil unit control methods.

Okay, okay. Point taken, already! Two-stage heating and cooling is better, in terms of space temperature control, than one-stage heating and cooling, and still falls under the realm of simplicity. For staged control of both heating and cooling, thus, a simple multistage thermostat (programmable or non-programmable) will do the trick, in terms of controlling the fan, one or two stages of heating, and one or two stages of cooling. But what about the economizer controls? You have this mixing box on the back end of your fan coil unit, and you don't know what to do with it. Okay, let's take a look at a few options, that remain within the realm of fan coil unit simplicity.

First and foremost, there needs to be a damper actuator. How that actuator is controlled is dependent upon the particular method of economizer control chosen. One method, that has been around for quite a long time, is to use an integrated economizer control package, or system (see Figure 10-1). The system consists of several components, the sum of which make up a package that can directly and seamlessly interface with the thermostat. These components consist of the damper actuator, an outside air enthalpy sensor (return air sensor as well, for comparative enthalpy sensing), a discharge air temperature sensor, and an economizer control module. The module is a solid state controller that is wired out to the actuator, the sensors, and also the thermostat. A central point of control, the economizer control module monitors the enthalpy sensor, determines if the outside air is suitable for free cooling, and if so, presses economizer operation into action if the space thermostat calls for cooling, thus modulating open the outside air damper to maintain a suitable discharge air temperature. If the outside air is not suitable for economizer operation, then the control module maintains the outside air damper at minimum position, as set by a manual potentiometer either on board the module or remotely located and wired back to it. Upon a subsequent call for cooling, the damper stays put, and thermostat control is directed, through the economizer control module, straight to the "energy using" means of cooling (DX or chilled water).

While this is a most effective method of implementing economizer operation, it does have its drawbacks. First, that's a lot of components there, and the cost of those components adds up quickly! We could minimally eliminate the economizer control module, if the thermostat chosen for the application had the ability and the "smarts" to handle economizer control. The thermostat, in essence, must first be electronic. It must also have inputs for outside air and discharge air sensing. Last, it must have a proportional output to control the mixing box damper actuator. That's asking a lot from a simple thermostat. Yet, if the thermo-

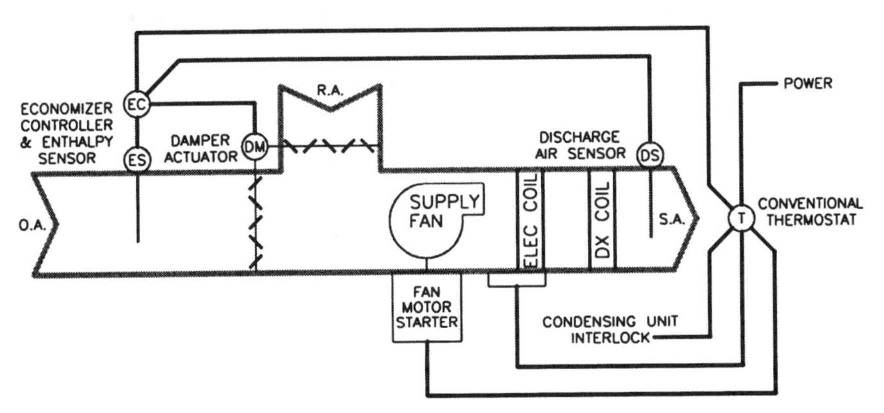

Figure 10-1. Fan coil unit with staged electric heating, staged DX cooling, and integrated economizer control package, operated by a conventional electromechanical or electronic thermostat.

stat has the capability, it dramatically simplifies things, from both an installation and an operational standpoint. Giving the thermostat complete control over all facets of fan coil unit operation sounds like a pretty good idea, and it is. Of course, a thermostat with these capabilities has to 1) exist, and 2) be affordable. In this day and age, that's not too tall of an order to fill, as thermostat manufacturers are realizing this and are building economizer control into their offering of programmable thermostats. Advancements in technology and the ever-decreasing physical size of electronic components are the reason for this, and we are to be thankful for that! The only thing we may give up with this method, is decision making based on outside air enthalpy. We are typically restricted to straight temperature sensing of the outdoor air, and have to base our decision to economize on dry bulb temperature, with no regard for humidity. If the thermostat has an analog input for outside air temperature sensing, then economizer changeover setpoint is likely set via the thermostat keypad, and is in units of degrees. However, if the thermostat can accept a contact closure for economizer changeover, then an enthalpy controller can make the decision to economize, with the controller's contacts feeding the binary input of the thermostat.

A third method that we should touch upon, one that is within the realm of simplicity yet is not necessarily achievable using "off-the-shelf" components, is DDC control. An "**application specific**" controller can provide the means for controlling the heating, cooling, fan, and economizer section of a built up fan coil unit. With this method, the controller is the central point of control; all sensors and end devices are wired back to the controller, which likely resides local to the fan coil unit itself. Instead of a thermostat in the space, there is a space sensor. The sensor may be equipped with a setpoint adjuster, and not much else. The DDC controller is a configurable type controller. The term "application specific" implies that the controller is designed and manufactured to fulfill a specific application. A "**unit level**" DDC controller, one specifically designed for a fan coil unit application, must be configured, or tailored, however, to fit the application at hand. This requires someone with the tools and the know-how to access the controller, via a human interface, and configure it. Once properly configured, the DDC controller will control all facets of fan coil unit operation (heating, cooling, fan, and economizer) as one integrated control system. While this is an attractive and cost-effective alternative to thermostatic control of a fan coil unit, it is not one that is capable of being implemented by the average, non-DDC oriented designer or contractor.

To be capable of pulling off this method of control, the installing and commissioning contractor must have a familiarity with the DDC product, in terms of hardware and software, and a skill level in the design, installation, and commissioning of a system utilizing such a product.

In summary, to control a fan coil unit with staged heating and cooling, and economizer, at least three methods exist, that stay within the realm of simple fan coil unit controls. The first method that we discussed is to use a simple, conventional style heating-cooling thermostat, in conjunction with a separate economizer control package. While this method is quite acceptable, it isn't necessarily the most affordable or the simplest. Component costs, and installation and wiring of the system, tend to offset the attractiveness of this method. The second method is to use a specialized electronic thermostat that can take direct control of the economizer section. With this method comes simplicity, in terms of installation and operation, and with the never-ending advancements in technology, should become the trend among electronic thermostat manufacturers. The third method of fan coil unit control is to use a DDC controller specifically designed to fit the application. And while this method may be the slickest and most cost-effective, it does require a knowledge of DDC and a skill level with the chosen controller.

This whole section demonstrates the added complexity of a small air handler when you include economizer control. Without an economizer section, a simple thermostat will do everything you require the fan coil unit to do. Yet with an economizer, another dimension is thrown into the mix, that substantially complicates things. With the added complexity of system operating requirements comes the added complexity in the control system operating the unit. And of course, added costs, for the components themselves, for the installation of them, and for the commissioning of the control system as a whole. In light of all of this, we may be tempted to step back from our application at hand, take a good objective look at it, and question whether economizer is worth the added hassle. Perhaps more so with a project with not just one or two, but several or even dozens of fan coil units. To decide whether or not to economize takes on great importance on such a project. The wrong decision up front becomes a real problem in the end, as you can imagine. To save money up front by eliminating economizer operation, only to find out upon completion of the project that a real need for it exists, is a scenario that is to be avoided, and something that must be considered on a job-to-job basis.

Staged Heating and Proportional Cooling (or Vice Versa)

This section builds upon the concepts and methods discussed in the previous section. Since we covered the aspects of economizer control and options, we need not go into too much detail here. Instead, we focus our attention on the integration of all system control, into one working system, for fan coil units that require proportional control for one of its two possible modes (heating or cooling).

By way of example, consider a fan coil unit with electric heating and chilled water cooling (Figure 10-2). A decision must be made, up front, as to whether to provide two-position or proportional control on the chilled water side. The deciding party may brainstorm and initially conclude that, hey, the heating is only one stage, why do anything more elaborate with the cooling? This being the case, we're back to the relatively simple control methods discussed last section. Yet somebody in the brainstorming party suddenly raises an issue, and favors proportional control for the chilled water. Additionally, it's pointed out that not one, but two stages of control are warranted on the heating side. The final decision is made, and the fan coil unit will have two-stage heating and proportional cooling (and economizer!).

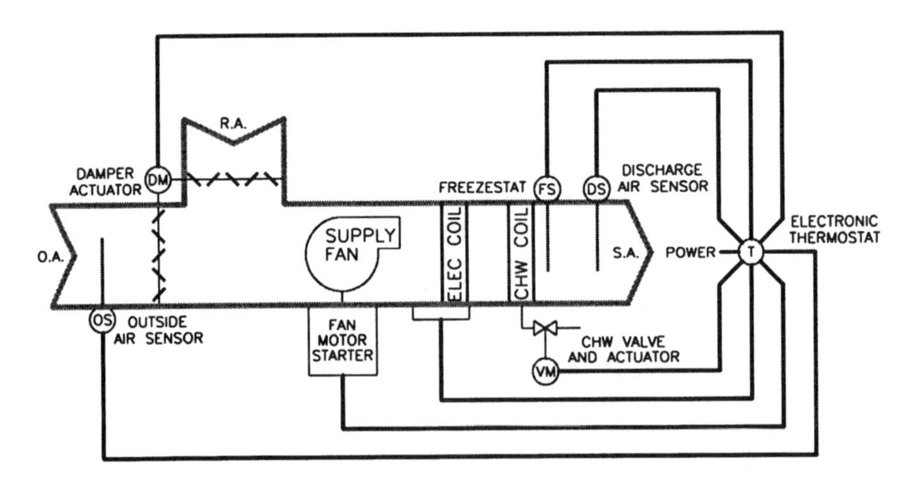

Figure 10-2. Fan coil unit with staged electric heating, proportional chilled water cooling, and economizer, operated by an electronic thermostat with modulating capabilities.

At this point, we are basically out of the realm of the simple heating-cooling thermostat, yet not necessarily out of the realm of fan coil unit simplicity. Defining our requirements, we need a control system that can operate the fan, two stages of heating, a modulating chilled water control valve, and the economizer section. The most probable solution to this challenging control scenario is to employ a specialized electronic thermostat. The thermostat needs two staged heating outputs, a proportional output for the chilled water valve, and a proportional output for the economizer actuator. Sounds like a tall order to fill, yet thermostats do exist that can perform all of these functions. Caution must be exercised, however, and the requirements for the application must be fully addressed and understood. With an application demanding this much control, the application has been deemed to be at least somewhat critical (by the decision to go with proportional control), and the overall capabilities of the electronic thermostat could fall short of the application's requirements. For most fan coil unit applications, this type of control should suffice. The shear simplicity of it makes it an attractive option.

The other option here that falls within the realm of simplicity is DDC. A controller specifically designed for the application can quite easily handle the job. Another proportional (or floating) type output (chilled water valve in addition to economizer) is really no problem for such a controller. The capability is there. However, as we stated in the last section, there must be some know-how; some skill level in regard to how to properly apply, configure, and commission the controller in such an application.

To summarize our "simple" controls options for this section, we have two: use a specialized electronic thermostat or go the DDC route. Any other means of handling a control system complexity of this level would likely merit a "higher end" control system, one that is normally reserved for built up air handlers, and has the sticker price to boot!

One final note before we leave this section. Take away the economizer and the staged heating here, and you dramatically simplify things. You still need to proportionally control the chilled water valve and operate the supply fan, and for this you may still opt for a specialized electronic thermostat, especially if you require occupied/unoccupied mode programmability. However, you can feasibly save some money by simply controlling the valve with a proportional space temperature controller, and then somehow integrate fan operation and control. You may conclude that any cost savings in components is gobbled up in extra design and installation time, and thus choose to pay the premium for that thermostat. The point is that, by simplifying the system, other options open up; you aren't necessarily locked into the options discussed above, and you can still remain within that "realm of simplicity."

Proportional Heating and Cooling

So now you want to modulate two valves and control an economizer section all with one simple control system, huh? Well, to put it bluntly, this is easier said than done. Depending upon the application, you may be forced to "cross that bridge" from the realm of simplicity to the realm of "**equipment level**" control systems. At any rate, let's analyze the situation, and see where our options lie...

If we forget about economizer operation for a minute, what then, are our requirements in terms of control? We need to proportionally control both a hot water and a chilled water valve, and also control the operation of the supply fan. Can we ask a thermostat to do all of that? Well, we can ask, but... Actually, electronic thermostats do exist that offer two proportional outputs and fan control. We're getting close to that bridge, however, and depending upon what else we want to throw into the mix here, control-wise, we may actually end up crossing it, into the land of "air handler controls."

By throwing economizer into the mix, we can virtually rule out any chance of simple thermostatic control of the fan coil unit. By adding the requirement for a third proportional output, we've basically surpassed the capabilities of a "current state" electronic thermostat. For some reason, that third proportional control output is difficult to cram into a thermostat, whether it be due to physical and technical limitations, or due to added complexity. At any rate, we are forced to look at other alternatives, at least until technology can pack the whole ball of wax into a single package, and manufacturers can offer it inexpensively and attractively.

Until that day comes, we need to consider the other options here, for control of a fan coil unit with proportional hot and chilled water, and economizer. The first option is DDC. A unit level controller

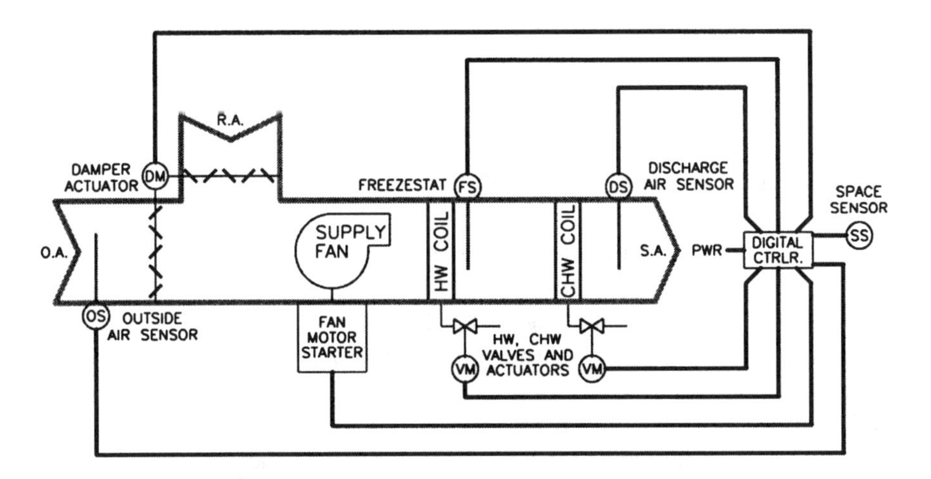

Figure 10-3. Fan coil unit with proportional hot water heating and chilled water cooling, and economizer, operated by an "application specific" digital controller.

can handle this, and if you are a DDC aficionado, then this is exactly the method you would implement (Figure 10-3). Again, if you don't have the requisite tools of DDC skill and know-how, then this method isn't very feasible.

Other methods exist, but I'm afraid that we now cross the line into air handler complexity. To design a control system that incorporates all of these different control schemes into one seamless control system is not a simple task. Yet it is possible for a control system designer, using "off-the-shelf" controls components, to come up with a working control system that incorporates fan control, hot and chilled water valve control, and economizer control as well, into one package. In fact, certain manufacturers have for years and years now offered electronic "control modules" that integrate mod heating and cooling with economizer operation. These modules serve as a "building block" for this type of system, and require other components to complete the package. All components would typically reside in a common control panel, custom-built for the application. At this point, though, we've definitely crossed that line, and are talking about an air handler control system. Cost, complexity, component count, and installation time all just went up considerably. Unless the application is extremely critical, applying such a system to a small fan coil unit delivering only a few hundred cfm may be more than anyone cares to invest in, and as such a compromise has to be made. Specifically, lose one of the proportional control loops and get back over that bridge, into the realm of simple fan coil unit controls.

In summary, proportional control of fan coil unit heating and cooling, and economizer as well, approaches the complex, and isn't something that can necessarily be accomplished with a single electronic thermostat. Not that it's unthinkable, for sure, and is something to definitely look out for. Nevertheless, even with the impending introduction of electronic thermostats with three proportional outputs (Has it happened already? What year is it?), we will still need to exercise caution, and be certain that the overall capabilities of the thermostat don't fall short of the requirements of the application. We may be otherwise forced to "cross that bridge," whether we like it or not, and equip a unit of "fan coil unit size" with air handler controls.

Dual-Temp or Two-Pipe Systems

One more type of fan coil unit that we need to talk about is one that enjoys a lot of popularity these days. "Dual-temp" or "two-pipe" systems are heating/cooling systems using only a single piping/pumping system. A boiler plant produces hot water during the heating season, and a chiller plant produces chilled water during the cooling season. Only one set of pipes carries water to and from the equipment connected to the system. This set of pipes is connected to the boiler plant in the winter, and to the chiller plant in the summer. Thus, depending upon the season, the temperature of the water delivered to the equipment is either hot or cold. Hence the term "dual-temp."

Dual-temp or two-pipe fan coil units have only a single coil. The coil can provide heating in the wintertime, when there is hot water in the pipe connected to it, and can provide cooling in the summertime, when chilled water is pumped to it. It's either one or the other at any one time, folks. A space thermostat controlling a dual-temp fan coil unit, calling for heating during the summer mode, is basically "outta luck." Likewise for calls for cooling during winter modes (what about economizer?).

Why do this to begin with? Well, the price is right! First cost of installation and equipment is a deciding factor. Physical space, the limitation of it, is another reason. Third, year-round system requirements also weigh in. What is meant by that? It means that, even with a "four-pipe" system serving fan coil units with two coils, the hot and chilled water plants will not (necessarily) be in operation year-round. In other words, hot water is typically only available in the colder months, and chilled water in the warmer months. So by having four pipes instead of two, what is really gained? Well, with a four-pipe system, there is some overlap during those "in-between" months (spring and fall), in which there may be needs for both heating and cooling, and therefore both the hot and chilled water plants can simultaneously be in operation. With a two-pipe system, you give up the luxury of having both hot and chilled water potentially available during those in-between months.

With that said, we now focus on the utilization of a fan coil unit designed to operate in conjunction with such a system. We mentioned that a dual-temp fan coil unit will have only one coil, and depending upon the mode of the dual-temp system, will accept hot or chilled water flow through it. If we count our "components of control," we have fan, proportional hot/chilled water valve, and economizer. The same options that we have at our disposal for fan coil units discussed back in section 2, we have here as well, for we only have the requirement for two proportional outputs; valve and economizer. Refer to Figure 10-4 for the dual-temp fan coil unit and its required components.

The controller of choice in this scenario needs to know the temperature of the water in the pipes, so that it can operate the control valve accordingly. For instance, if there is hot water in the pipes, then the controller must operate the valve only in a heating mode; calls for cooling would thus be prohibited, and the control valve would remain closed to the coil upon such calls. Same goes if there is chilled water in the pipe; a call for heating during this mode would need to be prohibited, and the control valve would need to remain shut.

This is typically performed by an "aquastat" that senses the water temperature and conveys this information to the controller. With this knowledge, the controller knows how to operate the valve; it knows to modulate it upon calls for the current mode, and keep it closed to the coil upon calls for the opposite mode. The aquastat is sometimes directly immersed into the pipe, yet more often is

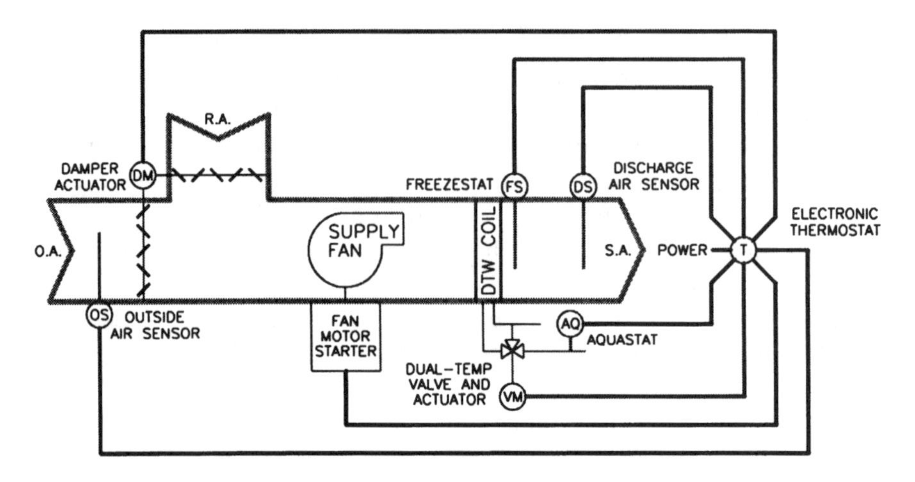

Figure 10-4. Fan coil unit with a single coil for both proportional hot water heating and chilled water cooling, and with economizer, operated by an electronic thermostat with modulating capabilities.

"strapped on" to the pipe. Precise temperature sensing is not required of the aquastat. The aquastat simply needs to determine if the water in the pipes is hot or cold. During the heating mode, the temperature of the water is typically 160-180 degrees, and during the cooling mode, the water temperature is 45 degrees or so. For this reason, the aquastat is normally set to "change over" at around room temperature, or 70 degrees.

Placement of the aquastat is critical. The aquastat must be able to sense water that is continuously flowing. If sensing water that is sitting in the pipe and not flowing, the water can potentially warm up (or cool down) to the changeover setpoint of the aquastat, thus erroneously registering the wrong mode. A three-way valve makes good sense here, whether required or not for other reasons. With a three-way valve, there is always flow through the common port of the valve, and the aquastat can be installed at the piping connected to this port.

For fan coil unit applications requiring two-way valves, the next best thing to do is to install the aquastat back at the mains, where water is continuously flowing. A third option is to install a very small "continuous flow bypass" right before the two-way valve, from supply to return; so small that it won't have much of an impact on the temperature control process, yet large enough for an aquastat to be feasibly and functionally mounted to. This is the method favored by manufacturers of packaged fan coils.

The importance of changeover should be apparent. The valve must receive the proper control based upon the mode of the dual-temp system. It is interesting to ponder the absence of changeover in a system like this. Consider for example that the dual-temp system is in a heating mode, with hot water being produced and distributed via the two-pipe system. A particular fan coil unit is operating to maintain the temperature of the space that it's serving. This particular space requires heating, and the control valve is modulating to allow the required amount of hot water to flow through the coil. All is well with the temperature control process. Until... the mode of the dual-temp system changes over from heating to cooling. If the fan coil unit's temperature controller does not know that this changeover has taken place, and the space still requires heating, then the control valve will continue to modulate, yet to allow chilled water through the coil. This creates a vicious cycle; the colder it gets in the space, the more the control valve modulates open to the coil, and the more the valve opens, the colder it gets in the space. The controller thinks that it's heating by allowing hot water to flow through the coil, but it's not hot water in the pipes, it's cold water! The ultimate result of this situation is that the control valve modulates fully open, allowing full flow through the coil, and the space becomes "very chilly."

If a dual-temp system is operating in a cooling mode, then the fan coils associated with the system have no means of providing heating. Any spaces served by any fan coils, that require heating, are basically outta-luck. Turn this around and we see that the opposite is not true, if the units are equipped with economizers. If the dual-temp system is in a heating mode, it is true that the fan coils cannot use the control valve to provide cooling. They can use the economizer, though, and they will! The point made here is that dual-temp fan coils can only provide cooling during dual-temp system cooling modes, yet can provide both heating and cooling during system heating modes, if they are equipped with economizers.

To sum up this section, dual-temp fan coil units have only one coil, and are to be connected to a dual-temp piping system. The coil can provide heating and cooling, just not at the same time; what it can provide of course depends upon the temperature of the water supplied to it by the piping system. The temperature of that water is usually either a function of the outside air temperature, or a seasonal function. Control methods for this type of fan coil unit that remain within the realm of simplicity are very similar to those of which are outlined in section 2 (and even section 1) of this chapter. The fan coil unit controller needs to know the mode of the dual-temp system, so that it can operate the control valve appropriately. This is normally accomplished with an aquastat that senses the temperature of the water in the pipe. Finally, a fan coil unit can only provide cooling (and not heating) when the dual-temp system is in a cooling mode, yet can provide both heating and cooling when the dual-temp system is in a heating mode, if equipped with an economizer.

Chapter 11

BUILT UP AIR HANDLING UNITS

The term "air handling unit" is a generalization, used to describe everything from single zone packaged rooftop units and fan coils to full blown VAV air handlers. Technically, anything that "blows air" can be called an air handler. For our purposes however, we reserve the term for the types of equipment discussed in this chapter.

This is the author's favorite topic, and also probably the most involved subject of this book. It is the last chapter to actually be written. This single chapter cannot possibly cover all there is to know on the topic of air handlers, but it will attempt to give the reader a clear overview of common applications. The entire chapter will be based upon the use of digital controllers operating the air handling equipment, with digital logic performing the various control sequences described herein. Remember from our discussions on fan coil unit controls, that if the control system becomes too complex, we need to "cross over" into the land of air handler controls. We will assume that, for all practical purposes, equipment of this level of complexity will be most appropriately controlled by an "equipment level" digital controller, with all switches, sensors, transmitters, and end devices wired to the controller, and the controller programmed to perform the required sequence of operation.

The first step here is to define, or reiterate, what is meant by the term air handling unit. To do this, we make a distinction (or two) between air handlers and built up fan coil units, and also between air handlers and built up make-up air units. Like fan coil units, air handlers are made up of a series of singular components. All air handlers will of course have a fan, and at least one means of heating or cooling, and usually both. Built up air handlers can range in size from a small "fan coil size" unit to a large central station VAV unit. They can serve a single zone, as can a fan coil unit, or they can serve multiple zones, as is the case for a VAV or reheat application. Large air handlers will always be referred to as such, whereas small, single zone air handlers can theoretically be referred to as fan coil units. If air handler control complexity exceeds that of what would merit unit level controls, then the unit is called an air handler, and given equipment level controls.

To distinguish the term "built up air handler" from the term "built up make-up air unit," consider that back in the chapter on make-up air units, in the sections on built up systems, we talked only about built up make-up air units that handle 100 percent outside air, with no return air connections or capabilities. We can say that built up make-up air units are built up air handlers that handle nothing but outside air. The moment we put a return air connection on a built up make-up air unit, we change the designation of the unit from "make-up air unit" to "air handling unit." At least for the purposes of this writing.

The approach of this chapter, as with most of the chapters in this book, is to lay down the basics, and build upon them. To that end, we will first discuss the basic mechanical components that can make up an air handler. Then we move on to talk in detail about controls components commonly utilized in air handling unit control systems. From there we engage in a discussion on formulas and analysis tools that help to provide insight into how an air handler has been designed to operate, and how it is intended to be controlled. This is a very important section, and serves to equip the control systems engineer with an arsenal of tools that will help him/her understand the mechanical design issues, and how they relate to air handler control.

The next few sections will cover the various air handler applications, specifically, single zone systems, reheat systems, and VAV systems. And following that is an entire section devoted to ventilation strategies. With the subject of Indoor Air Quality being at the forefront of our industry, this section is "required reading," not only for the control systems engineer, but for anyone in this business involved in the design, installation, or maintenance of ventilation systems.

Are you ready? Okay then, here we go! We start by taking a real good look at all of the singular mechanical components that could possibly make up an air handler. Air handler configuration ranges from the simple to the complex, as we start adding on and building more features and functions into it. Without any further banter, let's move on and see what this is all about!

Basic Components

The construction of built up air handlers typically consists of multiple "modules" all fitted together in such a manner as to physically and functionally fit the given application (refer to Figure 11-1). At the heart of all air handlers is, what else, the supply fan! Other fans can make up the air handler as well, as we see ahead.

Fan Modules

The supply fan takes its place in the built up air handler either up front, ahead of everything else (draw-through configuration), or somewhere in the middle of the unit, perhaps behind the coils (blow-through configuration). The fan is properly sized to overcome the static losses of the air handler itself (coils, filter section, etc.), and also the static losses associated with the duct distribution system. A fan of this size will most always be powered with a three-phase motor.

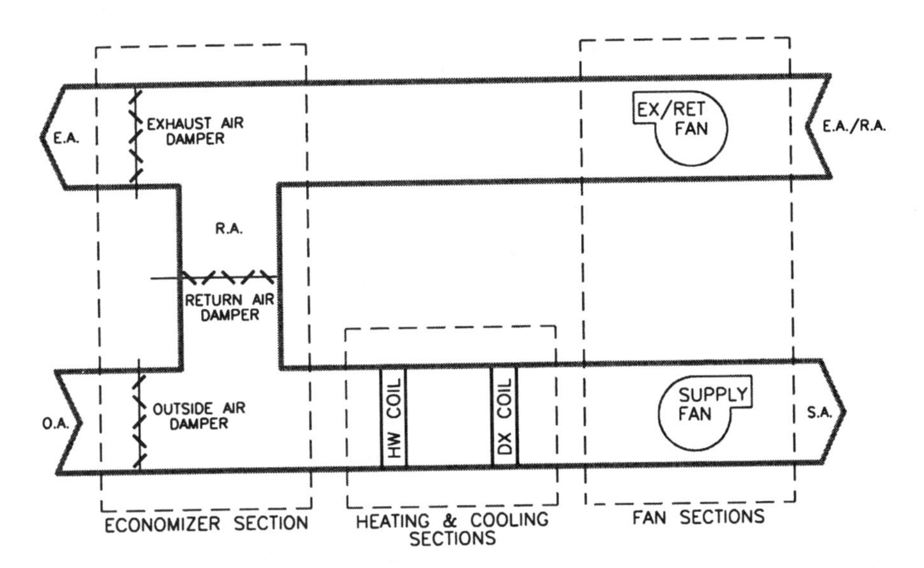

Figure 11-1. Built up air handling unit showing the various sections or "modules" that can comprise such a unit.

An air handler may be equipped with an exhaust/return fan, which is positioned to assist the air to get from the space served back to the unit. Depending upon the position of the economizer dampers at any point in time, the fan may actually deliver more air to the exhaust damper than back to the unit for another go-around. If the outside and exhaust air dampers are fully open and the return air damper is fully closed, then the fan returns no air to the unit. Hence the dual description. In an air handler application requiring an exhaust/return fan, you can bet that this fan will be driven by a three-phase motor.

An air handling system may also consist of an exhaust fan (in lieu of an exhaust/return fan). This fan is not situated to help air return to the unit. It is located after the return of the unit. If the economizer dampers are positioned to allow only a minimal amount of outside air to be introduced, then this fan is off, and the majority of the air is returned to the unit. If the outside air being brought in exceeds some critical point, then the exhaust fan is energized.

Heating Modules

The heating section of an air handler can take the form of a steam, hot water, or electric heating coil. Steam coils are generally reserved for make-up air units, so we won't consider these types of coils for air handling unit applications, which is not to say that they can't be utilized.

Hot water coils will be fitted with either two-position or proportional control valves, depending upon their required function. We will see why this is, when we get to talking about air handler applications. More and more, a proportional control valve will be specified for all air handler hot water coil applications. The premise is that two-position control can easily be done with a proportional control valve, and since a digital controller can be programmed to perform either type of control on the valve, why not have the added flexibility of proportional control, even if it's not specifically required.

Electric heating coils will either have enough stages required for the given application, from one to a million (give or take!), or will be equipped with an SCR (see Chapter 9, section 3). They will also be factory equipped with a thermal cutout and an air proving switch. If staged control of the heating coil is specified, then the air handler's main digital controller can feasibly take over the staging duties. However, if the number of stages is rather large, then a step controller can be specified to come as factory mounted, and the air handler's main controller can simply send a control signal to the step controller, which in turn handles the staging duties. In the case of SCR control, the SCR will be capable of accepting a signal from the digital controller, and this is how control of the heating coil should be implemented.

Gas-fired furnaces or heating modules are available, depending upon the application, and also on the manufacturer of the air handler. In other words, does the particular manufacturer even offer such a product for integration with the air handling unit? Other applications may consist of adding an indirect-fired duct furnace downstream of the air handler, and incorporate control of the furnace into the air handler's control system. These types of applications are the exception rather than the norm, so we will focus our concentration mainly on those applications that utilize either hot water or electricity as the means to heat.

Cooling Modules

The cooling section of an air handling unit can generally take on only one of two forms: chilled water coil or DX coil.

A chilled water coil will be fitted with a proportional chilled water control valve, always! Remember, we're talking about air handler controls here, and their associated level of complexity. There is no reason to ever go with two-position control on the chilled water coil of a built up air handler. Another way to put it is, if the application at hand can get away with simple two-position control on the chilled water side, then call it a fan coil unit! At least in terms of controls requirements. For any sizable air handling unit, its application will undoubtedly demand proportional control of the chilled water coil.

A DX coil produces cooling by means of refrigeration. A remote air cooled condensing unit is located outdoors and refrigeration piping connects the condensing unit with the DX coil. This type of configuration, with components of the refrigeration cycle located both indoors and outdoors, and refrigeration piping between them, classifies as a "split system." Cooling is implemented by staging compressors back at the outdoor unit, and thus the air handler's controller must be wired out to the unit. Two stages of control is typical for single zone applications, whereas more stages of capacity control are generally required for reheat and VAV applications, where control is performed to maintain discharge air temperature setpoint, rather than space setpoint. As with electric heating coils, staging of the condensing unit can be performed by a factory furnished step controller or sequencer, wherein the air handler's main controller simply sends temperature and setpoint information (in the form of a control signal) to the sequencer, and the sequencer itself handles the staging of the compressors.

Economizer Module

The economizer section of a built up air handling unit will minimally consist of an outside air damper and a return air damper. Unlike with the small built up fan coil unit, these dampers will typically not be mechanically linked together, and will need to be controlled by separate damper actuators.

Most often an exhaust air damper will be part of the economizer package, set up to operate in unison with the outside air damper. Though this is most often the case when the air handler is equipped with an exhaust/return fan, it is not *necessarily* the case when the air handler is equipped with an exhaust fan. Often in this application, the exhaust air damper will be controlled independently, to maintain space static pressure. This is normally the case if the exhaust fan is not equipped with any type of air quantity control. If the fan is equipped with air quantity control, then the damper actuator becomes a two-position actuator, energized whenever the exhaust fan is on. Sometimes a simple non-motorized backdraft damper will serve as the exhaust air damper, depending upon the application.

The purpose of the economizer section of a built up air handling unit is two-fold. First, it is utilized to take advantage of outside air as a source of free cooling, outdoor conditions permitting. Second, and perhaps more importantly in the eyes of engineers and indoor air quality proponents, the economizer section of an air handler allows the introduction of outside air at a rate required for the proper and adequate ventilation of the spaces served by it. The operation of the economizer section with respect to indoor air quality will be dealt with in detail in the upcoming section on ventilation strategies.

Other Modules

Various other modules can make up the construction of a built up air handling unit. At least one filter section will be part of any air handler. Sound attenuators, air mixers, and access sections are just a few of the additional components that can make up an air handler. We will restrict our discussions to those types of modules that rely upon controls in some form or other. Suffice it to say that other components do exist, and could be required for any given application.

Common Controls

As pointed out in the introduction of this chapter, we will base our air handler control systems on the use of digital controls, as opposed to built up electromechanical/electronic control systems. This is not to say that air handler control schemes can't be handled without the use of microprocessor-based controls. The complexity of air handler control sequences simply lends it to be more easily handled by digital controllers, whether they operate as stand-alone, or whether they be fully networked with other controllers, thereby making up a Building Automation System (BAS).

With that said, we begin this section by introducing the "main controller," its physical description, its functions, etc. Next we cover sensors and controllers that are typically required for air handler control. Following that we explore the various end devices that enable air handler operation and control. Finally, we discuss the various safety devices that an air handler will normally be equipped with.

Main Controller

Figure 11-2 shows the air handling unit main controller as an equipment level digital controller with inputs and outputs, enough required for the given application. Analog inputs will accept signals from sensors and transmitters, and binary inputs will accept contact closures from external switches and two-position controllers. Analog outputs will serve end devices that can accept a proportional type signal, such as modulating valve and damper actuators, VFDs, and equipment embedded sequencers, and binary outputs will serve to control two-position end devices, such as two-position valve and damper actuators, fan starters, and miscellaneous field relays. A virtual brain exists between the inputs and outputs. The brain of course must be programmed to perform the required equipment-wide sequence of operation. In essence, the controller must process all sensor values and two-position inputs, coordinate all output values, and tie everything together as a single sequence.

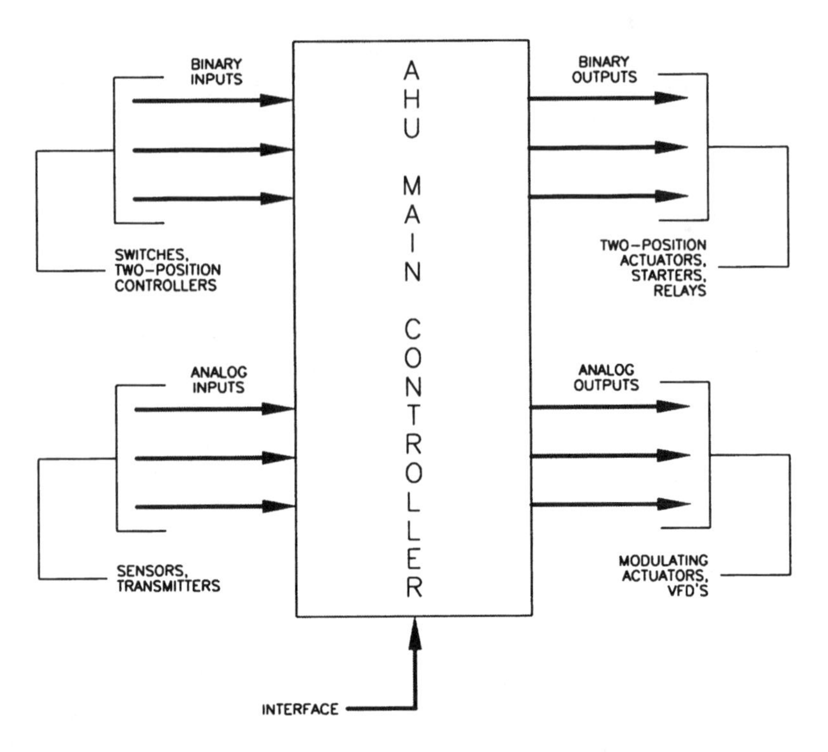

Figure 11-2. Equipment level digital controller, serving as the main controller for a built up air handler. The main controller processes all inputs and coordinates all output values in alliance with the Sequence of Operation. Interface to the controller is required for initial programming, commissioning, and adjustment of operating parameters (as necessary).

In addition to its obvious duties of controlling all outputs in a manner beneficial to the overall operation of the air handler, the main controller also has many other functions, some of which are discussed here.

Time-of-day scheduling is typically performed by the main controller. A weekly schedule is set up within the controller, which establishes occupied and unoccupied modes. The controller will be programmed to operate the air handler differently between the two modes, depending upon the requirements of the application. This typically includes fan control (continuous during occupied, intermittent during unoccupied), outside air damper control (closed during unoccupied), and setpoint control.

Speaking of setpoint control, perhaps the main controller's most important function is the implementation and the enforcement of this. Occupied and unoccupied temperature setpoints, as well as pressure and humidity setpoints (as the application requires), are typically programmed into the controller, and the controller operates the equipment in accordance with these setpoints. Other setpoints that can be programmed into the controller are: outside air damper minimum position, economizer enable or changeover setpoint, outside air temperature "lockout" setpoints (of heating and/or cooling modes), etc.

Note that all of these settings need to be set at the main controller, via some means of human interface. The programmer will have access to the controller, for initial program download and for commissioning purposes, and he will establish initial values for these various setpoints. However, if the end user needs to change or adjust any parameter, he must also have a means of interface to the controller, as well as some requisite knowledge. With a networked system, this is normally accomplished through the front end, which is a desktop computer hooked into the control network and running the appropriate software. The end user is trained on how to access changeable parameters, and is typically locked out of the actual programming of the controllers on his network. In other words, he is given just enough access to where he can perform time-of-day scheduling and change setpoints, but not enough to get into the nitty-gritty of the controller configurations.

With that said, what about the project that consists of a single air handler, controlled by a single stand-alone digital controller? The project hasn't been sold as "networked DDC," yet the complexity of the control schemes required to operate the air handler have driven the designer to utilize digital controls. There are a few ways to handle this. You can give the end user a software package, to run on his laptop or hand-held, that he can use as a means of interfacing to the digital controller. He must of course be willing and accepting of this method of interface. Another solution is to provide him with a "manufacturer specific" operator interface, which can be as simple as a Liquid Crystal Display (LCD) and a keypad, all in one device. This interface is set up to give the end user limited access to the controller, enough to allow the user to establish operating schedules and adjust important setpoints.

The third method consists of simply providing user adjustment with external devices. For time-of-day scheduling, a discreet time clock can serve as a binary input to the controller. All scheduling is done via the outboard time clock, and not via the digital controller, even though the controller has the capability. Temperature setpoints, and other settings requiring user adjustment, can be performed with external potentiometers, wired as analog inputs to the controller. These devices can be acknowledged by the digital controller as setpoint devices, provided that the controller is configured as such. Outside air damper minimum position can also be set by an external potentiometer, and the economizer enable setpoint can be performed by a two-position temperature or enthalpy controller serving as a binary input to the digital controller. In essence, just because an air handler is controlled by a digital controller, it doesn't mean that user access to the controller's "digital world" is mandatory. On the contrary, if enough adjustability is given to the user in the form of external devices, the user need not be concerned with the inner workings of the digital controller. Not in the least.

Sensors and Controllers

This section serves, in part, as a vehicle for discussing the various types of sensors and transmitters that are required for *digital* air handler control. The section starts out by discussing temperature sensor types, and then finishes by talking about other types of devices normally encountered in the control of air handlers, ranging from static pressure and humidity transmitters to airflow proving and clogged filter switches. For an illustration of a built up air handler equipped with some of these devices, please refer to Figure 11-3.

Electronic temperature sensing in HVAC applications is typically performed by a device called a **thermistor**, which is simply a variable resistance temperature sensor. As the temperature sensed by the thermistor changes, so too does its resistance value. This relationship of temperature versus resistance is exploited in the area of electronic temperature control, as we see a perfect example of in our digital air handler control system. Thermistors will serve as passive analog inputs to our digital controller, in which the wired resistance value can be converted into actual temperature information and processed by the controller's brain. The term "passive" implies that the device itself does not generate any voltage. A voltage is sent to the device via the digital controller, and a voltage is returned from the device, which corresponds to the resistance of the device, and therefore also to the temperature that the device is measuring.

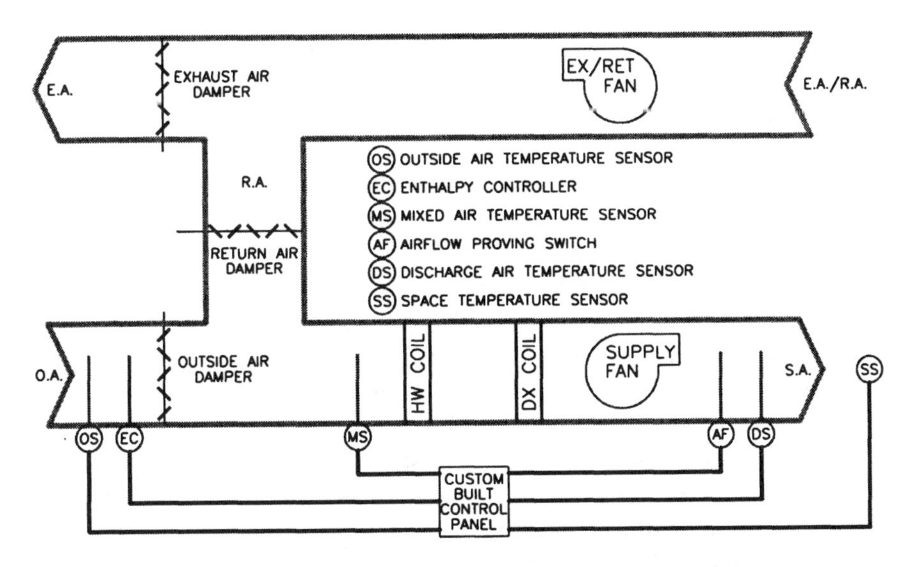

Figure 11-3. Built up air handler showing some of the various sensors and controllers that the air handler can be equipped with, all wired back to a common control panel. Control panel houses the main controller, as well as other devices.

First on the agenda is to talk about the **space sensor**. A space sensor is that which is mounted in the space served by the air handler. The thermistor takes the form of a small solid state electronic device, mounted on a printed circuit board. The printed circuit board is enclosed in a decorative cover, and the whole apparatus is mounted on an inside wall. The space sensor may also have a temperature adjustment, which simply allows the user some setpoint adjustability at the sensor. In addition, the sensor may have a button or two, that can typically be utilized for "timed override" (and cancel) from unoccupied mode to occupied mode.

A **discharge air sensor** will be mounted in the supply duct, downstream of the fan and any coils. The discharge air sensor will be constructed as a probe, that is inserted into the duct and secured. The length of the probe can vary, and sensor manufacturers will typically offer various discreet lengths. Depending upon the duct size, the length of your discharge air sensor may be as short as 4 inches, or as long as 18 inches or more.

A **mixed air sensor** will be installed in the mixed air chamber of the air handling unit. This sensor takes the form of a long averaging element, and should be installed such that the element traverses the cross section of the chamber, in a serpentine fashion. The mixed air chamber is the section of the air handler where outside air meets return air. The mixture of these two components is subject to stratification, which is a fancy term used to describe a mixture of air that is not uniform in temperature. This is typically the case with the mixed air section of an air handler. The mixture of outside and return air has not yet had a chance of blending thoroughly, and therefore stratification results. To accurately measure the mixture, an averaging type temperature sensor must be used, which when installed correctly, effectively senses the average temperature of the stratified mixture of outside and return air. Construction of the sensor actually consists of several thermistors, embedded along the length of the sensing element, and wired in a series/parallel configuration.

An **outside air sensor** will be mounted to a weatherproof electrical box, and equipped with a weather shield. The sensor should be installed outdoors, in a location indicative of outdoor conditions, and out of direct sunlight.

A **return air sensor** can be mounted in the return air duct of the air handling system. This sensor is identical to the discharge air sensor, in terms of construction. A simple probe type sensor, suitable for duct mounting.

The next few devices that we will talk about are generally found as "active devices." Unlike the thermistor type temperature sensors discussed above, these devices utilize active electronics to produce a control signal. The devices themselves require power, typically in the form of 24 volts DC, as provided by an external power supply. The devices measure the variable of which they were designed for, i.e., pressure or humidity. The variable is converted to an electronic signal, amplified by the device's active electronics, and transmitted as a voltage or current signal, back to an analog input at the digital controller.

Static pressure transmitters measure pressure, relative to some reference point, and transmit a signal in accordance with the pressure sensed. Duct static pressure transmitters, used in VAV applications, measure supply duct static pressure as compared to the static pressure outside of the duct. Space static pressure transmitters, used in both VAV and constant volume applications, measure building pressure with respect to the outdoors.

Humidity transmitters measure relative humidity, whether it be space, duct, or outdoor air. For our purposes, we will strictly limit humidity sensing, as used in this chapter, to outside air sensing. An outside air humidity transmitter can be used to aid in the decision of whether to economize or not.

The remaining devices discussed under the present heading are simple two-position controllers, used to handle various singular tasks in the overall control of an air handler.

An **enthalpy controller** is a two-position device that can be wired as a binary input and utilized as an external means of establishing economizer enablement. Rather than measuring outside air temperature with a thermistor and outside air humidity with a humidity transmitter, and then performing an enthalpy calculation with the digital controller, this device can establish economizer changeover externally, with a simple onboard adjustment.

An **air proving switch** determines whether the air handler supply fan is operating or not. This takes the form of a sail switch or, more preferably, a differential pressure switch. Current sensing switches are also employed as fan proving switches. They don't directly monitor airflow, yet they confirm fan operation by monitoring the electrical current draw of the fan motor.

A **clogged filter switch** is a differential pressure switch installed across the entering and leaving side of a filter section. This device does not serve any control purpose. Rather, it is employed as a means of monitoring the air handler's filter bank, and alerting maintenance personnel as to when the filters need to be changed out. The switch can be wired into a binary input of the digital controller, and a binary output can be used to light a "Change Filters" light. More commonly in standalone applications, the switch will bypass the digital controller altogether, and be hardwired directly to the light.

End Devices

Typical end devices found as components of a built up air handler are the topic of this heading (see Figure 11-4). We of course talked at great length about end devices back in Chapter 5, so we shan't go down that path again. This section simply serves to list out the devices as they apply to the air handler, and to cover some typical requirements in terms of the devices themselves and how they are to operate.

Fan starters will be required for all constant volume applications, and for all VAV applications that utilize inlet guide vanes as the means

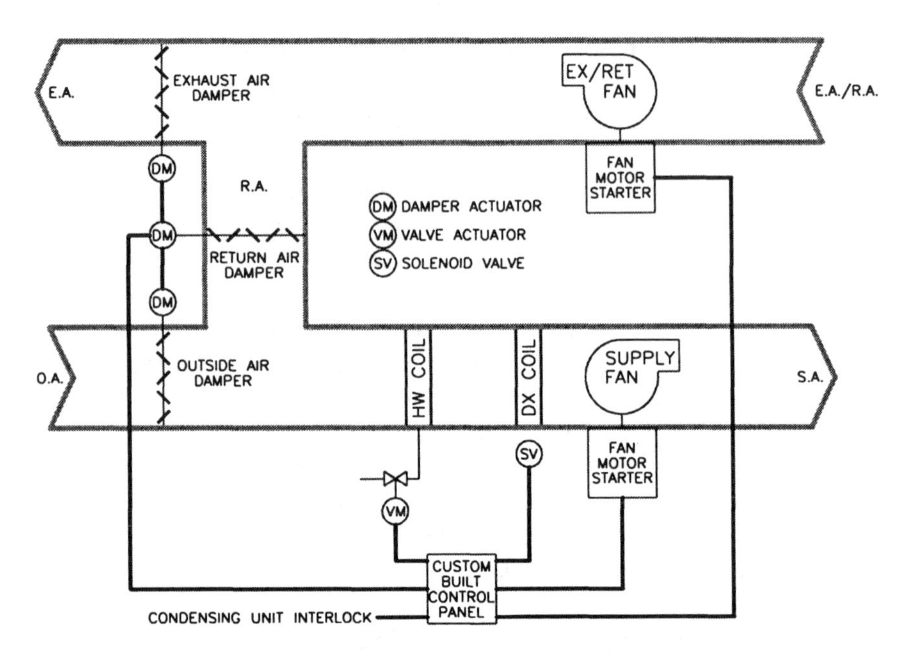

Figure 11-4. Built up air handler showing required end devices for a typical application.

of air quantity control. This statement assumes that the fans of a built up air handler will be three-phase. A foregone conclusion for air handlers of any considerable size. The starters will interface with the main controller via "interposing relays." The coils of the relays will be controlled by binary outputs off the main controller, and the contacts of the relays will tie in to the starters' control circuits, thus allowing for fan start/stop control via the main controller. On paper, the binary outputs can directly interface with the starters, yet in practice, it's wise to drive a relay with the binary output, and have the relay interface with the starter. This is primarily done to protect the sensitive binary outputs of the controller from starter control circuits whose current draws may be excessive or subject to large coil inrushes, or "bursts" of current.

Inlet guide vanes are a means of implementing air quantity control. Used on supply fans in VAV applications, and on exhaust fans for building pressure control, they will be operated by an IGV damper actuator. The damper actuator will typically be mounted so that it springs the vanes closed (minimal airflow) upon loss of power. This is the normal fail-safe position of the inlet guide vanes. The proportional actuator will be controlled by an analog output off the main controller.

Variable frequency drives will be utilized for VAV applications and building pressure control, as a high tech alternative to inlet guide vanes. The VFD will require two points of control from the main controller: a start/stop signal (binary output) and a proportional control signal (analog output). The start/stop command simply enables the VFD to operate, and the VFD varies the speed of the fan motor as a function of the proportional control signal received.

Outside and return air dampers will be fitted with proportional damper actuators, and will operate in unison, in opposite directions, meaning that as the outside air damper opens, the return air damper closes in direct physical proportion. The actuators will be mounted so that the outside air damper springs closed and the return air damper springs open upon removal of power. The actuators will be controlled as one, by a single analog output.

An **exhaust air damper** will normally be fitted with a proportional damper actuator. If the air handling system includes an exhaust/return fan, then the damper will be controlled in conjunction with the outside and return air dampers, by the same control signal. The damper will stroke open as the outside air damper strokes open, and will spring closed upon removal of power. If the air handling system includes an exhaust fan in lieu of an exhaust/return fan, then the damper may be controlled independently from the outside and return air dampers. If the exhaust fan is equipped with IGVs or a VFD (for building pressurization control), then the damper actuator will simply be a two-position device, energized whenever the exhaust fan is placed into operation. Otherwise, the damper itself may be employed for space pressure control. The actuator will be controlled proportionally, by a different analog output than the outside and return air damper actuators.

Hot and chilled water valves have been discussed in the first section of this chapter. The hot water control valve will be proportionally controlled in all single zone applications, yet may be a two-position valve in VAV and reheat applications. The chilled water control valve will always be proportionally controlled, regardless of the application. The use of two-way versus three-way valve bodies is project dependent. Generally, the hot water valve will spring open to the coil and the chilled water valve will spring closed to the coil, upon removal of power. This is traditionally what is called for, yet all applications should be uniquely evaluated, rules of thumb aside.

Solenoid valves are two-position refrigeration valves. These are normally closed devices, and are powered open, either by 24 volts or 120 volts. The requirement of solenoid valves with split system air handlers is generally dictated by the manufacturer of the remote air cooled condensing unit. Refrigeration piping is run from the outdoor condensing unit to the air handler's DX coil. Solenoid valves are required to be installed local to the DX coil, one per refrigeration circuit, as per the manufacturer's recommendations. For air handling units utilizing DX cooling, the condensing unit normally has control over the remote solenoid valves. In other words, the solenoid valves are wired back to the condensing unit, as per the unit wiring diagram. Staging is handled either directly by the main controller, or by the condensing unit's integral step controller (if equipped). The condensing unit's control system determines when to open and close the solenoid valves, in coordination with the number of cooling stages called into operation.

Safety Devices

Safety devices that built up air handlers are typically equipped with are discussed under this heading. Each particular device has its application and its function. Each device is of the manual reset variety, affecting a unit shutdown upon a trip of the device. Safety devices such as these are generally hardwired, in series with each other, so that if any single device trips, control power is interrupted, and operation of the entire unit is halted. Figure 11-5 shows our built up air handler equipped with some of these devices.

Not all of these devices are required for every air handler application. In fact, certain applications may call for the need for none of these.

A **freezestat** would require to be utilized in any application having water (or steam) coils. As stated in the chapter covering safety devices, the freezestat's sensing element is normally mounted to the leaving side of the air handler's coil. When an air handler has more than one coil, then it's up to the controls designer to determine the best spot for the freezestat. Coil configuration within the air handler provides the basis for this all-important decision, as does some insight into the overall operation of the air handler, and a suitable location should be chosen so as to avoid nuisance trips, and at the same time not compromise the purpose of the device.

A freezestat should also be installed in applications in which the

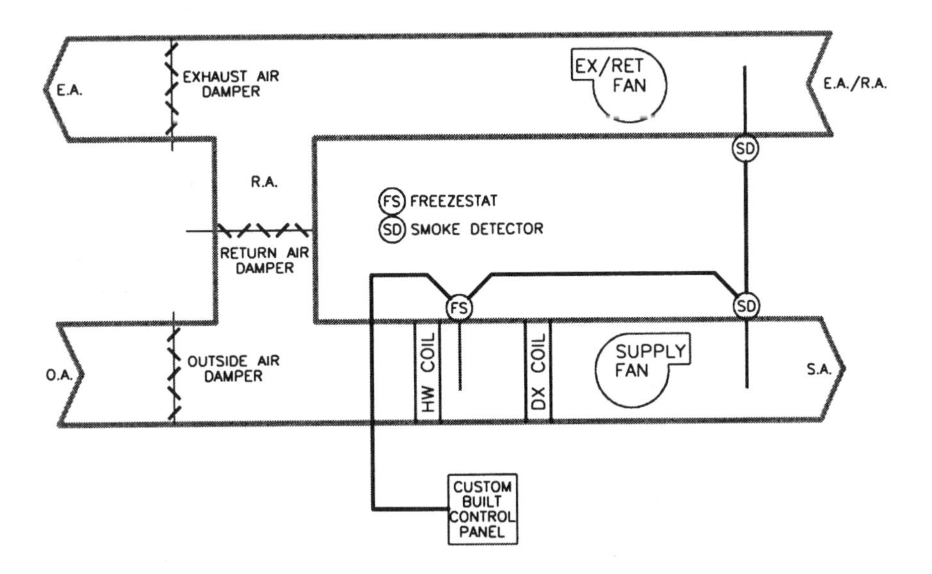

Figure 11-5. Built up air handler showing some typical safety devices. These devices are normally hardwired is series with each other, such that if any single device trips, the entire control system is shut down.

air handler serves downstream water coils, as would be typical for a VAV/reheat system. Even if the air handler itself does not have a water coil, it should be equipped with a freezestat, with the device's sensing element installed such that it "cross-sections" the discharge air duct.

It should be noted here that if there is no possible chance of coil freeze-up, then a freezestat need not be used. Applications such as this would include: units that bring in no outside air at all, units serving in climates that never get cold enough for one to be concerned with such potential problems, etc.

Smoke detectors are code-mandated safety devices, their requirement depending upon the physical size of the air handler, in terms of how much air they move, and also depending upon the code that's followed by the particular municipality. Generally mandated for any air handler moving more than 2000 cfm, the smoke detector will be installed in either the supply or the return air duct. Different schools of thought dictate which duct it goes in. On the one hand, installing the smoke detector in the supply air duct would make it easier for the detector to shut down the unit if the unit itself in some way caught fire. On the other hand, the smoke detector would be better suited to detect smoke in the actual space served by the air handler if the detector were installed in the return air duct. Yet if the overall air handling unit has no exhaust capabilities, then its return air duct may experience little or no airflow during periods of economizer operation, rendering the detector useless during these cycles. Eradicating this notion is the fact that many of the larger systems *will* have exhaust capabilities as part of the air handling unit, and the main return air trunk, which also serves as the exhaust path, will always have airflow through it, whether the system is economizing of not.

Regardless of where it's installed, the detector will require power, typically of the line voltage variety, and often specified to be from a different source, or circuit, than the air handler's control power source. You might ask why this needs to be. If the purpose of the smoke detector is to shut down the air handler upon detection of smoke, then why would the smoke detector need to remain powered in the event of air handler control power failure, since loss of control power equates to air handler shutdown? The issue of concern is that the smoke detector often pulls "double duty," with its auxiliary set of contacts wired to a building level fire alarm or smoke detection system. A random trip of air handler control power should not kill power to the smoke detector, as this may effectively trip the detector as well, and issue a "trouble condition" alarm at the building level system.

The use of two duct mounted smoke detectors, one in the supply air duct and one in the return air duct, would be a requirement dictated by the size of the unit, or simply dictated by the engineer. In other words, there comes a point in size, in terms of how much air the unit handles, that codes will mandate the use of two detectors. In addition, some engineers will specify the use of two detectors, simply out of professional preference, regardless of the size of the air handler.

A **high limit static pressure switch** is required in variable air volume applications, in which the supply fan operates at intermediate levels of air delivery. This of course accomplished by either IGVs or a VFD. Failure of such equipment can result in the air delivery of the unit exceeding required volume, with no regard to system capacity. In the event of such a failure, the static pressure switch, with its high pressure port piped into the supply air duct, will prevent overpressurization of the duct, by affecting a unit shutdown upon detection of excessive pressure.

FORMULAS AND ANALYSIS TOOLS

Before we take a look at some air handler applications, we take the time here to develop a working knowledge of the processes that make up a typical air handling unit. Processes such as, how the outside air is mixed with return air, and what the resultant temperature of that air is, how this air is then delivered to the heating and cooling components of the air handler, and how it is conditioned by these components. Formulas describing these processes provide the controls designer with some idea as to how the system was designed to operate, and also provide invaluable insight as to how it should be controlled to meet that end. One of the best tools that a control systems engineer can have is an understanding of how to "work the numbers" to formulate his/her own theories on how the air handling system should operate. Instead of blindly designing a control system for a large multifunctional air handler, the controls designer must have insight into the various processes of the system. The designer will thus be better equipped to handle the overall design of the entire system, so that all processes are appropriately accounted for, and all variables are worked into the control system design.

Outside Air, Return Air, and Mixed Air

As stated earlier in this chapter, the economizer section of an air handling unit has two functions. One is to utilize outside air for free cooling, conditions permitting. And the other is to introduce the proper amount of outside air as required for adequate ventilation of the space(s) served by the air handler. The latter function brings about a world of complexity to an air handler, in terms of how the outside air component is introduced and conditioned. Saving the bulk of discussions on these complexities for a subsequent section of this chapter, we focus here on the mixture of outside air and return air, and how these air paths are related, mathematically.

Let us explore that relationship, by way of example, and take a look at a typical application in which a constant volume air handler serves one large zone. From the ventilation and equipment schedules on the engineer's mechanical plans, we gather that the air handler is sized to deliver 8,000 cfm. Also from the schedules, we see that the total required amount of outside air is tallied to be 1,200 cfm. In this example, the outside air damper must be positioned to allow this quantity of outside air to be introduced, as a minimum, for all times that the air handler operates in an occupied mode.

The first thing that we need to determine is what percentage of the total cfm needs to be outside air? The percentage is simply found by dividing the outside air value by the total cfm value, and multiplying the result by 100. So in this case, the percentage of outside air is found to be (1,200/8,000) × 100, or 15 percent. The return air, therefore, is 85 percent.

So what's the big deal? Well, by knowing the ratio of outside air to return air, and also knowing the temperature of the outside and return air, we can calculate the temperature of the mixed air, using the following formula:

$$MAT = \{(RAT \times \%RA) + (OAT \times \%OA)\}/100$$

A useful formula indeed! How so? Well, from this formula a table can be generated, showing mixed air temperature values for all outside air temperatures. The only thing missing from the formula is the return air temperature. For the sake of generating such a table of values, let us assume that the temperature of the air returning from the space to the unit is 74 degrees. This is a fairly safe assumption, assuming that the space is being maintained at room temperature.

By plugging in values for outside air temperature (***OAT***), at 5-degree increments, using Chicago design days to define our range, and holding the return air temperature (***RAT***) constant, Table 11-1 is generated.

So what does this tell us about the air handling system, and how it's been specified? Well, maybe nothing yet, but when we start looking at the heating and cooling components of the air handler, we begin to understand the relevance of the above table. The table on its own does provide some insight into what the air handler is doing if it's neither heating nor cooling at any point in time. In such a scenario, it's basically serving up mixed air as the main course! In other words, the temperature of the air delivered into the space(s) served by the air handler during periods of no heating or cooling is what the temperature of the mixed air is, save for picking up a degree or two from the heat generated by the operation of the supply fan.

Incidentally, another good little rule of thumb derived from this equation (that we've brushed upon in a previous chapter) is that, for a return air temperature of 72 degrees, an outside air percentage of 20, and

Table 11-1.

Outside Air:	15%
Return Air Temp:	74 deg.
Outside Air Temperature	**Mixed Air Temperature**
–10	61.4
–5	62.2
0	62.9
5	63.7
10	64.4
15	65.2
20	65.9
25	66.7
30	67.4
35	68.2
40	68.9
45	69.7
50	70.4
55	71.2
60	71.9
65	72.7
70	73.4
75	74.2
80	74.9
85	75.7
90	76.4

an outside air temperature of –10 degrees, the mixed air temperature equates to 55 degrees. We will see further on why this is important, in climates where the temperature of the air outdoors can get down this low, and lower.

Now, what else can we learn from this formula? If we take it and twist it around, we can use it to solve for $\%OA$. Whoa, wait a minute! Isn't that fixed at 15 percent, from the engineer's ventilation schedule? Yes, the *minimum acceptable percentage* of outside air is extracted from the mechanical drawings. But during an economizer cycle, the economizer section of an air handler will bring in varying amounts of outside air, from the minimum acceptable value, up to 100 percent! This depending upon how cold it is outside. You see, the economizer cycle normally strives to maintain a 55-degree air temperature (or thereabouts) at either the mixed air chamber or at the discharge of the unit. It does so by modulating the outside and return air dampers in unison, in opposing directions. So as the outside air damper opens from minimum position toward fully open, the return air damper closes accordingly. Proper control of the economizer will result in a steady mixed air temperature of around 55 degrees, by modulating the outside and return air dampers and mixing the two air streams in the correct proportions.

So back to the equation. If we solve for *%OA*, then the formula ends up looking like this:

$$\%OA = \{(MAT - RAT)/(OAT - RAT)\} \times 100$$

Holding the mixed air temperature constant at 55 degrees, and holding the return air temperature constant again at 74 degrees, we generate Table 11-2.

Reworking the formula in this manner to come up with the preceding table may or may not be that helpful in terms of air handler analysis. Still, it's a good way of illustrating what happens with "the mix" during economizer cycles. Notice of course that the table only covers outside air temperatures of up to 65 degrees. For temperatures any higher than this, the outside air doesn't have much value as far as trying to use it for free cooling. The table assumes that the decision to economize is based on outside air dry bulb temperature (humidity unaccounted for), with changeover set at 65 degrees. For any outside air temperatures below this, the economizer is enabled, and for outside air temperatures above this, the economizer is disabled and the outside air damper resides at minimum position. Of course, with an outside air temperature of 55 degrees and an economizer cycle striving to maintain a 55-degree mix of outside and return air, the outside air damper will be driven to its fully open position, with the return air damper fully closed. During economizer cycles when the outside air is between 55 and 65 degrees, the outside air damper will be fully open. The economizer cycle will do what it can to maintain mixed air temperature setpoint. Once the outside air damper is modulated fully open, the economizer has done all that it can,

Table 11-2.

Mixed Air Temp:	55 deg.
Return Air Temp:	74 deg.
Outside Air Temperature	**Percent of Outside Air**
–10	23%
–5	24%
0	26%
5	28%
10	30%
15	32%
20	35%
25	39%
30	43%
35	49%
40	56%
45	66%
50	79%
55	100%
60	100%
65	100%
(70)	(minimum)
(75)	(minimum)
(80)	(minimum)
(85)	(minimum)
(90)	(minimum)

and can't do any more. The mixed air is what it is: all outside air at whatever temperature. Assuming that the return air is coming back in excess of 65 degrees, however, then the outside air still has value over the return air, at least in terms of temperature.

Either of the preceding tables can be generated, for any values, by properly manipulating the equation. A computer program can even be written to perform this function. The 21st century solution is to use an automatic spreadsheet to calculate the values and build the tables. Software may even be available, in the form of vendor software or mechanical design software, that has this feature. Of course the old-fashioned way works just fine, and for this the equation is of the utmost value (don't leave home without it!).

With all of this junk under our belts, let's briefly touch upon other uses for these tables. The first table assumes a fixed percentage of outside air. What if that value changes (aside from economizer cycles)? For instance, what if the outside air damper "minimum position" is varied or reset, based on some other variable, say, CO_2 level in the space served? What if the minimum position setting is simply intervened with manually? With these scenarios, the table's use comes in the form of being able to calculate mixed air temperatures for differing percentages of outside air. This comes in handy, especially if you're verifying the limits of your heating or cooling coil, to see just how much you can realistically reset your minimum position based on CO_2, on that worst case day (more on this to come). Another use for that first table is to analyze a VAV air handler whose outside air damper is operated to maintain a "fixed" cfm value, regardless of supply air quantity demand. Take that same air handler that we opened up this section with, only now consider it to be a VAV air handler. The maximum deliverable volume of air is 8,000 cfm. The amount of outside air is fixed at 1,200, for all operating modes. When the air handler is delivering its maximum cfm, then the amount of outside air is 15 percent. However, as the volume of air delivered by the unit decreases, the percentage of this that is outside air increases. Up to 100 percent, if the capacity drops off to 1,200 cfm! Utilization of the table may help define the lower boundary of deliverable air volume, and also provide insight into how the heating coil is sized, and how it should be controlled.

Now that we've learned how to analyze mixed air, for a number of different scenarios, it's off to the coils! The next two sections take a look at how the mixed air meets the coils, how it's conditioned by the coils, and how it leaves the coils.

Cooling Coil Capacity

When the mixed air hits the cooling coil, it's safe to assume that, if it is to be conditioned by it in any way, the temperature of the air will be in excess of 55 degrees. A cooling coil will generally be sized to take the mixed air down to a temperature of 55 degrees, and not much less. At mixed air temperatures below this, the cooling coil serves little or no purpose in the air handling process. Whether it's the economizer doing all the work on a demand for cooling, or whether the amount of outside air required to be brought in results in mixed air temperatures below 55, the cooling coil is basically out of a job in these scenarios, which I'm sure it's okay with!

Time to introduce a couple more terms used in the description of air handler processes. We speak of the role of the cooling coil as taking the temperature of the air entering the coil down to 55 degrees leaving the coil. **EAT** and **LAT** are acronyms that describe the temperature of the air entering and leaving the coil. Standard LAT (Leaving Air Temperature) for a cooling coil, though varying slightly from application to application depending upon psychrometrics (don't ask), is 55 degrees. The coil is sized to take such and such cfm of air at an EAT (Entering Air Temperature) of so and so, down to an LAT of 55 degrees. Typical Midwest design practices dictate that the coil be sized for a dry bulb EAT of 80 degrees. With an EAT of 80 degrees and an LAT of 55 degrees, we gather that the coil is sized to impart a 25-degree delta T to the full amount of air that it's capable of handling.

Where does the EAT of 80 come from? We can gain insight into the answer for this by again looking at the table relating mixed air temperature to outside air temperature, keeping return air temperature and percentage of outside air constant (Table 11-1). If we fix the return air temperature at 76 degrees and the percentage of outside air at 20, the mixed air temperature, or the EAT of the cooling coil, approaches 80 degrees as the outside air temperature approaches (summer) design day. A return air temperature of 76 degrees in the midst of summer is pretty realistic, especially if you account for the fact that the air picks up a couple of extra degrees after it leaves the conditioned space(s), on its way back to the air handler.

If the cooling coil is properly sized for these criteria, and the criteria are correct and adhered to, then the coil will be able to perform its job of cooling and dehumidification, for even the worst case days. If the variables are altered, for example, if the amount of outside air is increased upward from 20 percent, then the coil will not be able to perform up to spec on design days, and the comfort needs of the space(s) served by the unit will be compromised.

If the EAT and LAT are not available from the equipment schedule on the mechanical plans, then the following formula can be used to determine these values:

$$Btuh = 1.08 \times cfm \times delta\ T$$

This formula is the "airside" equation for heat transfer. The cooling capacity of the coil will be given, in Btuh (British thermal units per hour) or in MBH (thousands of Btuh). The cfm of the air handler will also be given. By using the above formula to solve for delta T, and assuming an LAT of 55 degrees, the EAT can be found.

Determining the delta T of a coil, whether it be a cooling or heating coil, can give us important information about air handler design criteria, if such information is not given. For instance, without explicitly being given information about entering air temperature or minimum percentage of outside air, these values can actually be found, by using our analysis tools. Consider that the cooling coil of an air handler is found to have a delta T of 27 degrees. Assuming that the design LAT is 55 degrees, then the EAT is 82 degrees (82 – 27 = 55). This in fact is the design mixed air temperature. Now, by using the formula that relates mixed air temperature with percent outside air, we can solve for the minimum acceptable percentage of outside air, assuming a return air temperature of 76 degrees and a worst case outside air temperature (summer design day) of 90 degrees. The result is an outside air percentage of approximately 43 percent. This tells us that the cooling coil was sized, after the ventilation requirements were figured out, to handle up to 43 percent of the total air handler volume, on worst case cooling days.

If you are given this information on the mechanical equipment and ventilation schedules, the above exercise at least serves as a check. If the values are validated, then you're ready to design your control system. However, if the values are in contrast, then there may be a problem somewhere, perhaps with the engineer's calculations. More likely, there may just be more to the design criteria of the air handler than is depicted in the design documents. For instance, if the coil seems to be oversized, the engineer may have built in some safety factor, or is perhaps concerned about dehumidification requirements. Likewise, if the coil appears undersized, the engineer may have a reason for that as well. At any rate, these exercises help to align the control systems engineer's thought processes with the mechanical design engineer's performance criteria. If things seem out of kilter after performing the exercises, then maybe it's time for the controls engineer to establish communication with the mechanical engineer, and get the design scoop "straight from the horse's mouth!"

Did you know that you can roughly size a chilled water valve by simply knowing the cfm handled by the chilled water coil, with virtually no other information on hand? It's true, with the aid of yet another analysis tool. We've introduced the airside equation for heat transfer. Now let's introduce the "waterside" equation:

$$Btuh = 500 \times gpm \times delta\ T$$

where gpm is the rate of flow through the coil, and delta T is the difference in temperature between the water entering the coil and the water leaving the coil. **EWT** and **LWT** are acronyms used to describe these values. We know that, if the gpm is known, we can size a proportional control valve. Most often the gpm will be given, as well as the EWT and LWT, normally shown on the equipment schedule for the air handling unit. However, even without these values given to us, we can still come up with a general idea for the size of the control valve, if we can just find the required gpm. That's where the above equation comes in handy. By being given or by calculating the Btuh value for the air passing through the coil, we can take that same value and plug it in to the waterside equation, solving for gpm. For a chilled water application, we assume that the design EWT is 45 degrees, and the design LWT is 55 degrees (pretty standard). So the waterside delta T is 55-45, or 10 degrees. Work the formula appropriately, and you have your gpm value! And hence your control valve size, by following the procedure laid out in the section of this book devoted to valve sizing.

As mentioned, cooling coil specifications (Btuh, gpm, and EWT & LWT) will normally be found on the equipment schedule, but the above exercise still serves as a check, and further aids the control systems designer in understanding the mechanical design concepts behind the numbers.

Heating Coil Capacity

Unlike with the cooling coil, we can say that the temperature of the air entering the heating coil can pretty much be anything, from room temperature all the way down to winter design day temperature. A heating coil in a simple single zone application may be sized to take the mixed air up to a temperature suitable for heating purposes. For such a system, the worst case mixed air temperature may be around 55 degrees or so, based upon a return air temperature of 72 degrees and an outside air requirement of 20 percent. The heating coil, therefore, may be sized for a delta T of 40 to 50 degrees, to take 55-degree air up to as much as

95 degrees or more. In systems in which the air handler is required to maintain a constant discharge air temperature of 55 degrees (VAV and reheat systems), a heating coil may simply serve the function of night setback and morning warm-up. In this case, the mixed air is 100 percent return air (no outside air during these cycles!), at a temperature somewhere around 60 degrees. So the heating coil may be sized for a delta T of only around 20 to 30 degrees, if that. Finally, in systems that are required to bring in substantial amounts of outside air, the delta T of the heating coil can approach 80 degrees or more! We initially touched upon this when we discussed the heating sections of make-up air units.

We see from the above paragraph that the EAT, LAT, and delta T of the air handler's heating coil are not as "standard" as with the unit's cooling coil. Like with the cooling coil, the capacity of the heating coil will be given, in Btuh (or in MBH) and the delta T can be found. Yet simply knowing or finding the delta T of the coil does not give all the answers, and some introspect into the mechanical design of the system is warranted. Asking questions about outside air requirements, discharge air temperature requirements, and space heating requirements will provide some insight and put the control systems engineer in a better position to make the right choices of heating coil control, and integration of its control into the overall operation of the air handling unit.

When an air handler's heating coil is electric, its capacity will be expressed not in Btuh or MBH, but in the electrical term of kilowatts (kW), or "thousands of watts." The watt is a unit of electrical power, which is a function of both voltage and current. The formula relating kW to Btuh is as follows:

$$Btuh = kW \times 3410$$

The constant in the above formula has been rounded off to a whole number, for the sake of practicality. Anyway, we see that there's an extra step involved in determining the delta T of an electric heating coil: finding the capacity in Btuh. By way of example, let's check the heating capacity of a heating coil that might be used in a simple single zone application. From the section on Cooling Coil Capacity, we were able to roughly determine, in a roundabout way, the design outside air requirement of an handling unit, in percent, by analyzing the characteristics of the cooling coil. Let's assume that an electric heating coil is also a part of this air handler, and is sized to handle space heating requirements on a winter design day.

We found that the design outside air requirement was approximately 13 percent. Assuming that this is the minimum requirement of outside air for all times, we can then use this information to find the EAT of the heating coil. For this calculation, we'll use an outside air temperature of –10 degrees, and a return air temperature of 72 degrees, since we are solving for winter design criteria. Plugging in these values into the equation that relates outside and return air ratios and temperatures, we solve for mixed air temperature, and come up with 37 degrees. This is the "assumed" design EAT of the electric heating coil. If the coil capacity is given, then at this point the delta T and the LAT can be found. Imagine for a second that the coil capacity is not known, and assume that the heating coil should be sized to take the EAT up to a temperature suitable for heating purposes, say, 95 degrees. This yields a delta T of 58 degrees (95-37). To solve for capacity, one thing is missing here: cfm. Okay, for the sake of the example, this air handler delivers a constant volume of 3,000 cfm of air. Now the capacity of the heating coil can be calculated, and is found to be 1.08 × cfm × delta T, or 1.08 × 3,000 × 58, which equals 187,920 Btuh. Since the heater is electric, the capacity will be expressed as such, or as 55.1 kW.

Now check the equipment schedule for the actual capacity of the heating coil. If it's (close to) what you came up with, then your assumptions are probably accurate. If not, then you may be missing something, pertaining to how the unit is designed to function. For instance, if the heating capacity is substantially greater than what you've calculated using the above procedure, then maybe the outside air requirement is not constant, but to be reset upward as a function of occupancy. How much it can be reset can be figured out using the analysis tools that have been presented thus far. What if the heating coil capacity is less than what you've calculated? This can mean a number of things. It is most likely that the space heating requirement isn't what you might think it is. There may be heat generated by lighting and other internal loads, of which the engineer has figured for, that translate to a heating requirement that can be handled with supply air temperatures lower than 95 degrees. In fact, there may even be enough heat generated in the space that the heating coil is sized only to handle the outside air load. In this case, the heater will be sized for an LAT of 70 degrees or so. What this means is that the space isn't relying on the air handler for heat; that the space generates enough heat on its own. Of course, since there is an appreciable requirement for outside air, the heating coil must be there, and have the ability to "temper" the mixed air on colder days.

As the last issue to discuss on the topic of heating coils, we give equal time to hot water valve sizing based upon coil characteristics (as was done with chilled water). Consider that the design temperature of the water entering an air handling unit's hot water coil is typically 180 degrees, and the design temperature of the water departing the coil is 160 degrees. As with the chilled water coil, by knowing or calculating the airside Btuh value for the air blowing through the coil, we can determine gpm by plugging Btuh (and the waterside delta T of 20 degrees) into the waterside equation for heat transfer, and ultimately arrive at a proportional control valve size. Again, the coil data will likely be readily available from the equipment schedule, thereby eliminating any guesswork or speculation as pertaining to proper control valve sizing.

We are finally ready to pull everything together and start talking about air handling unit systems in their entirety. We start simple, by looking at single zone systems, then move on to constant volume reheat systems, and finish off with VAV systems. These next three sections attempt to cover the "basics" of air handler operation and control, and won't necessarily get into each type of system too deeply. There are a great many variations on each topic, too many to cover here. Still, the sections should do a pretty decent job in conveying the basics, and the section following these on ventilation strategies will expound on the basics and provide a nice finish to this all-important chapter.

Single Zone Systems

A single zone system is basically an air handler that is designed to operate at a constant volume of air and serve a single zone, with the temperature sensing/control point within the zone. The air handler's fan, heating coil, cooling coil, and economizer will all operate to meet the comfort needs of the zone served by the unit. In Figure 11-6, a single zone air handler is portrayed, with all the sensors, controllers, end devices, and safeties that might be required for a particular application.

Fan Operation

The supply fan of a single zone air handler will normally run continuously during occupied modes, at a constant speed, or air delivery

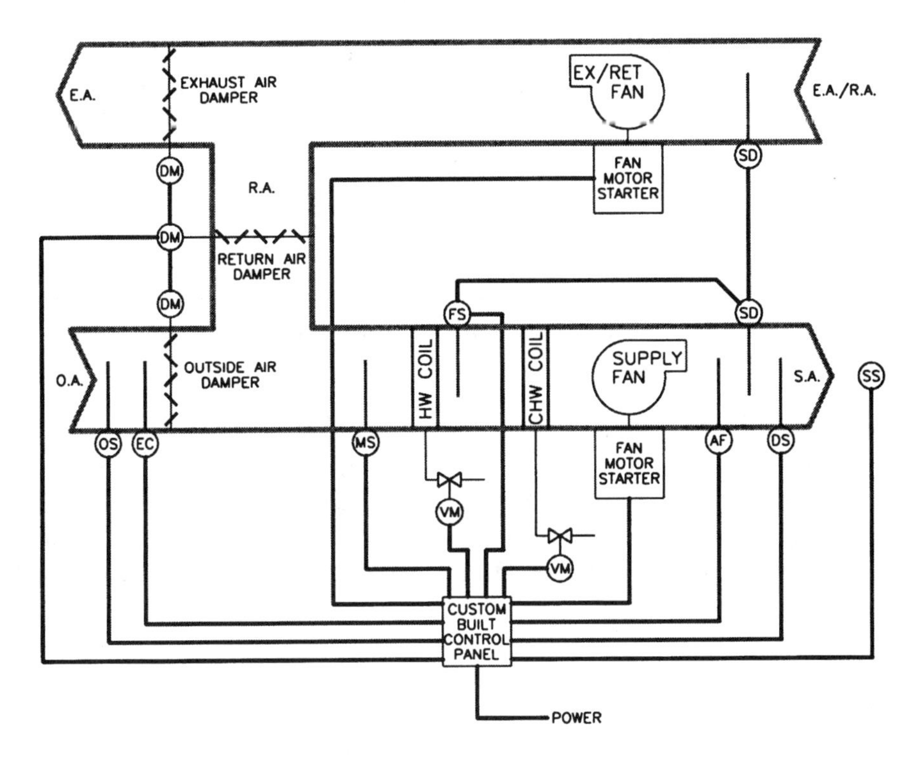

Figure 11-6. Single zone constant volume air handling unit, hot water heating and chilled water cooling, fully equipped.

rate. Operation of the fan, during occupied modes, *can* be intermittent (cycling upon calls for heating and cooling), though for most commercial and industrial applications, the fan must run continuously, whether or not there's a demand for space heating or cooling. This of course is because of the fresh air requirement (usually mandated by code) during the occupied mode of operation.

If the single zone air handling system is equipped with an exhaust/return fan, then this fan simply operates in unison with the supply fan. If the system is equipped with an exhaust fan, then the fan will be enabled whenever the supply fan is running, yet other circumstances might dictate the actual operation of it. The most popular mode of operation for such a fan is to simply enable it whenever the amount of outside air reaches some value, this generally done via an end switch on the outside air damper actuator. A step fancier is to impose space pressure control on the exhaust air damper; when the exhaust fan is engaged by the end switch, then proportionally control the exhaust air damper to maintain space pressure setpoint. The fanciest mode of operation by far is to have the exhaust fan operate via a variable frequency drive, and control the speed of the fan to maintain space static pressure. This is often more than anyone cares to invest in, at least for small single zone systems, but it is an option that can be considered and applied. The reason for the need for space pressure control stems from the nature and the position of this fan in the air handling system. An exhaust/return fan is not subject to the potential requirement for space static pressure control, in single zone air handling systems.

Heating and Cooling Operation

For this section, we consider two combinations of heating and cooling coils: electric heating/DX cooling (staged control), and hot water heating/chilled water cooling (proportional control). We consider both combinations aside from economizer operation, which we will cover in the next section.

For staged control of heating and cooling in a single zone system, the more stages the merrier! Actually, there is a reasonable cutoff to that statement. On the heating side, the electric heating coil will be sized to take the "worst case" mixed air up to a temperature suitable for the heating needs of the space. This can be a difference of as much as 40 to 50 degrees (or more). There is an oft used rule of thumb stating that, for mediocre staged space temperature control, the delta T per stage should be in the range of 6-12 degrees. For a 40-degree total temperature rise through an electric heating coil, this equates to three to six stages of temperature control, three being the low end of course, and six being the high end of precision control. If an electric heater for such an application is scheduled to only have one or two stages, then chances for acceptable heating temperature control are slim.

Control of the heating stages will be performed either directly by the air handling unit's main controller, or by a factory furnished step controller, which can accept a control signal from the main controller and stage the heater accordingly.

On the cooling side, staged control pertains to DX cooling, in which a remote air cooled condensing unit is controlled in accordance with system requirements. The DX coil, as stated in the section on Cooling Coil Capacity, is generally sized for a delta T of 25 degrees or so. For a single zone constant volume application, two or three stages of control are adequate. Two-stage control of a 25-degree delta T split system falls on the upper cusp of the above-mentioned rule of thumb. Yet three-stage control falls nicely within the rule. Four-stage control is better yet, though considered by some to be more of a luxury for single zone control.

Control of the stages of cooling, on smaller lower end systems, will be performed directly by the air handler's main controller, as the air cooled condensing unit will not be equipped with a factory furnished sequencer, and will need to be fed with external contact closures (binary outputs from the main controller) for staging of compressors. With larger, more sophisticated condensing units, a sequencer capable of receiving a control signal from an external source (the main controller) may be standard equipment.

For proportional control of heating and cooling in a single zone system, the hot and chilled water valves will generally be controlled directly as a function of space temperature and setpoint. P+I control can be applied if the application warrants it, since the digital controller has the capability. A deadband will be established, to separate the heating and cooling setpoints. As space temperature falls toward heating setpoint, the hot water valve is modulated open. Likewise, as space temperature rises toward cooling setpoint, the chilled water valve is modulated open. The deadband can be as little as two degrees, though care must be exercised so that the throttling ranges of the two control loops don't overlap. For example, if the deadband is set for two degrees, and the throttling ranges for the heating and cooling control loops are centered about their respective setpoints, then the throttling ranges for both loops should be no more than two degrees (±1 degree from setpoint). An additional safeguard is to prohibit cooling if there is a pre-existing call for heating, and vice versa.

Before moving on to the next heading, we must discuss the concept of low limit control. We have mentioned twice already, once in this chapter and once in the chapter on Rooftop Units, that on a winter design day of –10 degrees, with a return air temperature of 72 degrees, and an outside air requirement of 20 percent, the mixed air temperature is 55 degrees. To reiterate what was expressed in the rooftop units chapter, this is fine for a reheat or VAV system air handler (one that is designed to blow 55-degree air for all occupied modes), and is borderline acceptable for a single zone constant volume air handler. Yet for any mixed air temperatures lower than this, the air must be tempered.

For a single zone air handler, one that operates in response to the needs of the zone that it serves, low limit control should be incorporated into the heating coil control scheme, such that the discharge air is at a suitable temperature during periods of no heating demand from the space. In other words, if the space served by the air handler at any point in time is not calling for heat, then without low limit control, the air delivered to the space would be the temperature of the mixed air. At mixed air temperatures below 55 degrees, this is typically not acceptable. Imposing low limit control ensures that the heating coil, whether hot water or electric, maintains the discharge air temperature at 55 degrees or greater, during periods of no space heating demand. The debate that follows this rule is that, if the outside air is so cold that the mixed air temperature is less than 55 degrees, then the space will continually be calling for heat, at least to some extent, and the heating coil will always be active because of this. While this may be true, it's still wise to implement low limit control, especially when using digital controls. It's easy enough to do; just monitor the discharge air temperature during periods of no demand (heating or cooling), and operate the heating coil accordingly.

The implementation of low limit control is more difficult with electric heat than it is with hot water heat. In the case of the hot water heating coil, the proportional hot water control valve is simply modulated to maintain the low limit temperature setpoint. For the electric heating coil (one that is not equipped with SCR control), the coil must be staged in order to maintain this low limit setpoint. Easier said than done, since we're talking discreet stages of control. The more stages that the electric heating coil is equipped with, the easier this becomes. To maintain precise low limit setpoint, the "leading" stage of electric heating would need to be cycled rapidly. The alternative to that is to control to within an acceptable range, and not short-cycle the heater. The more stages of heat, the narrower the range can be. The good news is that maintaining a precise low limit setpoint is not that critical, and an acceptable range of control can usually be achieved without an extravagant number of stages.

Economizer Operation

If outdoor air conditions permit, then the economizer is the "first response" to a call for cooling with single zone systems. We have spoken quite a bit about economizer operation thus far. Now we discuss it in terms of built up air handler operation. To restate the definition of economizer operation, it is the process of using outside air for free cooling, conditions permitting, by modulating open the outside air damper.

Economizer operation, when considered apart from mechanical cooling, is not that difficult to understand. It is when you throw DX or chilled water cooling into the mix that things become complicated. Integration of economizer with "mechanical" cooling is what we're talking about here. DX cooling with economizer is considered first.

Let's back up a step, and assume that the outside air is not suitable for free cooling. This being the case, the outside air damper resides at minimum position. Any call for cooling by the space temperature sensor must be handled with mechanical cooling. For DX cooling, this means staged cooling: the further from setpoint, the more stages engaged. Now consider that the outside air is suitable for free cooling. What happens upon a call for cooling? We could set up the control loop to be similar to what is done with a packaged single zone rooftop unit. To give a nutshell retread of that, upon a rise in space temperature from setpoint, the economizer is engaged, and outside and return air dampers are modulated in an effort to maintain a suitable discharge air temperature of around 55 degrees. Upon a further rise in space temperature above setpoint, the first stage of DX cooling is engaged as well. For a more detailed description of economizer operation as it pertains to a packaged rooftop unit, please refer to the appropriate section in the beginning of Chapter 8.

This method of integrating economizer operation with DX cooling is acceptable, and definitely time-tested and well-proven. Yet it does leave a bit to be desired, at least for a built up system of this caliber. Consider that, before the space temperature rises to the point of calling for cooling, the temperature of the discharge air is that of the mixed air. On milder days, this temperature could approach 70 degrees. Now when the space initiates a demand for cooling, the economizer is pressed into operation, and the discharge air temperature drops toward 55 degrees. That's close to a 15-degree delta T between "no call for cooling" and "call for cooling." For a rooftop unit controlled by a conventional thermostat, this may be the best that can be done. Yet for a built up air handler utilizing digital controls, we should be able to do better.

Another approach is to actually proportionally control the economizer as a direct function of space temperature and setpoint. For instance, as the space temperature rises to and through occupied mode space cooling setpoint, then proportionally open the outside air damper. If the space temperature continues to rise, then the outside air damper will reach its fully open position, and the space will call for a stage of DX cooling. Regard must be given though to mixed air or discharge air temperature, so that the outside air damper is not allowed to stroke fully open if the outside air temperature is below 50 degrees or so. This is typically handled by imposing low limit control, which overrides space temperature control of the economizer.

Both methods described herein of economizer control integrated with DX cooling are acceptable, and either method gets the job done. The differences between the two are as follows: the first method is straight out, space-initiated, staged temperature control. The second method actually gives the space the ability to employ proportional control for its comfort needs, at least when economizer operation is permitted. As can be surmised by the preceding paragraphs, the second method is technically and theoretically the superior method, and for this caliber of system, should be the method chosen and implemented.

For economizer operation integrated with chilled water cooling, the most straightforward approach is to modulate both components (economizer dampers and chilled water valve) in sequence, as a function of space temperature and setpoint. This is "staged proportional control," with economizer being the first stage, and the chilled water valve being the second stage. For each control loop, low limit control must be imposed, which overrides space initiated control. For the economizer, the mixed air sensor will read mixed air temperature, and be utilized to impose mixed air low limit setpoint control, typically around the 55-degree mark. For the chilled water valve, the discharge air sensor will be used, and a discharge air low limit setpoint will be enforced, which can be a couple of degrees below the mixed air low limit setpoint, or around 52-53 degrees.

The "economizer changeover setpoint," that point at which economizer operation is allowed or disallowed, can be based on dry bulb outside air temperature, or on outside air enthalpy. For dry bulb changeover, the energy utilizing means of cooling can be locked out when the outside air temperature is less than the dry bulb changeover setpoint, provided that this setpoint is 55 degrees or less, and assuming that 55-degree air is sufficient to satisfy the space cooling requirements. With this scenario, there is no overlap, or periods of which both economizer operation and mechanical cooling may be engaged. If the dry bulb economizer changeover setpoint is specified to be greater than 55 degrees, then the mechanical means of cooling should be allowed for outside air temperatures below the changeover setpoint, at least down to the 55-degree mark. Given this situation, there is opportunity for periods of overlap of the two processes (economizer operation and mechanical cooling). Raising the dry bulb changeover setpoint allows for more opportunities for free cooling, albeit with no regard to humidity.

For enthalpy changeover of economizer operation, humidity is taken into account, and is a more refined method of determining whether or not to economize. The outside air "lockout setpoint" for mechanical cooling should be set relatively low, to accommodate for those "cool yet wet" days (52 degrees and raining!), when economizer operation is not allowed. Overlap of the two processes is probable, especially on those "warm yet dry" days in which the outside air has value in terms of air conditioning, yet is not cool enough to entirely satisfy the needs of the space.

It is customary to lock out mechanical cooling for all outside air temperatures below 50 degrees, as it is a realistic assumption that economizer operation will be permitted and will be sufficient to satisfy the cooling needs of the space served by the air handling unit.

Overlap of economizer operation and mechanical cooling is not only possible, but often desirable. If there were no benefits in allowing for possible overlap between the two processes, then life would be easy and the control of each process would remain separate. It is because of the fact that there are scenarios in which the overall system benefits from allowing both processes to work simultaneously, that the above material was presented in the fashion that it was.

One final note before moving on. The term "mechanical cooling" has been kicked around quite a bit now, basically on the assumption that the reader has heard and is familiar with the term. To clarify, mechanical cooling, in regards as to how this chapter uses the term (whether technically correct or not), refers to the "energy utilizing means of cooling," be it DX cooling or chilled water cooling.

Unoccupied Mode Operation

If the single zone air handling unit is not to run 24/7, then the unit will have an unoccupied mode of operation. Unoccupied mode simply refers to unit operation when the space served by the unit does not have any occupants. For these periods of time, there is no requirement for outside air, and no requirement for precision comfort control. Thus, the fan can shut down and rest upon transition to this mode, and the outside air damper can go fully shut.

Unoccupied space heating and cooling setpoints are typically established, spread out to where unit operation is not engaged unless the temperature in the space strays out of the range defined by these setpoints. Default unoccupied heating and cooling setpoints are 60 and 80 degrees, respectively. Upon transition to the unoccupied mode, the unit will shut down, and the temperature in the space will potentially begin to deviate from the occupied mode setpoints. For instance, if it's cold out, the temperature in the space served by the air handling unit will fall from room temperature, toward the unoccupied heating setpoint. Likewise, if the temperature outdoors is high, then the space temperature may tend to rise, toward the unoccupied cooling setpoint. As long as the temperature in the space is above the unoccupied heating setpoint and below the unoccupied cooling setpoint, then the air handling unit remains shut down.

If the temperature in the space strays out of the unoccupied mode "deadband," then the rooftop unit will come to life: the fan(s) will start, and the appropriate mode (heating or cooling) will be pressed into operation. The outside air damper should remain shut during such a cycle, since there is no occupant driven requirement for outside air, and conditioning the outside air is costly.

For whichever mode is called for (heating or cooling), the means to heat or cool may be implemented at full throttle. The purpose of an unoccupied heating/cooling cycle is to quickly bring the space temperature back into the range defined by the unoccupied heating and cooling setpoints. A modest differential should be incorporated, so that the unit doesn't cycle frequently in the unoccupied mode. For instance, if the unit goes into an unoccupied heating cycle, the unit should not come out of this cycle until the space temperature rises to at least a couple degrees above the unoccupied heating setpoint. Same goes for cooling. If the unit enters an unoccupied cooling cycle, then the unit should be allowed to run and blow cool air until the temperature in the space comes down in temperature to a couple degrees below the unoccupied cooling setpoint.

If there are any concerns about "limiting" the discharge air temperature in either an unoccupied heating or an unoccupied cooling cycle, then this can be addressed, via the main controller, by monitoring the discharge air sensor and establishing limits. A concern may stem from the fact that the heating or cooling coil is sized for the outside air required to be brought in during the occupied modes, and when the unit is strictly recirculating air from the space (as it does in an unoccupied heating or cooling cycle), the coil is in essence oversized, and excessive discharge air temperatures (high or low) can result. This generally will not be a concern with such a system, though discharge air temperature limiting during unoccupied cycles is something that can be implemented, if the concern does exist.

When the single zone air handling system transitions from the unoccupied mode to the occupied mode, does it first go through a "morning warm-up cycle?" When we discussed the terms night setback and morning warm-up in the chapter on Common Control Schemes, we noted that the term "morning warm-up" does not really apply to single zone systems, at least not to thermostatically controlled rooftop units. Upon transition to the occupied mode of operation, the single zone rooftop unit begins to control to its occupied mode setpoints, as established via the thermostat. If the temperature in the space has fallen from (occupied heating mode) setpoint during the unoccupied mode, then as soon as the unit transitions into the occupied mode, it will go into a heating mode of operation, automatically, without any special control cycle required.

For a built up, single zone air handling unit, this is also the case. There needn't be any special cycle implemented to provide "first thing heating." However, we *can* keep the outside air damper shut for a while, so that we don't unnecessarily expend energy on heating the outside air. Say, upon transition to the occupied mode (which should occur a good half-hour to an hour before anyone actually arrives to occupy the space), if the space temperature is more than 2 degrees below the occupied space heating setpoint, then do not position the outside air damper for its minimum, not until the space temperature rises to within 2 degrees of space heating setpoint. With a single zone rooftop unit controlled by a conventional space thermostat, the thermostat has no control over the outside air damper, so this is simply not possible. At least not without adding more hardware and implementing more complicated control schemes. Yet with a built up air handling unit controlled by a digital controller, this is a simple task to accomplish. And worth the effort on the programmer's part, especially if the minimum requirement for outside air is substantial.

Reheat Systems

An air handler serving a system of nothing but reheat coils is the topic of this section. Like the single zone system air handler, the reheat system air handler will operate at a constant volume of air. Unlike the single zone air handler, a reheat system air handling unit will serve multiple zones of comfort. The concept is simple; provide cool air at a constant volume to all reheat coils. The zone controllers for those coils can then choose whether the cool air delivered into the zones should be heated or not, and if so, to what degree. The notion of cooling down air and heating it back up, for comfort control purposes, tends to be a compromise on the energy side of things. Still, this is a fairly precise method of individual zone temperature control, and continues to be a popular alternative (code and application permitting), for serving multiple zones with a single air handler. Figure 11-7 diagrammatically illustrates the basic reheat system air handler. The illustration doesn't look all that different from Figure 11-6 (single zone air handler). Both types of systems have many of the same components. The differences between the two systems is in how they are designed, and how they are set up to operate.

Fan Operation

The operation of the supply fan of a reheat system air handler must be continuous during occupied modes. There is no opportunity for intermittent operation of the fan during these modes, as there might be for the single zone air handler. The fan is serving many zones, all with their own unique comfort needs. It is unlikely that all zones will be satisfied at any given point in time, and even if that were the case, chances are good that that situation wouldn't last too long. To let the air handler supply fan rest during a period of zero demand from the zones is not a scheme too often applied, or even thought about for that matter. Of course code requirements for fresh air all but wipe out the idea altogether. The consensus is to have it run continuously.

Operation of any related exhaust or exhaust/return fan of a reheat air handling system is identical to that of a single zone system. No need to reiterate; just refer back to the appropriate heading in the last section.

Heating and Cooling Operation

To understand the operation of the heating and cooling sections of the reheat system air handler, we must first think in general terms about

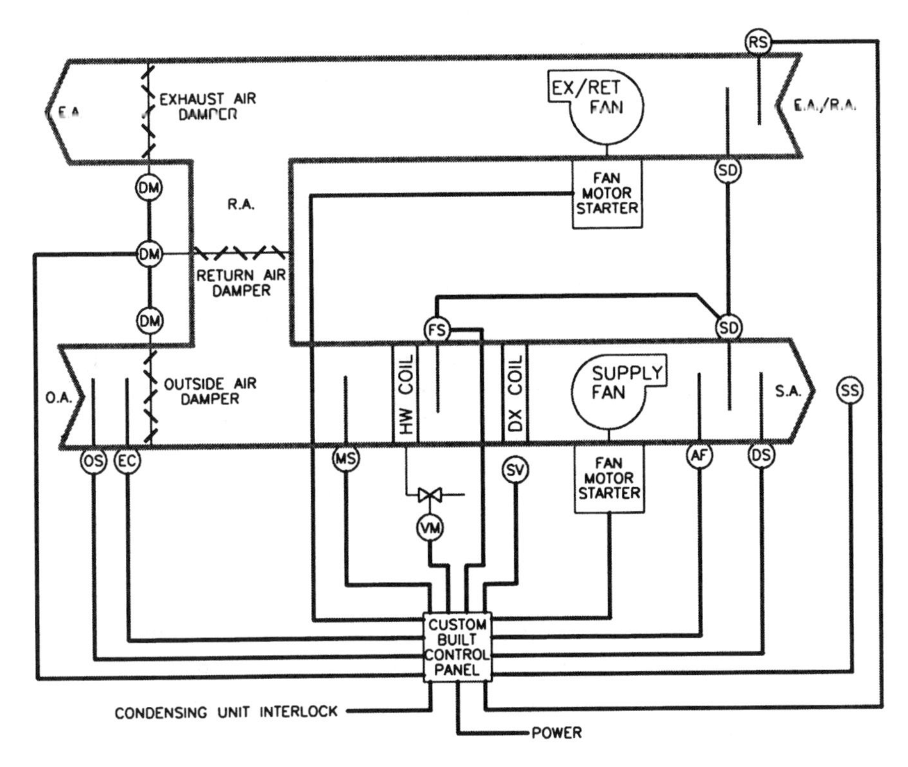

Figure 11-7. Reheat system air handling unit, hot water heating and DX cooling, with all of the required controls components.

the overall function of this unit. The unit is employed to blow air at a constant volume and constant temperature. The temperature of the air that it delivers should be in the range of 52-58 degrees. The unit will utilize heating and cooling as required in order to accomplish this. That is, discharge air temperature control, with no regard to what the temperature is in the spaces being served.

Why would a reheat system air handling unit need to be equipped with heat? Maybe it's there for night setback and morning warm-up cycles. But there's heat at the zone level, that can be utilized for these cycles. So it really doesn't need to be there for that. For outside air requirements of 20 percent or less, there would also be no need for heat in the unit. Remember that, for a winter design day of –10 degrees, a return air temperature of 72 degrees, and an outside air requirement of 20 percent, the mixed air temperature will be 55 degrees, which is smack dab in the middle of that above-mentioned range of acceptable discharge air temperature. During occupied modes, never would there be a need to utilize air handler heating to maintain discharge air setpoint.

It is for outside air requirements greater than 20 percent that the unit will require a heater, employed for discharge air temperature control. We tend to think of a unit designed to serve a system of reheat coils as one that cools the air delivered down to a suitable discharge air temperature (of around 55 degrees). Yet when the unit must bring in substantial amounts of fresh outside air, the unit must be able to heat to discharge air temperature setpoint, in addition to cooling to it. Control of the heating section will likely need to be more precise than what two-position control can offer. For a hot water coil, the control valve should be a proportional control valve, utilizing a P+I control algorithm. For an electric heater, the number of stages may need to be more than one, the quantity being a function of just how much outside air the air handling unit is specified to bring in.

Since we can relate percentage of outside air to mixed air temperature, we can talk in terms of the heater's entering air temperature (EAT), in order to find out the acceptable number of stages for the heater. Now, since we're trying to maintain reasonably precise temperature control of the discharge air, we should be able to establish a guideline for the number of stages required for such an application. As always, the more stages the merrier, but by looking at what the EAT will be, worst case, we can determine delta T and decide upon a reasonable number of stages. To go out on a limb here and say that the amount of outside air ever required to be brought in by a built up air handler during winter design conditions will never be more than 50 percent (a substantial amount!), we can find the worst case EAT to be 31 degrees (chilly!). That yields a heater delta T of 55 – 31, or 24 degrees.

In the chapter on make-up air units, we stated that, for mediocre discharge air temperature control, the number of stages of electric heat should not exceed 10. But that was a different application, perhaps not calling for the degree of precision that we are looking for here. For space-initiated staged temperature control, we mentioned last section that the delta T per stage should be in the range of 6-12 degrees. For discharge air temperature control of electric heat, in this application, the delta T per stage should probably be even less, more like 4-6 degrees per stage. For a total delta T of 24 degrees, the high end of this range requires six stages of control, and the low end (of precision control) requires four stages.

From the above assumptions we can conclude that the most stages typically required for an electric heater used in this type of application will be no more than six. The stages can be controlled directly by the main controller, with binary outputs, or by a factory mounted step controller that can receive a control signal from an analog output off the main controller. This works out pretty well, for if a factory furnished step controller is specified, many manufacturers will equip their heater with a discreet step controller, one that will have six stage capability, whether or not all six are used. In other words, if you specify a four-stage heater and factory mounted step controller, you will receive a four-stage heater, the four stages of control operated by a step controller capable of operating up to six stages. Of course if your application requires more than six stages, then specifying a factory furnished step controller will actually get you two step controllers, one to handle the first six stages, and one to handle the additional stages.

The decision to heat to discharge air setpoint, rather then to cool to it, must be based upon a couple of things. First, there must be zero demand for cooling. If there is no cooling demand, then the outside air damper will reside at minimum position, for economizer operation will not be engaged. With no demand for cooling and the outside air damper at minimum position, the mixed air temperature can then be used to determine if heating needs to be engaged. If the mixed air sensor registers a value below, say, 52 degrees, then it's safe to say that the percentage of outside air being brought in by the outside air damper, coupled with the temperature of it, is such that the air handler must now heat to maintain discharge air temperature setpoint. Under these circumstances, it is important to prohibit "opposite mode" operation (cooling). It is also important to build in a differential of a few degrees into the mixed air changeover setpoint (which is 52 degrees in this case), so that the air handler remains in this mode until the mixed air comes up in temperature above the changeover setpoint *plus* the differential (or above 55 degrees). Once this takes place, then the air handler can return to its "normal" mode of operation: discharge air cooling.

The mixed air changeover setpoint, that which determines whether to heat or to cool to maintain discharge air temperature, is closely related to the desired discharge air temperature setpoint. In the above scenario, the discharge air setpoint was 55 degrees, and the mixed air changeover setpoint was 52 with a 3-degree differential. If the discharge air setpoint were different, then so would the mixed air changeover setpoint need to be. For instance, if the air handling unit were employed to maintain a 60-degree discharge air setpoint, then the mixed air changeover setpoint would need to be more like 57 degrees, again with a few degrees of differential built into it.

For all mixed air temperatures above "changeover," the air handling unit must use its means to cool in order to maintain discharge air temperature setpoint. Economizer operation aside, the unit's main controller will either stage a remote condensing unit (DX cooling) or modulate a control valve (chilled water cooling).

For DX cooling, a condensing unit will be staged in order to maintain discharge air temperature setpoint. The same rule of thumb applies here, as it does to discharge air control of electric heat in this type of application. That is, a delta T of 4-6 degrees per stage. With entering air temperatures of the DX coil approaching 80 degrees, the total delta T of the coil will be in the area of 25 degrees. Four stages of control would be the *minimum* requirement for the condensing unit, using this criterion. Six stages is more appropriate. The stages of control can be handled with binary outputs off the main controller, or a control signal can be sent to the condensing unit, if the unit is equipped with a factory furnished sequencer.

For chilled water cooling, as for hot water heating in this application, the valve should be controlled using a P+I control algorithm. Straight proportional control may or may not be adequate for this type of application, and since we're trying to perform precision discharge air control using a digital controller, the controller's capabilities should be exploited in order to optimize the performance of this control loop.

Since we're talking about discharge air temperature control of an air handling unit, I suppose that we should at least touch upon the concept of reset, as it can apply to this type of system. Reset control of discharge air temperature setpoint can be based upon either outside air or space temperature. If based upon outside air temperature, then a decrease in outside air temperature results in an increase of discharge air temperature setpoint. We can set up a reset schedule so that, as the outside air temperature falls toward –10 degrees, the discharge air temperature setpoint will be reset upward, toward 65 degrees. For space temperature reset of discharge air setpoint, a temperature decrease in the general space served by the air handling unit will result in the discharge air setpoint being increased, or reset upward. A space temperature sensor, mounted in some general location, can be used to perform this operation.

Reset control should not be routinely incorporated as an initial design feature of the control system (unless explicitly called for by the mechanical engineer, as part of his design criteria). It is more of a possible "remedy" to a potential problem down the road. For instance, it might be found upon colder outside air temperatures that certain zones of reheat aren't performing up to spec, and as a consequence these zones are too cold. If the discharge air of the main air handler were to be kicked up in temperature, then these zones have a better chance of maintaining comfort levels. Care must be taken so as not to compromise the comfort levels of other zones, simply to satisfy these problem zones. In other words, there may be certain zones that need every bit of that 55-degree air at certain times, and if it's not available, these zones can tend to overheat. So reset control isn't the ultimate solution to zone temperature control problems during the winter, yet if applied carefully, it can at least potentially minimize the problems that may occur during extreme conditions.

Economizer Operation

Integration of DX and chilled water cooling with economizer operation, for air handlers that are controlled to maintain discharge air temperature (reheat *and* VAV systems), is what is discussed under this heading. DX cooling with economizer is considered first.

If the outside air temperature is not suitable for free cooling, then economizer operation is prohibited and DX cooling is implemented to maintain discharge air temperature setpoint. If the outside air temperature is below the lockout point for the condensing unit, then economizer operation is the only means of cooling. What if the outside air is suitable for free cooling, and also above the mechanical cooling lockout point? How is the control of the economizer and the mechanical cooling integrated?

The goal is to maintain a 55-degree discharge air temperature setpoint. If the outside air is suitable for free cooling, then the economizer should be the first step toward achieving that goal. If the outside air damper modulates fully open, without being able to reach that goal, then it's time for DX cooling to step in and help! A simple way to do this is to engage a stage of cooling, and initiate a time period, of just a few minutes. During this period of time, watch the discharge air temperature, and when it begins to drop, then modulate the economizer to maintain discharge air temperature. In other words, back off on the amount of outside air, in order to maintain the proper mixture of economizer and DX cooling. After the time period times out, then release the stage of cooling, and let the economizer continue to go about its business. The outside air damper will likely once again go fully open. After another few minutes, again engage DX cooling, and perform the same process. And so on, and so on.

During an "on" cycle, the combination of economizer and one stage of DX cooling *could* result in a discharge air temperature that stabilizes above the desired setpoint. If discharge air temperature setpoint cannot be maintained with one stage of DX cooling active and the outside air damper fully open, then one stage of DX cooling apparently isn't enough to satisfy the setpoint needs. If this is still the case at the end of this cycle, then instead of cycling off the current stage of DX cooling, bring on another stage! Start another time period, and watch the discharge air. With two stages of DX cooling engaged, the discharge air temperature *should* fall, and the outside air damper can back off. At the end of this cycle, simply turn off the second stage of DX cooling, leaving the first stage active, and start another cycle.

As an example of this process, consider the following givens: 55-degree setpoint, 5-degree delta T per stage of DX cooling, and an *initial* outside air temperature of 58 degrees. The starting point of this example is immediately prior to a DX "on" cycle; the outside air damper is fully open, and the discharge air temperature is ...58 degrees. Now, the on cycle is initiated, the first stage of DX cooling is activated, and the discharge air temperature will potentially fall 5 degrees. However, as the discharge air temperature drops to and through setpoint, the outside air damper modulates toward is closed position, in an effort to maintain setpoint. To maintain 55, the outside air damper will reposition itself to introduce, to the DX coil, a mixture of outside and return air that is roughly 60 degrees in temperature. With the outside air temperature being 58 degrees, and assuming that the return air temperature is 74 degrees, the outside air damper will position itself to bring in around 90% of outside air.

After a few minutes, the stage of DX cooling is released, and the discharge air temperature begins to rise. The outside air damper counteracts by driving open. Once fully open, the discharge air temperature is again the outside air temperature: 58 degrees.

A few more minutes lapse, and the first stage of DX cooling is once again engaged. Assuming that it's morning, the outside air temperature is on its daily upswing, and by this time has risen to 60 degrees. Engaging one stage of DX cooling will result in a 55-degree discharge air temperature, without having to back off on the amount of outside air. During this on cycle, the temperature outdoors continues to rise, and so does the discharge air temperature, above setpoint. At the end of this on cycle, rather than cycling off the first stage of DX cooling, the second stage will be brought on. With a total DX delta T of 10 degrees now, the discharge air temperature setpoint will be attainable, and the outside air damper will modulate closed in order to achieve and maintain it.

This method of cycling stages of DX cooling in conjunction with economizer operation can be done with as many stages as needed, and is a very well-proven technique for discharge air temperature control. The time periods that make up the on-off cycles should be long enough to prevent short-cycling of compressors. A few minutes is fine for the off cycles. For the on cycles, the time periods can be longer, for it is during these periods that the discharge air temperature is being maintained precisely at setpoint, by backing off the economizer. It is also during these periods that the most energy is being used; we're not getting everything we can from the outside air, so we're actually sacrificing some amount of available free cooling in order to maintain precise setpoint.

For integration of economizer with chilled water cooling, the process is more straightforward. The accepted method is to control the chilled water valve to maintain discharge air temperature setpoint, and control the economizer to maintain a mixed air temperature setpoint, before the chilled water coil, of a couple of degrees below discharge air setpoint. What happens is, the economizer will do its thing, and if mixed air setpoint can be maintained, then the chilled water valve will remain closed, since the discharge air temperature is a couple of degrees below setpoint. However, if the outside air damper modulates fully open, and mixed air setpoint is not realized, then the discharge air temperature will wander into the control band of the chilled water valve, and the valve will begin to modulate, to maintain discharge air setpoint.

To illustrate with real values, consider that the discharge air temperature setpoint is 55 degrees. This is the value that will try to be maintained by modulating the chilled water valve. The mixed air temperature setpoint, then, will be 2 degrees below this, or 53 degrees. If economizer operation alone cannot satisfy the demand for cooling, then the valve will modulate, and 55 degrees will be maintained at the discharge of the unit. If the economizer can maintain (mixed air) setpoint, then the valve remains closed, and 53 degrees will be maintained at the discharge of the unit.

In terms of the economizer changeover setpoint, the same issues apply here as with single zone systems. To reiterate, changeover setpoint is that point at which economizer operation is allowed or disallowed, and is based on either dry bulb outside air temperature or outside air enthalpy. For dry bulb changeover, if the specified dry bulb changeover setpoint is less than or equal to the discharge air temperature setpoint, then mechanical cooling can be locked out at and below the changeover setpoint.

That was kind of a mouthful, so let's try to visualize this with some real numbers. Assume that the dry bulb changeover setpoint is 55 degrees, and the discharge air temperature setpoint is also 55 degrees. For outside air temperatures above 55, economizer operation is disallowed and mechanical cooling is employed, and for outside air temperatures below 55, economizer operation is allowed and mechanical cooling *can* be locked out. This is because the economizer will be able to satisfy discharge air temperature setpoint, without help from mechanical cooling. Now assume that the dry bulb changeover setpoint is 50 degrees, keeping the discharge air temperature setpoint at 55 degrees. In this case, the economizer is disabled for outside air temperatures above 50, and mechanical cooling must "fill the bill." For outside air temperatures below 50, economizer operation is permitted, and mechanical cooling can be locked out, because the economizer will be able to achieve and maintain discharge air temperature setpoint. Finally, consider what the case is if the dry bulb changeover setpoint is *higher* than the discharge air temperature setpoint, say, 60 degrees. For outside air temperatures above 60, mechanical cooling is the only alternative. For outside air temperatures between the discharge air setpoint and the changeover setpoint (between 55 and 60), the economizer is enabled, yet economizer operation alone is not sufficient to satisfy discharge air temperature setpoint. So mechanical cooling must still be allowed, for this range of temperatures, and overlap of economizer operation and mechanical cooling is likely to take place. Of course, once the outside air temperature drops below the discharge air temperature setpoint, then mechanical cooling can be locked out, as economizer operation will handle the situation.

If the preceding material was difficult to absorb, then I know exactly how you feel! What all of it really means is that, with dry bulb changeover of economizer operation, there can be allowable periods of overlap, when both economizer and mechanical cooling may be simultaneously implemented. It's basically a matter of trying to get more out of your outside air, in terms of "free cooling," by raising the economizer changeover setpoint. Simply put, the economizer alone may not be able to entirely satisfy discharge air temperature setpoint, but if it can at least help in getting there, why not use it? The mechanical means to cool can take it the rest of the way!

As far as enthalpy changeover goes, there is nothing new to add here that wasn't already spoken of when we discussed single zone systems. To avoid the risk of being repetitive, we will simply state that enthalpy changeover takes outside air humidity into account in the decision to economize, and is a more refined approach to making that determination. Overlap of economizer operation and mechanical cooling is highly probable, when you consider that warm dry days are candidates for economizer operation.

As it was in the economizer section of Single Zone Systems, it is appropriate to note here that the whole concept of overlap is what makes this topic as complex as it is. If economizer operation and mechanical cooling were mutually exclusive, then each process would be treated individually and things would be simple. Yet benefits are gained by allowing such overlap, and thus integration of the two processes must be performed, control-wise, in order to realize the benefits.

Unoccupied Mode Operation

Unoccupied mode operation of a reheat system air handler is similar to that of a single zone air handler, with some noticeable differences. Upon transition to this mode, the unit will shut down, and temperature control will revert to the maintenance of unoccupied space heating and cooling setpoints, of 60 and 80 respectively. A space temperature sensor is required for the enforcement of these setpoints. At night, the system in essence becomes a single zone system, operating as required to maintain the general space temperature between unoccupied heating and cooling setpoints.

Let's take a look at unoccupied cooling first. If the general space temperature rises to the unoccupied cooling setpoint, then the unit will come to life; the fan(s) start, and the unit will do what it needs to do to maintain discharge air temperature setpoint. If economizer operation is not permitted (which will likely be the case), then the outside air damper will remain fully shut. If the general space temperature, as read by the centrally located space temperature sensor, has risen to the unoccupied mode cooling setpoint, it is a fairly safe assumption that all zones are at temperature levels in excess of their respective zone setpoints. Therefore, no amount of reheat will take place at the zone level. The unit will deliver cool air to and through all zone reheat coils, and into the spaces served by the air handler. The general space temperature will fall, and once it drops a couple of degrees below unoccupied cooling setpoint, the air handling unit will shut down. Until needed again.

Now for heating. Even if the unit has no form of heat in it, an unoccupied heating cycle can still be implemented. Upon a drop in temperature to the unoccupied mode heating setpoint, the unit comes to life, with cooling disabled, and the outside air damper fully shut. So the unit just recirculates air from the space. If the general space temperature has fallen to the unoccupied mode heating setpoint, it's likely that all zones served by the unit are at temperature levels below their respective zone setpoints. So all reheat zones, once enabled upon engagement of the supply fan, will call for full heat. The unit will deliver air, at the current space temperature of 60 degrees (or thereabouts), to all zone reheat coils. The reheat coils do what they do best, and the general space temperature begins its upswing. Once the space temperature has risen to a couple of degrees above unoccupied heating setpoint, then the air handler shuts down, and the reheat zones are disabled.

For morning warm-up, the same thing essentially happens: the air handler starts up, in a recirculation mode, and the reheat coils are engaged. The space warms up, and the cycle is terminated, and not again allowed until the after the next unoccupied cycle. Termination of the morning warm-up cycle can be done by monitoring space temperature via the space sensor, or by monitoring return air temperature, via a return air temperature sensor. Return air monitoring is the preferred method of determining when to end the morning warm-up cycle, as this temperature is a more accurate representation of the average space temperature.

Upon termination of this cycle, the system enters the normal occupied mode of operation: discharge air cooling. The purpose of the morning warm-up cycle is to get the general space served by the air handling unit up to room temperature as quickly as possible, before entering the occupied mode of operation. Morning warm-up cycles are commonly employed in reheat and VAV systems alike, for it gives the system a chance to warm up the space and take the chill out of the air, before going into discharge air control mode.

If the main air handler has heat in it, perhaps because the unit is expected to handle substantial amounts of outside air, then this heater can be utilized for unoccupied mode heating and/or morning warm-up. Care must be exercised, so that the temperature of the air delivered to the reheat coils is not too great. Why? Well, by using the unit's heater and using the reheat coils for a heating cycle, the air is being "twice heated": once by the unit, and once at the zone level. This twice-heated air can be at excessive temperatures, and concerns may exist about sending air this hot into the zones. With a reheat system consisting of electric reheat coils, the temperature of the air leaving the coils might even reach the trip point of the coils' integral high limit thermal cutout switches, which, if they are manual reset devices, would effectively terminate the operation of the heaters.

If the reheat system air handler's heat is to be used in a heating cycle, discharge air temperature high limit control can be imposed on it, so that the temperature of this air does not pose a problem at the zone level.

VAV Systems

A Variable Air Volume (VAV) air handling unit is an air handler designed to serve a system of VAV and/or fan powered boxes. Like the reheat system air handler, the unit is to operate to provide cool air at a constant temperature. Unlike the reheat system air handler, the unit provides air not at a constant volume, but at a volume that varies as a function of zone level demand. We touched upon VAV box operation in the chapter on Rooftop Units, in order to explain the operation of a packaged VAV rooftop unit. Suffice it here to say that VAV boxes served by a VAV air handling unit (packaged or otherwise) will modulate their dampers in response to zone comfort levels. The VAV air handler will operate as a function of this demand; the more of a demand for cool air at the zone levels, the more air delivered by the air handler. So the VAV air handler can be thought of as an air handling unit, one that serves multiple zones, and operates at a constant temperature and variable volume of air delivery. A step up from the reheat system, the VAV system as a whole is an efficient and economical means (first cost aside) of providing multizone temperature control with a single air handling unit. Figure 11-8 illustrates the typical "fully equipped" VAV air handler.

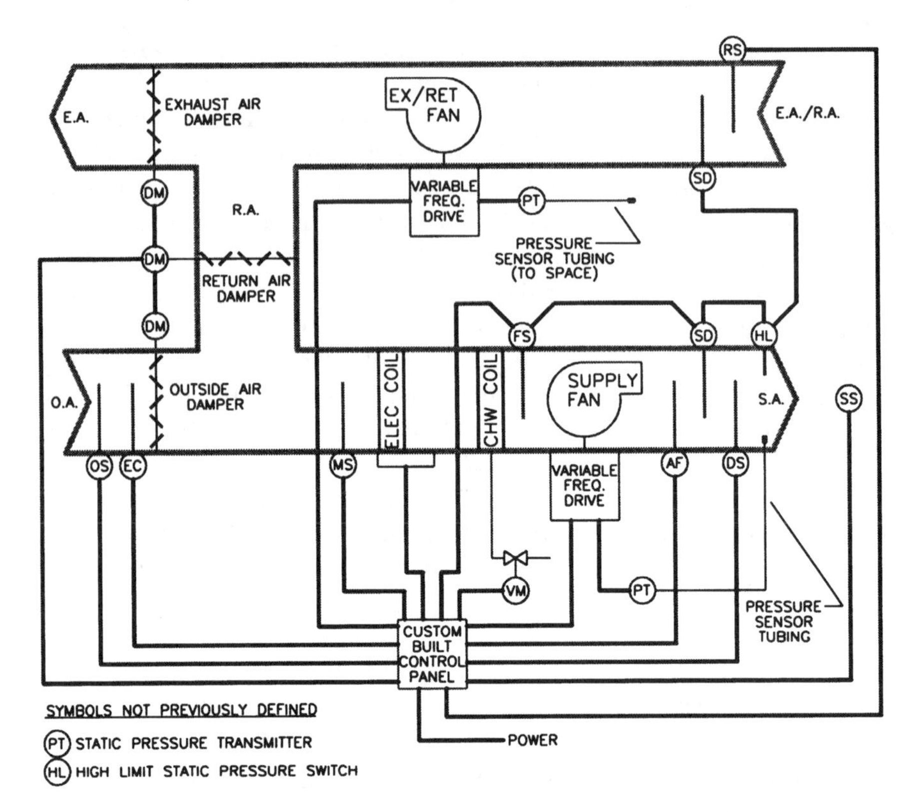

Figure 11-8. VAV air handling unit, electric heating and chilled water cooling, nicely equipped with all the fixings required to blow cold air at a variable rate.

Supply Fan Operation

Perhaps the single thing that most differentiates the VAV air handler from others is how the air handler's supply fan operates. The supply fan will operate continuously during all occupied modes. Yet unlike the single zone and reheat system air handlers, the VAV air handler must deliver air not at a constant volume, but at a variable rate of delivery. The supply fan must be able to "back off" from full capacity, all the way down to some acceptable minimum rate, as a direct function of zone level demand. In other words, as VAV box dampers served by the unit close from their fully open positions down toward their minimum positions, the VAV air handler's supply fan must be able to accommodate this, by reducing it's rate of air delivery. This is normally accomplished by either inlet guide vanes or by a variable frequency drive.

Supply duct static pressure control will be employed as the means to control the air delivery of the supply fan. A static pressure transmitter will measure duct pressure with respect to the pressure outside of the duct, and will transmit this information to the main controller, which receives it in the form of an analog input. A setpoint is established via the controller, typically around the area of 1.5" W.C., and an appropriate control signal is generated at an analog output of the main controller. The control signal feeds either a damper actuator, in the case of IGVs, or it directly feeds into the control board of a VFD. With either method of air quantity control that the supply fan is equipped with, air delivery is controlled by maintaining a constant supply duct static pressure.

The system of ductwork that connects the VAV air handler with the terminal units (VAV and/or fan powered boxes), is called the **medium pressure system**. It is in this duct distribution system that pressures are maintained at an inch and a half of static pressure, and the ductwork is constructed to a higher pressure class than what is typical with single zone and reheat air handler duct systems. In simple terms, the terminal units require a certain level of static pressure at their inlets for their proper operation, and the medium pressure ductwork must be able to withstand this pressure level, without blowing out and springing leaks.

Placement of the pressure transmitter is important, for systems that have straight (unlooped) duct runouts. The high port of the static pressure transmitter should be installed in the main trunk, at least two-thirds of the way down the line, if not further. This assures that the most remote VAV boxes on the system get the pressure that they need, when the system is maintained at setpoint. Of course, by placing the static pressure transmitter in such a location, and controlling the air delivery of the fan to maintain duct static pressure at this point in the duct, the duct static pressure maintained closer to the air handler will actually be higher than setpoint, due to static losses inherent in such a ducted system. Care must be taken so that this pressure doesn't reach or exceed the setting of the requisite high limit static pressure switch, which is located right at the discharge of the air handling unit. Generally this will not be an issue; the high limit controller is typically set at 3.5 to 4 inches of static pressure, quite a bit higher than setpoint, relatively speaking.

For looped duct systems, placement of the controller is more arbitrary. Say what? Take a look at Figure 11-9 for a visual, to go along with the following written explanation. A "main loop" system, common in

VAV applications, is a ducted system that is fed from one point, and goes out in two directions. The two branches form a loop by meeting up with each other at the opposite end of the building. In this type of a system, the ductwork may be all the same size, and VAV boxes can tap off the loop anywhere at all, with little regard to location. Static pressure within the loop, as maintained by proper control of the supply fan, should be relatively uniform and consistent throughout the loop, and proper placement of the static pressure transmitter becomes more of a "theoretically correct" thing. In other words, a location for the transmitter can be selected by reviewing the mechanical plans and gaining some insight into the air handling system as a whole. Realistically speaking though, you can't really go too wrong in the placement of this controller (at least with the smaller scale systems). Just don't put it at the immediate discharge of the air handler! And get it in a straight section of duct, away from transitions and elbows, which can cause turbulence and make it difficult to take a consistent measurement of static pressure.

The use of a single static pressure transmitter is suitable for smaller systems, however very large systems may require multiple transmitters, whose locations are chosen by reviewing the mechanical plans and determining proper sensing points. The signals of these transmitters can be

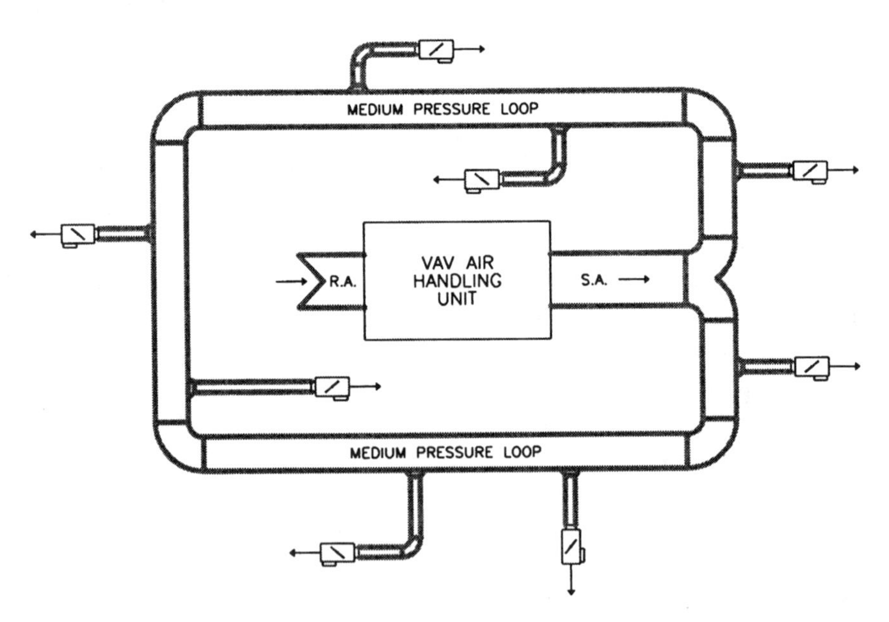

Figure 11-9. VAV looped duct, or "main loop" system.

averaged; each control signal serves as an analog input to the main controller, and the controller performs an averaging calculation in order to determine the proper course of action on the supply fan. The use of more than one pressure transmitter can also be a remedy in the event of a static pressure control problem in a system using a single transmitter. In this case, an additional transmitter is installed and wired in (provided that the main controller has an available analog input), and programming is performed to acknowledge and utilize the pressure signal of the new transmitter.

Exhaust/Return Fan Operation

In a VAV air handling system, an exhaust/return fan is often standard equipment. A common approach to exhaust/return fan control is to regulate air quantity to maintain building static pressure. In this scenario, the fan operates independently from the supply fan. Building static pressure is sensed by a space static pressure transmitter, and the fan's air quantity is regulated by its IGVs or VFD to maintain a slightly positive building static pressure setpoint of typically .1" W.C.

This is the most straightforward method of controlling building static pressure in a VAV system; measure the building static pressure, and control the exhaust/return fan accordingly. However, problems can arise with this method. If this fan is the only form of exhaust in the building, then the fan will virtually "track" the operation of the supply fan, and the system maintains a nice balance. Most often though, this will not be the only form of exhaust in the building. There may be several other exhaust fans serving the building: toilet exhaust, janitor's closets, and electrical and elevator machine rooms, to name a few. How and when these fans operate directly impacts building pressure. Toilet exhaust fans are either on a time-of-day schedule, or perhaps on a wall switch. Fans serving electrical rooms and elevator equipment rooms are normally operated from a thermostat, and thus turn on and off arbitrarily.

It would appear to make even more sense, then, to control the exhaust/return fan via building static pressure, the thought being that this fan could maintain building static pressure, regardless of what the other fans are doing. For instance, if the elevator machine room fan kicked on via its thermostat, then the building static pressure would instantaneously drop, and the exhaust/return fan would respond by "slowing down," thereby bringing the building static pressure back up to setpoint.

Upon closer analysis, a couple of potential problems are unveiled, with the potential of occurrence of these problems (and magnitude) increasing with the amount of additional exhaust in the building. For instance, consider the following scenario:

A VAV air handling system delivers a maximum air quantity of 10,000 cfm. The minimum required outside air is 30 percent. In addition, miscellaneous exhaust systems throughout the building total 2,000 cfm. Assume that economizer operation is not in effect, so that the outside and exhaust air dampers are at minimum position, and the return air damper is substantially open. Assume also that the supply air demand is such that the supply fan is operating at 40 percent of its full capacity, or 4,000 cfm. The outside air quantity is 30 percent of 4,000 cfm, which is 1,200 cfm. Finally, assume that all building exhaust systems are in operation.

The building static pressure is negative, the result of exhausting 2,000 cfm via the building exhaust systems, and only *actively* making up 1,200 of it via the air handler. If that isn't a problem in itself, the exhaust/return fan, which is trying to operate to maintain a slightly positive building static pressure, is therefore operating at its minimum "allowable" capacity, which is likely less than the supply fan's operating point of 40 percent. The scenario described here is commonly referred to as "starving the return"; the chamber between the supply fan and the exhaust/return fan undergoes a substantially negative static pressure as a result of the difference in the operating points of the two fans. In this scenario, typically what actually happens is that the supply fan will pull more air via "the path of least resistance," being the outside air intake damper (and possibly even the exhaust air damper), and pull less air through the idling exhaust/return fan. This can lead to low temperature problems during the winter months, and high humidity problems during the summer months.

It is important to note that there are two potential problems that exist here: negative building static pressure, and negative return air chamber static pressure. The increase in demand for cooling from all of the VAV boxes, and the resulting increase in supply air, impacts these two problems in different ways. As supply air increases, so does the outside air being introduced into the building. While this tends to alleviate the building negative pressure problem, the return air chamber negative pressure problem actually increases, and is maximized at the point where the outside air being brought in by the air handler equals the air being exhausted by the building exhaust systems. With the above example, this would happen when the air handler supply fan is operating at about 66 percent.

The reason for this is that the building remains in a negative up until the outside air quantity equals the exhaust air quantity. Up until that point, the exhaust/return fan remains operating at minimum capacity, while the supply fan's operating point is increasing. The result is that the difference in the two fans' operating points increases up to that point, and therefore so does the return chamber's "negative" static pressure.

So what are some of the potential solutions to these problems? Up front, the VAV air handling system should not necessarily be counted on to make up the total quantity of air exhausted by these miscellaneous building exhaust systems. For systems in which the position of the outside air damper is fixed for all times (except during economizer operation), the quantity of outside air introduced into the building by the air handler is a function of demand. The air handler's heating and cooling capacities *can* be sized for the additional outside air required to make up the exhaust, and the outside air damper minimum position *can* be set higher. Yet unless there is some means of keeping the outside air quantity at a constant cfm, then the potential for problems remains. If the air handler is operating at less than half capacity, then the outside air component of the supply air may still not be large enough to alleviate the building negative pressure problems. A better up front solution might be to include a separate make-up air system for the building's exhaust systems.

A common fix to the "starving the return" problem, when discovered upon start-up, is to have the exhaust/return fan "track" the supply fan, instead of operating to maintain building static pressure. With this method, the control signal sent from the main controller to the supply fan's IGVs or VFD would be relayed to the exhaust/return fan's IGVs or VFD as well, so that the operating points of the two fans would be close to equal for the whole range of operation. Another "solution" that is along the same lines is to monitor the return air chamber static pressure, and simply control the exhaust/return fan via this signal, to maintain the chamber at a neutral static pressure. While both of these methods would take care of the return air chamber negative pressure problem, they tend to compound the building pressure problem.

The above "fodder" begs the question "For VAV buildings with little or no miscellaneous exhaust, why not have the exhaust/return fan track the supply fan to begin with?" With no exhaust other than the exhaust/return fan, it would seem that this fan would essentially track the supply fan, even *if* controlled via building static pressure. Then why go through the trouble of installing a building static pressure transmitter, and all of the associated pressure sensing tubing, and then go through the hassle of setting up the control loop correctly? The problem with sending both fans the same control signal is that there is then no regard for building pressure. The supply fan and the exhaust/return fan usually differ in size and in horsepower. The exhaust/return fan is normally smaller than the supply fan, because of the lesser amount of static pressure and friction loss it has to overcome. Given the same control signal, these two fans may operate at different capacities over the entire signal range, causing the building static pressure to vary.

Suffice it to say that there's "more to it than meets the eye" when trying to control exhaust/return fans in VAV air handling systems. The purpose of an exhaust/return fan is to return air from the space, and exhaust it and/or return it to the air handler, as dictated by the position of the economizer dampers. It should be controlled to maintain building static pressure, and thus should operate independently from the supply fan and economizer dampers. If there is a substantial amount of additional building exhaust from other systems, then the potential problems discussed herein, most notably the "starving the return" problem, could become a reality. If this is the case, then a better solution might be to not use an exhaust/return fan, but strictly an exhaust fan, either as part of the air handling system, or perhaps completely separate from it. The fan would run only when needed, i.e., only when the building pressure has the potential of becoming excessively positive, and would then operate to maintain the building slightly positive. This is how the manufacturers of packaged VAV rooftop units do it. Yet though it eliminates the potential of "starving the return," it doesn't do much to alleviate possible building negative pressure problems. Application is the key here, and a well thought through, well designed mechanical system will serve to minimize the wide array of potential problems that can be brought to light upon system start-up and commissioning. This section serves in part to show that the problems that surface with mechanical systems upon start-up are not always attributable to the controls and to the control system, and controls cannot always provide solutions to problems that arise due to subpar mechanical system designs.

Heating and Cooling Operation

Control of the heating and cooling sections of a VAV air handler is quite similar to that of a reheat system air handler, with subtle differences. Like with the reheat system air handler, if the unit is to bring in substantial amounts of outside air, then there will be a requirement for a heating coil, and it will need to be controlled to maintain discharge air temperature setpoint during the more extreme outside air temperatures. However, even if the unit is not required to bring in a lot of fresh air, there still may be a heating coil, specifically for the purpose of night setback and morning warm-up cycles. With a reheat system, all zones will be equipped with heat, for that is the nature of the system, and the air handler itself need not be equipped with heat if night setback and morning warm-up cycles can be accomplished with the zone level heaters. For a VAV system, this is not necessarily the case, for there may be VAV boxes serving interior spaces that aren't equipped with heat. So depending upon the particular VAV system and its unoccupied mode heating requirements, the engineer may specify the main air handler to have a heating coil, strictly for these cycles.

Not to get sidetracked here, but it's worth noting that for a VAV system designed to bring in substantial amounts of outside air, a main air handling unit heating coil may be done without, if all terminal units are equipped with heating coils. I have actually seen a VAV project in which the air handler was scheduled to bring in 40 percent of outside air, yet the unit had no heating coil. The unit served a system of VAV boxes, all with proportional hot water reheat. At outside air temperatures approaching –10 degrees, the discharge air temperature of the air handler approaches 40 degrees. The design entering air temperature of the VAV reheat coils was actually scheduled as 38 degrees! The engineer had figured that the heat at the zone level would take care of the substantial outside air requirement, for the more extreme conditions. In theory, this is probably quite acceptable. Yet in practice, it leaves a lot to be desired. Instead of noting all of the potential disadvantages and drawbacks to this type of setup, I leave it up to you to think it through and come up with your own theories on why this may not be such a good application. I'll just say that it is definitely the exception rather than the norm.

Night setback and morning warm-up will be covered under the

Unoccupied Mode Operation heading of this section. We digress, and consider here that, if the unit is equipped with heat, it's due to the requirement for large outside air quantities. That being the case, then it should be specified and controlled in the same manner as if it were a coil in a unit serving a reheat system; staged control if an electric heating coil, and proportional control if a hot water heating coil.

In regard to cooling coil control, operation is also the same as with a reheat system. The VAV air handler, during occupied modes, should operate to maintain a constant discharge air temperature setpoint, and will control the cooling means, whether DX or chilled water, to that end. A notable difference between discharge air control of a constant volume air handler and a variable volume air handler has to do with the volume of air being conditioned by the cooling coil. The coil is sized for worst case conditions, at full delivery of air. Yet, when the air handler operates at periods of reduced air volume (which is almost always!), the cooling coil gains "relative capacity." A cooling coil sized for a 25-degree delta T at maximum air volume can (in theory) impart a temperature change of 50 degrees to the air passing through it, at half volume! For a VAV air handling unit, the operating point of the supply fan may be closer to half capacity than to full capacity, more often than not. So even though the cooling coil is sized for worst case, the coil must be able to handle reduced capacities of air and still maintain precise discharge air temperature setpoint. If this is a chilled water coil, the proportional control valve (controlled by the main controller in P+I fashion) shouldn't have a problem with this. Of course, for the DX coil, the remote air cooled condensing unit should have enough stages of control, not just for full tilt cooling, but also for partial load conditions, when the coil is essentially oversized. To account for this, the number of stages might need to be selected as a function of the delta T across the coil during these conditions. Using the rules of thumb from previous sections, and selecting the number of stages based upon a delta T of 50 degrees, the condensing unit would need to be equipped with upwards of 8 stages of control! Good luck finding such an animal! Actually, the larger condensing units should be able to be specified with that many stages, or at least with mechanical unloaders, in addition to the electrical stages of control. The systems with six stages of control (and less) are subject to less than precise temperature control during mechanical cooling at reduced loads. Luckily, this will not be a year-round situation. During the height of the cooling season, the unit will thankfully operate at or near full capacity, and during milder times, the economizer will help out matters, or even completely relieve the condensing unit of its duties.

Incidentally, the same concerns may exist with the heating coil, if the coil is there for occupied mode discharge air temperature control. In other words, a heating coil sized for the full cfm value of the air handler gains relative capacity for air volumes of any less than the maximum. If the heating coil *is* sized for the full cfm, then it's probably oversized to begin with, for all modes! Consider that the coil is there for extreme outdoor air conditions, and you can see why this may be. Put simply, when it's so cold out that the unit's heater has to be pressed into operation, then demand at the zone level will generally be for heating. As such, the terminal unit dampers will have modulated down to their minimum positions and be sitting there. VAV boxes with reheat coils may have a separate "heating position" that they will jump to, which is typically set to be upwards of 50 percent or so. What all of this means is that the air handling unit's supply fan will be operating closer to half capacity than to full capacity. The heating coil, in all actuality, can be sized for half the total possible air volume, or less, if you're buying all of this. A "right-sized" electric heating coil, for this type of application, will have enough capacity to handle the above-mentioned operating conditions, and be equipped with the appropriate number of stages.

Consider what the scenario would be if an electric heater in this type of application was sized for the full cfm of the air handler. For example, if a 10,000 cfm unit was required to bring in an amount of outside air that would result in a worst case mixed air temperature of 45 degrees, and the electric heater was sized to take the full amount of deliverable air from 45 to 55, the heater would be approximately 30 kW. The delta T of this heater, at full cfm, is 10 degrees. However, if the air handler supply fan is operating at half capacity, delivering 5,000 cfm, the total delta T of the heater is 20 degrees. Furthermore, if the fan is operating at quarter capacity, delivering 2,500 cfm, the total delta T is 40 degrees. At this capacity, the heater is essentially 30 degrees oversized, which is okay if you can operate it at 1/4 capacity and also have adequate staging. Required stages for this heater would approach 10.

Would the heating coil in this type of application ever be purposefully sized for the full capacity of the air handler's supply fan, in light of what has been put forth in the preceding paragraphs? Maybe, if someone has a genuine concern about something. Consider a hypothetical situation, again with that same 10,000 cfm air handling unit, at worst case outside air conditions. Consider (hypothetically!) that 90 percent of the VAV boxes are calling for full cooling, and the other 10 percent are sitting at their minimums. The supply fan is operating near its maximum (greater than 90 percent). If the heater was not full-sized but rather sized for, say, 40 percent of the maximum deliverable cfm, then at this instant in time, the heater would be operating at full capacity, and the discharge air temperature would only be around 48 degrees. Of course in this unlikely scenario, the zones calling for cooling would quickly satisfy, the VAV boxes for those zones would close off, the air handler cfm would decrease, and the heater would gain relative capacity, resulting in an increase in discharge air temperature. Still, it's worth discussing this kind of a hypothetical situation, at least to get you thinking about things so that you can make your own evaluations on how mechanical systems are designed and how they should operate.

Last issue on the topic of heating and cooling control is the issue of reset control. Not much to say here. As with the reheat system air handler, reset control of discharge air temperature setpoint can be employed, though only at the engineer's discretion, or as a possible solution to a temperature control problem that arises down the road.

Economizer Operation

Hey, this section is easy! We already talked about this under Reheat Systems. Integration of DX and chilled water cooling with economizer operation, for VAV systems, is pretty much identical to that of reheat systems. Any differences have to do with the amount of air being delivered by the air handler, and for integration of economizer with DX or chilled water cooling, operational differences between VAV and reheat system air handlers are nil.

Unoccupied Mode Operation

Differences between unoccupied mode operation of reheat system air handlers and VAV air handlers are limited to fan operation and to the heating side of things. All else being equal, we focus strictly on these two issues.

As far as fan operation goes, the only difference between VAV and reheat system unoccupied cycles is that the VAV air handler supply fan has the ability to operate at variable rates of air delivery. During the unoccupied mode, this essentially translates to a subtle difference between the two types of air handlers. With the reheat system air handler, the fan turns on (at full speed), upon a call for unoccupied heating or cooling. For the VAV air handler, the fan will also turn on upon such a call. The only difference is that the supply fan will have duct static pressure control imposed upon it, so it may not (immediately) ramp up to full throttle. This of course is dependent upon what the VAV (and fan powered) boxes are doing at the time of supply fan initiation. The topic of unoccupied mode operation of a VAV system, specifically as it applies to the terminal units, will be covered thoroughly in the chapter on VAV and Fan Powered Boxes.

If the VAV air handling unit has a heating section that's required for occupied mode operation, i.e., because the unit is expected to bring in substantial amounts of outside air, then it can be utilized to perform night setback and morning warm-up cycles as well. As with the reheat system air handler, there may be a concern about excessive discharge air temperatures during these modes, especially if there are terminal units out there on the system that have electric reheat. To mitigate the concern, either lock out the terminal reheat at the zone levels, or as with the reheat system air handler, impose discharge air temperature high limit control on the air handler's heater.

If the VAV air handler has a heating coil that's there exclusively for night setback and morning warm-up duties, it may be explicitly "specified" to be single-stage operation, all or nothing. In the case of electric heat, the heater could be ordered with only one stage of control. In the case of hot water heat, the coil can be fitted with a simple two-position control valve. The provision of "all or nothing" heat in a VAV air handler, for night setback and morning warm-up purposes, is a tradition that has been born out of the notion that VAV air handler heating cycles should be quick and at full thrust, with no concern for precision comfort control. The heater is sized specifically for these cycles, and staging or variance of heat output normally need not be required.

If the air handler's heater is there exclusively for these heating cycles, then any (electric) heat out at the terminal units ought to be locked out. Let the air handler's heater do its job, for that's why the engineer specified it to be there. If locking out the heaters at the zone levels isn't in the design (or in the budget!), then you may have a problem. As was discussed in the last section on reheat system air handlers, the concern is regarding "twice heating" the air, once by the air handler's heater, and once by the terminal units' heaters. So what can you do, at the "equipment level," to evade this potential problem?

If the heater is electric, and it is furnished with only a single stage of control (on or off), then there really isn't much you can do, short of simply cycling it during a heating cycle so as to maintain a high limit discharge air temperature setpoint. What that is saying is, upon a heating cycle, turn the heater on, watch the discharge air temperature, and turn it off if the high limit setpoint is exceeded. The other alternative is to disable the heater altogether.

For a hot water coil, the control valve can become a proportional control valve, and high limit control can be imposed upon it during these cycles.

In either of the two preceding situations, the primary function of the air handling unit's heating coil is being compromised. The engineer sized it for these cycles, with the intention that the heater operates at full bore when pressed into operation. Imposing high limit control on a heating coil whose primary purpose in life is to perform night setback and morning warm-up cycles seems to defeat the purpose of the coil, at least to an extent. The right thing to do is to "bite the bullet," lock out those zone level heaters, and let that main heating coil do its thing when called upon.

Unoccupied mode VAV system heating strategies will be dealt with in more detail in the upcoming chapter on VAV and Fan Powered Boxes.

Typical Sequences of Operation

With all that has been covered in this chapter, you might question as to how on earth there can be such a thing as a "typical" sequence of operation for a built up air handling system. There isn't, and so this section is technically at fault in its title. What is presented here are "stripped down" control descriptions for the three types of systems that have been discussed in this chapter. These are all based on systems having hot and chilled water coils for heating and cooling, typical economizer sections, and exhaust/return fans in addition to the supply fans. The descriptions do not get into any great detail or offer any technical explanations of the various processes that govern the operation of the systems as a whole. Rather, they are presented here simply as a "sorbet," or in other words a palette cleanser, to wash down all of the preceding information. Read these with the knowledge that there is a lot going on behind the words to these descriptions. Oh, and feel free to paraphrase for your own needs. Just do so with an understanding of what's going on "behind the scenes," and of what's entailed to make the following descriptions reality.

Constant Volume, Single Zone System

During occupied modes, the air handling unit operates to maintain the occupied mode comfort level of the space served. The fans run continuously (at a constant rate), and the outside air damper is open to its minimum position. If the space temperature falls from heating setpoint, then the hot water valve is modulated open. If the outside air is not suitable for free cooling and the space temperature rises from cooling setpoint, then the chilled water valve is modulated open. If the outside air is suitable for free cooling, then the outside air damper is modulated open as the first means of cooling (economizer operation). If space cooling demand is such that economizer operation alone cannot satisfy the needs of the space, then the chilled water valve is modulated open as well.

During unoccupied modes, the air handling unit operates (as required) to maintain the space temperature between unoccupied heating and cooling setpoints (typically 60 and 80 degrees, respectively). If the space temperature strays out of this range, then the fans turn on, and the unit heats or cools as necessary in order to bring the space temperature back into the range defined by the unoccupied mode setpoints. Outside air damper remains closed during unoccupied mode operation.

Constant Volume Reheat System

During occupied modes, the air handling unit operates to maintain a constant discharge air temperature (typically 55 degrees). The fans run continuously (at a constant rate), and the outside air damper is open to its minimum position. If the outside air is not suitable for free cooling, then the chilled water valve is modulated to maintain discharge air temperature setpoint. If the outside air is suitable for free cooling, then the outside air damper is modulated open as the first means of cooling (economizer operation). If discharge air temperature setpoint cannot be attained by economizer operation alone, then the chilled water valve is modulated open as well. Hot water valve is modulated open to "heat" to discharge air temperature setpoint, if conditions are such that an extensive amount of cold outside air is being brought in via the outside air damper. Otherwise, hot water valve is closed during occupied modes. Reheat coils perform individual zone comfort control, by receiving cool air and reheating it as necessary in order to maintain zone setpoints.

During unoccupied modes, the air handling unit operates (as required) to maintain the space temperature between unoccupied heating and cooling setpoints (typically 60 and 80 degrees, respectively). If the temperature of the general space served by the air handling unit strays out of this range, then the fans turn on, and the unit heats or cools as necessary in order to bring the space temperature back into the range defined by the unoccupied mode setpoints. Outside air damper remains closed during unoccupied mode operation.

A morning warm-up cycle may be implemented, upon transition from unoccupied to occupied mode. Fans turn on, outside air damper remains closed, and hot water valve is driven fully open. Unit remains in this mode until the space or return air temperature reaches the morning warm-up cycle termination setpoint. Upon reaching this setpoint, the air handling unit enters its normal occupied mode of operation (discharge air temperature control).

Variable Air Volume (VAV) System

During occupied modes, the air handling unit operates to maintain a constant discharge air temperature (typically 55 degrees). The fans run continuously (at a variable rate), and the outside air damper is open to its minimum position. Supply fan operates to maintain a suitable duct static pressure, and exhaust/return fan operates to maintain a suitable space static pressure. If the outside air is not suitable for free cooling, then the chilled water valve is modulated to maintain discharge air temperature setpoint. If the outside air is suitable for free cooling, then the outside air damper is modulated open as the first means of cooling (economizer operation). If discharge air temperature setpoint cannot be attained by economizer operation alone, then the chilled water valve is modulated open as well. Hot water valve is modulated open to "heat" to discharge air temperature setpoint, if conditions are such that an extensive amount of cold outside air is being brought in via the outside air damper. Otherwise, hot water valve is closed during occupied modes. VAV boxes perform individual zone comfort control, by receiving cool air, and varying the rate at which it is delivered into the zones in order to maintain zone setpoints. Zone level reheat may apply as well.

During unoccupied modes, the air handling unit operates (as required) to maintain the space temperature between unoccupied heating and cooling setpoints (typically 60 and 80 degrees, respectively). If the temperature in the general space served by the air handling unit strays out of this range, then the fans turn on, and the unit heats or cools as necessary in order to bring the space temperature back into the range defined by the unoccupied mode setpoints. Outside air damper remains closed during unoccupied mode operation.

A morning warm-up cycle may be implemented, upon transition from unoccupied to occupied mode. Fans turn on, outside air damper remains closed, and hot water valve is driven fully open. Unit remains in this mode until the space or return air temperature reaches the morning warm-up mode termination setpoint. Upon reaching this setpoint, the air handling unit enters its normal occupied mode of operation (discharge air temperature control).

VENTILATION STRATEGIES

Nothing impacts the design and operational criteria of an air handling system more than the issue of Indoor Air Quality (IAQ). Well, almost nothing. Anyway, this subject seems to be at the top of every engineer's concerns list, for reasons of which are too broad to go into any great detail about here. Yet at least the overall concept should be easily understood. The quality of the air inside any building is of great importance. To maintain indoor air quality is (among other things) to ventilate; the more of an issue, the more ventilation required. Simply put, the more of a concern for IAQ, the more fresh, outside air must be brought in (by the air handler's outside air damper) to deal with the concern. Not that the air outdoors is particularly "fresh" (I guess that depends on where you live and work)!

So the requirement for outside air has substantial impact on the design and operation of an air handler. Why is that? Simple. The more outside air required, the larger the heating and cooling coils must be, and the more complicated the control system becomes. The first part of the preceding statement is easy to understand. The heating and cooling coils will be sized for particular entering air temperatures, as we saw previously. The outside air component of the mixed air affects the EATs of these coils. The more outside air, the larger these coils must be. The second part of the above statement, the part about the control system becoming more complex, is what this section will attempt to clear up. The following headings will cover some typical strategies relating to how outside air is introduced via the outside air damper of an air handler (economizer operation aside), and will illustrate the operational impacts each method has on control system complexity. Good luck! Get through this section and you're done with this chapter. And hopefully better equipped to handle air handler design, controls-wise and otherwise!

Fixed Outside Air Damper

When a mechanical system is conceived, the mechanical engineer will perform an evaluation, using a set of established guidelines, to come up with a total "worst case" outside air requirement for the space(s) that the system is serving. This is done by evaluating the requirements of each room, and tallying the total cfm of outside air. This information is found in the form of a ventilation schedule, which lists out all rooms that the air handler serves, and each room's "code-required" ventilation. For instance, storage rooms and rooms with little or no continuous occupancy may require no ventilation in the form of outside air. General areas such as hallways and corridors may have a "cfm per square foot" requirement. Office areas, kitchens, conference rooms, and other continuously occupied spaces will be figured for the maximum number of people that can occupy the space at the same time, hence a "cfm per person" requirement. However it's figured, the engineer adds up the outside air requirements of all rooms, to come up with the total requirement for the air handler serving the rooms. Depending upon the individual rooms and how they are to be utilized, this requirement for outside air can be as little as 10 percent, or as much as 50 percent (or greater) of the air handler's maximum deliverable air volume.

The above paragraph used the term "worst case" as it applies to figuring outside air requirements. Worst case basically means "full occupancy" of all rooms served by the air handler. Generally this is, if not impossible, at least highly improbable. If you think about it, in any typical office building for example, if most daily occupants of the building are at their desks working, then the break room and conference room will likely be empty, or at least not at maximum occupancy. Conversely, if the lunchroom is full, then chances are good that the office areas are relatively vacant. Be it as it may, the air handling system generally must be designed to handle the "sum" of the worst case conditions of all of the individual spaces served.

If you buy the notion that "full occupancy" of all rooms at once is an improbability, then you also will agree that the air handling unit will seldom need to bring in the total scheduled amount of outside air. Yet the outside air damper, by code, still needs to be capable of being positioned for this situation (and the unit's heating and cooling coils need to be sized for it). There are basically several ways to do this, as we see in these next few sections. The simplest by far is to set the outside air damper's minimum position to a fixed value, for the worst case requirement. This value is taken directly from the ventilation schedule, or from the air handler equipment schedule, as a cfm value, or in the form of " minimum percent of outside air." The damper position corresponding to this value can be simply programmed into the main controller, so that the damper holds this position, for all occupied modes of operation, regardless of actual occupancy.

For constant volume applications (single zone and reheat systems), setting the damper's minimum position to satisfy maximum occupancy results in excessive energy usage for all times that the spaces served by the air handler are at less than full occupancy.

For a VAV system, establishing the fixed position of the outside air damper is more challenging, at least in terms of satisfying code requirements. As an example, consider a 5,000 cfm VAV air handling unit, specified to have an outside air ventilation requirement of 500 cfm, which is 10 percent of the maximum deliverable air volume. If the outside air damper were fixed at a position to bring in 10 percent of the air that the supply fan delivers, then the air handler would introduce fresh air at a rate below 500 cfm for all times that the supply fan operates at less than maximum capacity. This does not satisfy code, for it's the actual cfm value that must be consistently delivered to the space to satisfy occupant requirements, and not the real-time percentage of the air being delivered by the supply fan.

One might argue that, in a VAV system, supply air demand is occupant driven; the fewer people in the zones served, the less of a demand for cooling, and the less air delivered by the air handler. The outside air component of the supply air decreases along with the demand for cooling, but this is okay because, theoretically, demand is a function of occupancy, and the less of a demand, the less people in the space. Hence a lesser fresh air requirement. Good argument. Unfortunately, this viewpoint doesn't hold much water, at least in the eyes of code mandates. And while it may be somewhat true during periods of greater demand (summer months), it's not quite the case during the colder months, when the zone level demand is likely to be more for heating than for cooling, regardless of the level of occupancy. Were that it be, the terminal unit dampers will be closer to their minimums, and the air delivered by the supply fan will be closer to half capacity. The notion that supply air demand is a direct function of building occupancy loses much of its validity when it's looked upon in this manner.

So for the VAV air handler that is the focus of our example, how can the outside air damper be set to satisfy code ventilation requirements? Simple. The damper will need to be set so that the outdoor air component of the supply air is at least 500 cfm for all operating points of the unit. To guarantee this, the damper must be set to deliver 500 cfm at the supply fan's minimum operating point, or rate of delivery. To carry on with the same example, assume that the air handler will never provide any less than 40 percent of its maximum deliverable amount of air. This may be determined by analyzing the terminal unit minimums; figure the minimum cfm values for each and every terminal unit, add 'em all up, take the ratio of this cfm value to the maximum supply fan cfm, and you have your (theoretical) minimum supply fan operating point, in percent. Okay, so the supply fan's minimum operating point in this example is found to be 2,000 cfm, which is 40 percent of 5,000. With the supply fan operating at this rate, the outside air damper shall be set to pass 500 cfm of air (which is 25 percent of 2,000), to satisfy code. Of course, when the supply fan operates at any rate higher than this minimum, the outdoor air component of the supply air is in excess of the requirement, and energy will be wasted in conditioning it. Consider that if the supply fan is operating at its maximum, then the amount of outside air brought in will be more like 25 percent of 5,000 cfm, or 1,250 cfm.

The practice of setting the outside air ventilation requirement by fixing the position of the outside air damper for all occupied operating modes, while definitely the simplest (and the method seemingly most often employed), also leaves the most to be desired. It's an energy hog, plain and simple. Conditioning outside air is costly, and hard on the equipment as well.

Fixed Amount of Outside Air

The preceding method of establishing ventilation requirements is applicable to all types of air handling systems, constant volume systems (single zone and reheat systems) and VAV systems alike. The method discussed under this heading is specific to VAV systems, in which the supply fan operates at varying air delivery rates.

Fixing the amount of outside air that passes through the outside air damper is a step up from simply fixing the position of the damper, and definitely a step in the right direction! Take that same 5,000 cfm VAV air handler from the previous discussion, and now assume that the rate of outside air introduced can be held constant regardless of how much air the supply fan is delivering. Specifically, consider that the minimum outside air requirement is (again) 500 cfm, and the outside air intake allows this much air to pass at a supply fan operating point of 40 percent.

Now picture that this cfm value is held constant, even as the air handler's supply fan ramps up in capacity. In the previous scenario of the fixed outside air damper, the VAV air handling unit will introduce upwards of 25 percent of 5,000 cfm, or approximately 1,250 cfm, when the air handler's supply fan operates at full capacity. However by fixing the outside air cfm value rather then the outside air damper position, the unit will only introduce 10 percent of outside air, at maximum air delivery of the unit's supply fan. Code is met, without having to introduce and condition excessive amounts of outside air. Overventilation of the spaces served by the VAV system is reduced, and performance is improved.

All right, already! How's it implemented? Easier said than done. Actually, this method of control should become less complicated and more popular as technology allows for more precise methods of measuring air velocities. This, alas, is the key to accurately implementing such a control scheme; in order to control the amount of outside air introduced, we must be able to accurately measure the amount of air introduced. An airflow measuring device is used to perform this all-important task, and is described in the following paragraph.

Basically the device is nothing more than a pitot tube or averaging tube that is used in conjunction with a differential pressure transmitter. The tube gets mounted in the ductwork, and has the ability to measure static pressure and total pressure. The static pressure is applied to the low port of the differential pressure transmitter, and the total pressure is applied to the high port of the transmitter. The difference in these two values is known as velocity pressure. The signal that the pressure transmitter generates is a function of velocity pressure.

An airflow measuring station is a more accurate approach to measuring airflow in a duct. The apparatus is basically a rectangular section of ductwork, complete with an airflow measuring device, which may consist of several averaging tubes traversing the section of the duct. The section may contain airflow "straighteners," so that a more precise measurement may be made of the air passing through it. A more expensive alternative to the simple flow tube, in terms of component cost and installation labor, the airflow measuring station will likely provide a more accurate measurement of velocity pressure, with less fidgeting and fussing around with.

The signal generated by the pressure transmitter will feed an analog input at the main controller. The signal must be converted to useful information via the controller. Velocity pressure information will not be sufficient, and the main controller must be programmed to convert this signal to air volume (cfm). A calculation is made that extracts the square root of the pressure signal, and multiplies it by the cross-sectional area of the duct. Now the value is in units of cfm, and viola, outside air damper position can be controlled as a function of it! Realistically, it's still not quite that simple. It's one thing to get an accurate and repeatable measurement of outside air cfm. It's another thing to control a standard outside air damper in order to maintain a constant rate of ventilation. It is these two issues that make this type of control scheme difficult (to say the least). Don't get me wrong, now. It's done all the time with considerable success. It's just not as cut-and-dried as one would expect it to be, that's all. Accurate airflow measurement is as much a function of the mechanics involved as it is of the electronics. Physical construction of the air handler plays a major part, as does damper construction and configuration. A specialized outside air damper assembly, one specifically designed with this type of control in mind, is a good investment, if available. An air handler manufacturer may even offer complete packages, including the outside air damper and airflow measuring device all in one module! If maintenance of outside air at a fixed value is in the game plan, then these avenues should be thoroughly explored, and the best practices ultimately be put in place.

Controlling the outside air damper assembly to maintain a fixed amount of outside air, when correctly implemented and properly maintained, is a darn good way to meet code fresh air requirements while reducing operating costs. Yet, there will still be periods of overventilation, regardless. Quite honestly, during any periods of reduced occupancy, the code-required "cfm per person" will be less than maximum. An air handler required to bring in 500 cfm of outside air for the worst case of maximum occupancy will technically not require this much fresh air if occupancy levels are less than the max. Figure that the fresh air value is calculated at 15 cfm per person, and you'll find that the fresh air requirement is based on an occupancy level of approximately 33 people. If the building is occupied to half that level, then with only 16 or so people occupying the spaces, the actual real-time requirement for ventilation is somewhere around half the maximum required ventilation rate, say around 250 cfm. But actually figuring out real-time occupancy levels is a challenge, and so this particular method of satisfying ventilation requirements, while performing the job that it's been hired for, will still allow for periods of overventilation. Without adding more complexities to the overall air handling system, this is typically the best that we can do, and so it continues to be a popular alternative. The next heading takes the concept of ventilation a step further, yet the requirement for more sophisticated controls and control schemes comes along with it as part of the package.

CO_2 Based Demand Controlled Ventilation

After reading through the material presented under the last couple of headings, doesn't it seem like there should be a better way to determine the required amount of ventilation? It seems that there should be a means of actually measuring occupancy, so that the position of the outside air damper can be adjusted as a function of how many people are actually occupying the space. If there are a dozen people in a building served by an air handler that's designed to introduce fresh air for a maximum of over a hundred occupants, then realistically the air handling unit (at most operating capacities) is likely bringing in too much outside air. The position of the outside air damper should be able to be reset downward in such a situation.

This heading discusses a method of measuring occupancy, which has come into its own as an acceptable means of establishing the ventilation requirement of an air handling system. Applicable to constant volume systems as well as VAV systems, it is the practice of measuring the carbon dioxide (CO_2) levels in the space served, correlating the measured values with an occupancy level, and setting the outside air quantity as a function of this level.

A little background first. It is a well-established notion that *indoor* CO_2 levels can be used as an accurate measure of human occupancy. This is due to the fact that humans are the prime source of CO_2 indoors, and will exhale CO_2 at a very predictable rate, given a presumed level of activity. For practical purposes, activity level is typified as would be expected in a normal working environment, be it office activity, light warehouse activity, etc. It is also recognized that the outdoor air CO_2 level, in most locales, is very stable, and in the range of 350 to 450 ppm (parts per million).

When the prescribed amount of outside air (per occupant) is being supplied into a building, the indoor CO_2 level will stabilize at around 700 ppm above the outdoor CO_2 level. This simple statement is quite powerful, in terms of what it means to acceptable indoor air quality. What it implies is that, if the indoor CO_2 level can be maintained at 700 ppm (or less) above the outdoor level, then all occupants in the space, on average, at any given time, will receive their code-required ventilation.

The above statement is founded on the premise that 15 cfm per person (of outdoor ventilation air) is adequate, and is generally accepted as the "minimum" ventilation rate for certain spaces (with the rate for other types of spaces being more like 20 cfm per person). It is interesting to note that, for a requirement of 20 cfm per person, the differential between indoor and outdoor CO_2 levels would need to be more like 500 ppm, which essentially translates into "more outside air."

To implement a control strategy based on the above, we would be required to measure both indoor and outdoor CO_2 levels, and modulate the air handler's outside air damper in an effort to continuously maintain a 700 ppm differential between the two measured values. As people occupy the space, the differential would tend to rise, and the outside air damper would have to be modulated open. Likewise, as people depart, the differential would decrease, and thus the outside air damper would be allowed to modulate closed. The term used to appropriately describe this process is Demand Controlled Ventilation (DCV). It is a fantastic concept on paper, and theoretically can all but resolve the conflict between maintaining acceptable ventilation rates and using excessive energy.

In practice, many issues must be dealt with and hurdles overcome, in order to effectively and consistently provide "code-required" ventilation using DCV. First, it's important to mention that the standards organizations writing ventilation codes do not unequivocally endorse or condone the method but do affirm that it can be used, provided that certain conditions are met. An understanding of those conditions must be had, and is beyond the scope of this book. Second, CO_2 sensor technology is a relatively recent development. First cost, sensor reliability, and calibration issues have been bones of contention (hopefully to be resolved in the not so distant future!). Finally, sensor location and placement is of great importance, in regard to measurement precision and acceptable maintenance of the established ventilation rates.

With that said, let's see how we go about this, using what current technologies have to offer us in the way of state-of-the-art controls and components. We consider constant volume systems first, as this is the easier topic to digest, and then we discuss DCV as it applies to VAV systems.

For a constant volume system, achieving DCV appears to be a pretty straightforward task. We need the sensors, of course, and we need a way to measure and control the amount of outside air being brought in via the outside air damper. We've already talked at length about how to measure and control outside air, so we've got that covered. The focus here is on CO_2 sensors, the potential requirement for more than one sensor, and the location of these sensors. To be technically correct, these devices aren't simple sensors, but transmitters, as they require power and utilize active electronics to generate a control signal relative to what they are sensing. For simplicity's sake, we'll just call 'em sensors in this section.

First, one could argue that, if it is known that the outdoor CO_2 level remains constant at a value, of 400 ppm for instance, then we would not require an outdoor CO_2 sensor. We can simply monitor indoor CO_2 levels, and control to maintain an absolute indoor level of 700 ppm above the assumed outdoor level (say 1100 ppm), instead of controlling to maintain an indoor/outdoor differential. It's a good argument, yet code conformance may dictate the requirement for both sensors, and the control of the differential value as opposed to control of an absolute indoor value.

For the constant volume system (single zone or reheat), where does the sensor go? Space or return air duct? For most applications, the sensor will be better off in the space served by the system. The sensor should be located in the room or area that would be deemed as the most critical, in terms of ventilation requirements. Return air sensing is suitable when the space served by the constant volume system is relatively consistent throughout, occupancy-wise. If there is diversity, then return air sensing is not a good bet. The same argument that works in favor of return air temperature sensing (for single zone systems) actually works against return air CO_2 level sensing. Basically, the return air tends to be an average of all areas served and conditioned by the air handling system. Controlling space temperature via return air temperature sensing is not a bad way to go (depending upon the application). Sure, some areas may be warmer or cooler than others, but the *average* space temperature will be maintained. For CO_2 level control, it is less important to control to an average level, and more important to maintain the required CO_2 level in the area deemed most critical. The sacrifice is that theoretically other areas will be "overventilated." Yet on the other hand, if even one area is underventilated (as could result from return air sensing and control), then the mechanical system is out of conformance, and the entire control system is of moot value.

For VAV systems, as with everything else pertaining to these systems, there is generally more complexity to DCV. Of course with VAV systems, we are not delivering a fixed amount of supply air, but a volume of air that continually and dynamically changes as a function of zone level demand. So that needs to be taken into consideration. Along with the question of where to put the indoor sensor. Well, to echo a previous sentiment, we'll assume that we have the best of what technology has to offer, in terms of controls. In short, a fully networked system, with digital VAV box controllers, and VAV zone sensors that not only sense temperature, but CO_2 level as well!

Sensors that combine temperature and CO_2 level sensing can and do exist, and can be utilized in a DCV control scheme. One thought is to modulate the terminal unit dampers as a function of CO_2 levels, in addition to temperatures. In other words, as more people occupy the zones, the terminal unit air flow's increase (per zone), until ventilation requirements are met. Of course, this can lead to "overcooling" of zones, so a maximum should be set, per zone. This is some of what can be done at the zone level. Now the next step is to develop an algorithm within the air handler's main controller, to utilize the zone level information, and operate the outside air damper in accordance with ventilation requirements. This is where it can get hairy. A couple of different things can be done. A critical zone can be chosen, that zone that is registering the highest CO_2 reading, and the outside air damper can be positioned to satisfy this particular zone's ventilation requirement. Also, as zone maximums are reached, the outside air damper can be opened further, to satisfy system-wide ventilation requirements. All in all, integration of zone level control with the operation of the outside air damper is a task left to the skilled control systems engineer/programmer, and is not for the average hacker. Understand that this is a very dynamic and complex control scheme, and requires just a bit of expertise in this area. Nonetheless it can be done, with the proper tools, components, and skill level, and should become a viable alternative for maintaining indoor air quality in the 21st century.

Using a networked DDC/VAV system to effectively perform DCV is definitely within the realm of practicality. The power of distributed digital control is exploited to its full extent, and the effectiveness of the control scheme is only limited by the abilities of the programmer and his knowledge and grasp of the concept. The engineering involved is extensive, no doubt about that. However, we may be fortunate enough to see, as time goes on, the makers of DDC products actually building in the algorithms required to implement DCV. A pre-engineered product, well-proven and thoroughly tested, will no doubt make Demand Controlled Ventilation more accessible, even to the entry-level DDC installer.

We now conclude this chapter by saying that the topic of air handling systems is vast and varied, as you know now by looking back at the shear size of this chapter. Hopefully you've gained some tools and some insight, not only relating to how air handlers are controlled, but also to how air handling systems are designed as a whole. The material presented in this chapter is vital to understanding HVAC systems in general, and the more that is known about them, the better equipped one will be to evaluate, design, and successfully install and commission these systems. We quietly bid farewell to this chapter, and forge ahead, with a clearer understanding and a better overall feel for HVAC "systems," and the role of controls in these systems.

Chapter 12

RTU Zoning Systems and Stand-alone Zone Dampers

This chapter would have a tendency to refer to the term "rooftop unit zoning system" quite a bit throughout, so rather than repeating this rather long term for every reference, why don't we come up with an acronym that we can use in its place (one that hasn't been taken). To avoid any kind of copyright infringement, let's use the term **RTZ** (**R**oof**T**op **Z**oning) as the defining term for a system that takes a single zone constant volume packaged rooftop unit, and turns it into a multiple zone heating and air conditioning system. I'll keep my fingers crossed and hope that this term hasn't been trademarked!

RTZ systems are a decent, relatively inexpensive means of providing multiple zoning to a space served by a single piece of packaged heating and air conditioning equipment. It works well when applied properly. This means that the zones must be similar in load. Interior and perimeter zones should not be served by the same RTZ system. It also means that the setpoints to be maintained in the zones are close in value, relative to one another. Contrary to what might be expected, an RTZ system will not maintain a constant 68 degrees in one zone, and 72 degrees in an adjacent zone. It is not realistic to expect precision temperature control from an RTZ system. The number of zones that an RTZ system can have is limited; the more zones a system has, the less of a chance to consistently and continually satisfy the individual zones. While the concept of RTZ is solid, it is not a panacea for all zoning design challenges. An understanding of what to expect from an RTZ system should be thoroughly possessed by the end user, so that there are no disillusions, and no disappointments when all is said and done. Sometimes called a "poor man's VAV system," the customer should know up front what he is getting for his money. Of course the alternatives are more costly, and the customer may be just fine with the poor man's system of zoning, even having the requisite comprehension of what it is. With that being the case, then it's RTZ all the way!

RTZ systems are microprocessor-based, typically supervised by a single "smart" controller. RTZ system zones are networked together, and back to the controller. The controller has the ability to "poll" the zones for their comfort needs, and also has the ability to broadcast information to the zones such as time-of-day and rooftop unit operating mode information. The system architecture described here is typical between manufacturers, though it can vary somewhat. The system controller is normally a digital controller. It is not programmable but configurable. The basic RTZ algorithms are already burned into the controller, and parameters need to be set by the commissioning technician. This is true with the "store-bought" version of RTZ. It is feasible, however, that a DDC contractor can use his product line to implement RTZ with non-specific DDC controllers. For this "road less traveled," the contractor would have to buy or build his own zone dampers, design his own control algorithms, create his own wiring and hookup diagrams, etc. Yet at the same time the contractor also avoids manufacturer price markups, and can feasibly save some money by going this route, and even offer the customer some "custom" features not offered by manufacturers' "canned" versions.

First on the agenda here is to take a look at the components that will typically make up an RTZ system. An understanding of each component shall be gained (hopefully), and then we will head on to "operational characteristics" of each of the components, and of the RTZ system as a whole. The basic methods of system control are well established; we're not re-inventing the wheel here! What we are trying to accomplish here is simply to gain a solid understanding of how RTZ systems work, and of some of the requirements in terms of controls.

System Components

The following components minimally make up an RTZ system (plus the rooftop unit, of course!); refer to Figure 12-1.

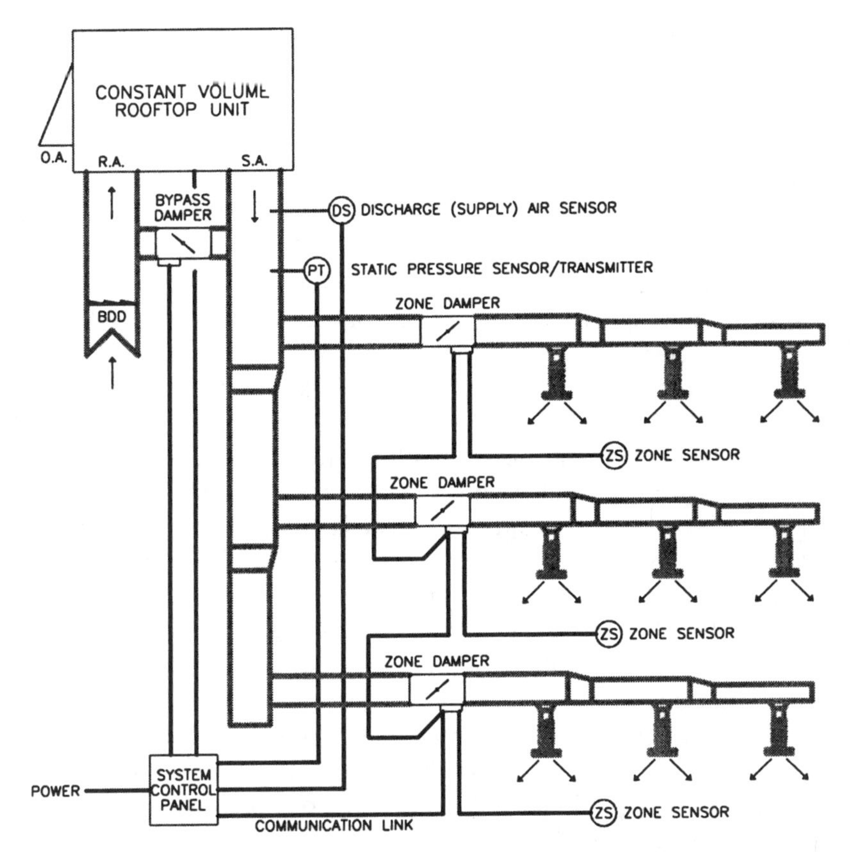

Figure 12-1. Constant volume rooftop unit serving an RTU zoning system.

- System Controller
- Zone Dampers
- Zone Sensors
- Supply Air Temperature Sensor
- Static Pressure Transmitter
- Bypass Damper
- Backdraft Damper

The **system controller** is the brains behind the operation. This is typically a microprocessor-based controller equipped with all of the capabilities required to supervise and implement the operation of an RTZ system. This includes software/firmware, control algorithms, and possibly even an internal time clock. The controller requires power, either 24 or 120 volts, and is typically mounted in an area suited to its presence, perhaps in a small telecomm room or electrical closet. A desktop computer can be connected to the controller for human interface, and thus the controller may be located in the office of the system's "supervisor." Of course there may be more than one RTZ system in any given facility, which makes seamless communication with all system controllers a bit more involved. A supervisory control system may be added, connecting all system controllers together via a network cable, for centralized access to all RTZ systems via a single location.

The last few sentences kind of got off track a little bit, so let's get back on track. The RTZ system controller's main function is to poll all zones on the system, and put the rooftop unit in the appropriate mode of operation. We are dealing with a standard, single zone, constant volume rooftop unit here; instead of a thermostat wired to the terminal strip of the unit, the system controller is wired to it. The controller therefore has full command of the rooftop unit's heating, cooling, and fan functions. On the other end, zone dampers are networked together and back to the system controller. It is over this network that the system controller communicates with the zones.

Zone dampers are nothing more than motorized butterfly dampers with controllers (Figure 12-2). The dampers themselves are round, and range up to 16 inches in diameter or so. The damper actuators are typically non-spring return, keeping costs per zone down. The controllers are electronic circuit boards, and are typically microprocessor-based, "application specific" controllers, meaning they are specifically designed to perform zone control. They are mounted to the zone damper, and enclosed in a sheetmetal box, along with the damper actuator. The whole assembly assumes its place in the duct distribution system, and therefore resides in the ceiling space. Ductwork connects the inlet side of the zone damper to the rooftop unit's main trunkline. Likewise, ductwork connects the outlet side to diffusers that serve the particular zone of comfort.

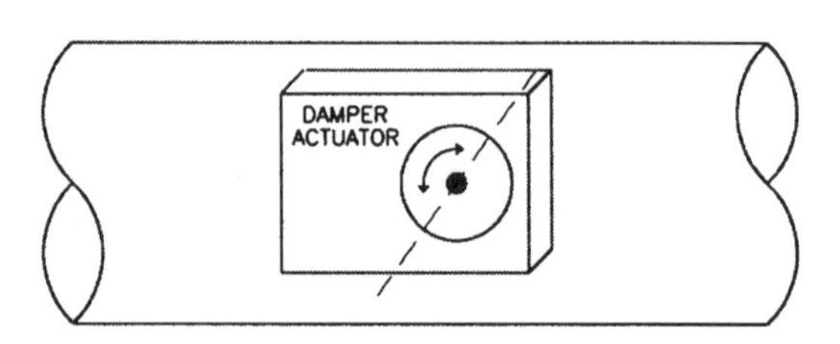

Figure 12-2. Round zone damper with actuator (controller not shown).

Zone sensors reside in the zones served by the zone dampers (one per zone). Each zone sensor is wired up to its respective zone damper, specifically, to the controller. The sensor monitors space temperature, and transmits this information to the controller. The sensor in addition may be equipped with an adjustment, allowing the individual user to set his or her own comfort level. The adjustment simply affects the temperature signal being sent back to the controller, so that it contains setpoint information as well as space temperature information.

The system **supply air sensor** is installed in the supply duct, downstream of the rooftop unit, and is wired back to the system controller. The controller uses this temperature information to validate rooftop unit operating modes, and to impose high and low limits, as will be discussed further on.

The system **static pressure transmitter** is installed in the supply duct as well, normally after the bypass damper and before the first takeoff. It is also wired back to the system controller. The controller uses this information to control the bypass damper, to maintain proper system pressure.

The previous paragraph jumps ahead a little bit, for we have not yet defined the term **bypass damper**. This is a motorized damper, either butterfly or bladed in style, that is installed between the rooftop unit's supply and return drops. The bypass damper's role is to maintain proper pressure in the supply duct, and at the same time provide adequate airflow through the rooftop unit, by bypassing supply air straight to the return as zone dampers close off and pressure begins to build. The damper actuator is normally non-spring return, though a spring return actuator can be used as well. Yet consideration is warranted in making the decision of which way the damper should spring return. Failing shut could cause excessive pressure in the system, as evidenced by excessive noise during periods of low demand. Failing open would cause reduced airflow capacity to the zone dampers, resulting in insufficient zone comfort control. Failing open can also cause rooftop unit operational problems, of which we will discuss in the next section.

The last component listed above as part of an RTZ system is not really a control component. Nevertheless, it is an important enough (and often enough overlooked) component to merit inclusion in this section. A **backdraft damper** is usually recommended for an RTZ system. The damper is installed in the return air duct, before the bypass damper. In other words, before the air returned from the spaces reaches the location of where the bypass damper meets the return air duct. Backdraft dampers only allow air to pass in one direction. In this application, the damper is situated to allow air (from the spaces) to return to the rooftop unit. During periods of light demand, the bypass damper may be bypassing the majority of the air that the rooftop unit's supply fan is delivering, straight to the return duct. The air may have a tendency to "take a left turn," and travel in the opposite direction of the rooftop unit, down the return air duct, out the return grilles, and into the spaces. Via the "path of least resistance," if you will. A backdraft damper prevents this scenario, and is a good idea on all RTZ system projects, even though certain duct system configurations may allow for the possible exclusion of such a damper.

Operational Characteristics

This section deals with the "ins and outs" of RTZ system operation, from the zone level to the equipment (rooftop unit) level, and everything in between. To accomplish this task, we have to begin somewhere, so why not right at the zone level?

Zone Control

To reiterate on what is meant by the term "zone," we recall in the chapter on rooftop units that a zone was defined as an area of temperature control. How can more than one zone of temperature control be served by a single piece of heating and air conditioning equipment? One way is to do it with an RTZ system. The rooftop unit serves not one, but several zones, with each zone consisting of a zone damper and a zone sensor. Forget for now about how the rooftop unit is controlled. Suffice it to say at this point that the unit will be in one of three possible modes: heating, cooling, or ventilation (neither heating nor cooling).

Each zone served by the rooftop unit has a zone sensor that transmits zone temperature (and setpoint) information back to its respective zone damper. The damper has the ability to modulate open and closed, to provide varying amounts of rooftop unit air into the zone served by the damper. Consider a single zone. If the temperature of the air that the rooftop unit is providing to the damper is conducive to the needs of the zone, then the proper amount of this air is allowed into the zone, via the damper. The further from setpoint that the zone is, the more air is allowed. The sensor measures the temperature of the zone, and relays this information to the zone damper's controller. The controller, in turn, modulates the zone damper to introduce the appropriate amount of conditioned air into the zone, in an effort to bring the zone back to setpoint.

Now, what if the temperature of the air delivered to the zone damper is not conducive to the comfort needs of the zone? The damper must close in this event, so as not to provide "opposite mode" air into the zone. For example, if the rooftop unit is in a cooling mode, and the temperature of a particular zone is below its setpoint, the damper serving this zone must close down, so as not to provide cool air into the zone. Typically, the zone damper will not be allowed to close completely, but only to some minimum position, so as to ensure ventilation of the zone at all times.

This in essence is the standard operation of an RTZ system zone damper. If the air received from the rooftop unit is conducive to the needs of the zone, allow it to pass (at the proper amount). If not, then don't allow it to pass. This is how a single piece of heating and air conditioning equipment can serve multiple zones of temperature control. It's not perfect (as we will see), but it's better than a single thermostat controlling the equipment.

The question raised by the previous paragraphs is "How do the zone dampers know what the temperature of the air is delivered to them?" The answer is "Because the system controller is telling them!" No kidding! The system controller is telling the rooftop unit to be in a certain mode of operation, through the unit's thermostat connection terminals, and therefore knows what the temperature of the air provided to the zone dampers should be (hot or cold). This information is broadcast to all of the zone damper controllers, via the network cable.

Rooftop Unit Mode Control

Okay, so what dictates the mode of operation of the rooftop unit? What drives the decision by the system controller to put the rooftop unit into a specific mode of operation? How does it know if the unit should be providing heating or cooling, or just ventilation? How? Because the zone damper controllers are telling it! The system controller "polls" the zones, to find out what is needed the most. As such, each zone gets to cast a "vote" for its required mode of operation. Zones whose temperatures are below their setpoints would vote for heating, and zones whose temperatures are above their setpoints would vote for cooling. Zones whose temperatures are pretty much at setpoint do not cast a vote. The zones that are the "majority vote" get their wish, and the system controller puts the rooftop unit into the appropriate mode. The dampers for those zones in the minority will close down, and will have to "sit and wait," until they become the majority, at which point the rooftop unit will switch to the opposite mode.

To illustrate this with an example, consider a six-zone RTZ system. At a given point in time, three of the zones are in need of heating, two are calling for cooling, and one is at setpoint. The zone at setpoint does not cast a vote. Therefore, it's three to two in favor heating, and the rooftop unit operates in a heating mode. The dampers serving the zones not calling for heating "sit tight" at their minimum positions, so that only a minimal amount of heating air is delivered into these zones. The dampers serving the zones that are calling for heating modulate as required to achieve their zone temperature setpoints. The further away from setpoint any particular one of these zones is, the more open its zone damper is. As these zones heat up toward their respective setpoints, their dampers modulate closed accordingly. The first zone reaching setpoint retracts its vote for heating, thus evening the score (two to two). In the event of a "tie" as illustrated here, the rooftop unit will likely remain in the same mode, until the score changes in favor of the opposite mode. That can happen in a number of ways. The deciding vote may be cast by the zone who, up until this point, has been satisfied. That zone's temperature may rise enough for it to cast a vote for cooling, at which point the score will change from two to three, in favor of cooling, and the rooftop unit changes over to the cooling mode. Of course now the dampers of those zones that are still in need of heating will have to close down to their minimum positions, whether they like it or not!

This is the basic principle of RTZ. Think of it as a democracy, where all members of the space served by the rooftop unit can "voice their opinion" and have a say-so in how the rooftop unit should operate. It's not a perfect system, as demonstrated in the above example, for a rooftop unit can only appease some of the zones some of the time. And possibly some zones never!

Certain specialized schemes can be incorporated into the overall voting process, that vary from manufacturer to manufacturer. A "priority zone" can be selected. This zone has "added pull" in the voting process, or perhaps even "executive authority" (good for the president's office!).

As such, this zone may get its way automatically, if the temperature in the zone strays enough from setpoint. Another variation of the voting process is "weighted voting." Higher priority zones can cast not one vote, but two or more, depending on how far out of whack they are. Low priority zones are stuck with only one possible vote.

In the event of ties, rather than sticking with the current mode of operation until the sides change, another method of determination is to look at the deviation from setpoints of each side, and go in favor of the mode with the greatest deviation.

In the unlikely event that all zones are satisfied at the same time, and no votes are cast for either heating or cooling, the rooftop unit simply operates in a ventilation mode. You can bet that won't last for long though, and soon enough one of those zones out there will call for something, be it heating or cooling, thus affecting all of the other zones, at least in a minimal fashion.

With all of the possible scenarios and different methods of control, RTZ systems are very dynamic, and quite complex in the manner in which they carry out their day-to-day operations. This complexity that makes it a very flexible and versatile system also makes it one that can be very difficult to set up and fine tune. A thorough understanding of the dynamics of the system must be had by the commissioning technician, to ensure that expectations are realized and callbacks are minimized.

System Controller and Bypass Damper

We have now talked at length about zones, zone dampers, and voting. The system controller is the "supreme and benevolent ruler" in an RTZ system. The controller polls its members, finds out the needs of the majority, and operates the rooftop unit in the required manner. The controller also has some other responsibilities, of which we now talk about. The first of these responsibilities is the control of the bypass damper.

The system controller continuously gathers information from a couple of sensors out on the system, one being the static pressure transmitter. This sensor measures the static pressure in the supply air duct, and transmits this information to the system controller. The system controller processes this information, compares it with the desired setpoint, and positions the bypass damper accordingly, to maintain setpoint. Consider an example to illustrate this. As before, we have a six-zone system. At a given point in time, say, first thing in the morning when the system comes to life, assume that all zones served by the system are below setpoint. When the system starts up, all of these zone dampers will cast a vote for heat, and will all modulate toward their fully open positions. The rooftop unit fan will start, and heating will be engaged. The demand for heat by the zones is such that all zone dampers go fully open, and the pressure in the supply air duct will be acceptable, even with the bypass damper fully closed. Now as zones come up to temperature, and their dampers modulate closed, the pressure in the supply air duct will tend to increase. The pressure transmitter senses the increase, and signals the system controller to do something about it. The system controller, in turn, begins to modulate open the bypass damper, to achieve and maintain acceptable duct static pressure and airflow.

Modulating open the bypass damper, to deliver supply air directly to the return air opening of the rooftop unit, seems like a pretty good way of maintaining acceptable duct pressure. There are consequences to this, however, that must be dealt with by the system controller. By reintroducing conditioned air to the rooftop unit for another go-around, excessively high or low air temperatures can be achieved. To illustrate again with an example, consider our six-zone system, right where we left off from before. The zones are voting for heat, the rooftop unit is in a heating mode, and the zone temperatures are coming up to setpoint. The zone dampers are modulating closed, and the bypass damper is modulating open. Assume that the system reaches a point to where all but one of the zones reach their setpoints and retract their votes for heat, with the remaining zone still voting for heat. The rooftop unit remains in a heating mode. All but one of the six zone dampers are at their minimum positions, and the remaining damper is modulating closed as its zone is coming up in temperature. The bypass damper is positioning to bypass the majority of the supply air directly to the return. The temperature of the air leaving the rooftop unit's heat exchanger, "first time through," is somewhere near 90 degrees. That air is delivered back to the return of the rooftop unit, for another go-around. The rooftop unit, still in a heating mode, and still capable of imparting heat to this air, does exactly that, and the supply air rises in temperature, possibly in excess of 110 degrees or more. This is excessive for such a system, and can lead to an array of problems, both operational and equipment related.

Consider the opposite scenario. The rooftop unit is in a cooling mode, only one zone voting for cooling, bypass damper substantially open. The DX coil of this particular rooftop unit is capable of imparting a

15-degree delta T to the air passing through it. The air delivered from the unit, first time through, may be around 55 degrees. As it returns for a second go around, the air mixes with the minimal outside and return air coming in to the unit. The mixture passes through the DX coil, and comes out the other side at a temperature of less than 50 degrees. Sustained operation of this kind will eventually result in the DX coil frosting up.

These scenarios must be avoided, or "nipped in the bud," before they have a chance to become a problem. The system controller must prevent these scenarios, and does so by imposing "high and low limits." The system controller continuously monitors the temperature information furnished by the supply air temperature sensor. If an extreme temperature condition occurs, the system controller simply terminates the cause of the condition, whether it be heating or cooling, for as long as need be. The controller may re-enable normal operation once the supply air temperature comes back in line, may require a mandatory time period of disablement, or may impose a combination of the two. Whichever method is taken, the supply fan still continues to run during the condition, and the system still receives its required ventilation.

Backdraft Damper

That about covers it for RTZ systems, at least as far as the basics go. The last thing on the agenda here is to discuss the need for a backdraft damper (refer to Figure 12-3). This is a non-motorized control damper that allows air to pass through it in only one direction. Depending upon the configuration of the ductwork distribution system, a backdraft damper may be required to be installed. As was stated earlier in this chapter, the damper is installed in the return air duct, upstream of the bypass damper. The damper allows air to return from the spaces to the rooftop unit, but does not allow air to pass the opposite way. Why is this important? Well, if the bypass damper is substantially open at some point in time, there may be a tendency for at least some of that air to travel in the opposite direction. Instead of returning to the unit, the "path of least resistance" may be down the return air duct and out of the return air grilles. This is especially true if the rooftop unit is in an economizer mode, when the return air damper is partially to substantially closed, and the unit supply fan is not pulling as much of a negative from the return air duct system. A backdraft damper prevents air from going the wrong way; the pressure will force the backdraft damper shut, and the air will have no choice but to return to the unit.

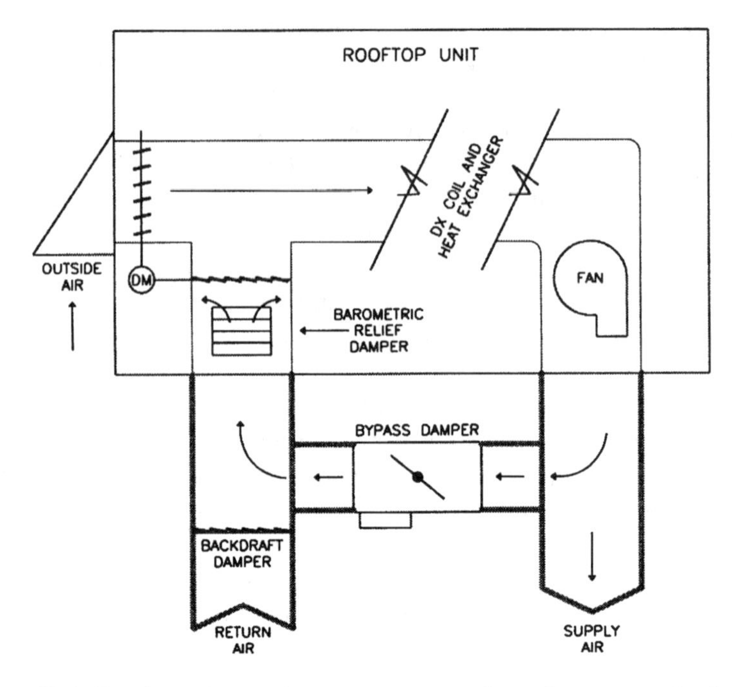

Figure 12-3. During economizer operation, when the return air damper is substantially closed, the supply fan will pull air mostly from the outside air intake. Air bypassed from the supply duct to the return duct will not be actively drawn by the fan, and may tend to travel in the opposite direction (down the return air duct). A backdraft damper prevents this scenario, and the bypassed air is forced out via the rooftop unit's relief damper.

In Closing...

The concept of RTZ is proven and well established. Its operating principles and methods of control have evolved and are widely accepted. There is no "reinventing the wheel" when it comes to these types of systems. However, their complexity does merit a certain degree of specialization, in terms of design, installation, and commissioning. Moreover, the RTZ system is not the end-all solution to all zoning dilemmas, and care must be taken so as not to oversell the concept. Yet, for what its worth, RTZ is an economical alternative to other methods of zoning, and as long as the end user knows what to expect from his investment up front, and knows of the alternatives and of their associated costs, he will likely be pleased with the performance of his system, and of the flexibility offered with it.

TYPICAL RTZ SYSTEM SEQUENCE OF OPERATION

The following is a brief generic explanation of RTZ system operation and control. It is not meant to reflect any particular manufacturer's methods, nor is it meant to encompass all facets of RTZ system control. What it does provide is a "nutshell" description of typical system operation, from the rooftop unit to the zone and bypass dampers. Take it for what its worth, and feel free to paraphrase it to fit your next rooftop unit zoning application.

Overall System Control

Rooftop unit mode (heating, cooling, or ventilation) is determined by the needs of the individual zones. For instance, if more zones are in need of cooling rather than heating, then the rooftop unit will operate in a cooling mode. Conversely, if more zones are in need of heating rather than cooling, then the rooftop unit will operate in a heating mode.

If there is no heating or cooling demand by any of the zones, then rooftop unit heating and cooling are disengaged, and the supply fan continues to run (ventilation mode).

Individual Zone Control

Once the mode of operation is established, then the zone dampers for those zones that are in need of the current mode (cooling or heating) will "modulate," or periodically adjust position, in order to maintain the desired zone temperature setpoints.

The zone dampers for those zones that aren't in need of the current mode will drive fully closed or to their minimum positions.

As setpoints are reached and consequently zone dampers close off, the system bypass damper will modulate open in order to maintain airflow through the rooftop unit, and also to prevent excessive pressure buildup in the ductwork.

STAND-ALONE ZONE DAMPERS

What happens if we take an ordinary RTZ system zone damper and install it in a single zone system, without a system controller, or anything else required of an RTZ system for that matter? By way of example, let's consider a single zone constant volume rooftop unit, serving an open office space, with the controlling thermostat located somewhere in the general area served. For installation and design reasons, an additional "zone" is served by this ductwork distribution system, in the form of a lunchroom or break room. The room is served by the rooftop unit, by way of a "stand-alone" zone damper. We call this zone a "subzone"; it has its own zone temperature controller, and is therefore at least partially qualified as a zone, according to our definition of zone in section 2. Yet this subzone has no say-so in what the heating and air conditioning equipment serving it should be doing. It simply can react to what mode the unit is in, and position itself appropriately.

Like a zone damper that is part of a full-blown RTZ system, this damper will modulate when the air provided to it by the rooftop unit is of benefit to its zone, and will close down to its minimum position when the air is of no benefit to the zone. At lunch time, for example, the break room may be filled, with people using the microwave to heat things up, coffee brewing, and heat being generated. As that being the case, the room would begin to warm up. This subzone is basically at the mercy of the rooftop unit, more specifically, at the mercy of the unit's controlling thermostat, which is located out in the open office area, on a column near the secretary's cubicle. At lunch time, with the area substantially vacated and at setpoint, the thermostat will likely have the rooftop unit in a ventilation mode. The break room, needing cooling, is outta luck. The only way this room will get any cooling is if someone went out to the rooftop unit's thermostat and turned it down a degree or two. Without that being done, the zone damper will sit at minimum position, allowing a minimal amount of ventilation air into the space.

You may ask yourself what good that damper does. Well, the way to look at this is to consider what would happen if it wasn't there. With no damper, the room would get its full scheduled cfm always, whether it was hot, cold, or otherwise, whether it would benefit from it or not. With the damper, no substantial improvement is made upon things when the rooftop unit is in a mode that is beneficial to the zone; the damper simply modulates open to allow air into the space. The benefit gained from having the damper is when the rooftop unit is in a mode opposite to what the room needs it to be in. At least in this scenario, the damper can close down to its minimum position, to allow only a minimal amount of "opposite mode air" into the room.

When a manufactured RTZ zone damper is applied in this manner, as a stand-alone entity, how does it know what mode the rooftop unit is in? There is no system controller to broadcast this information. How can it alternately get this info? Easy. Just install a sensor on the inlet side of the damper, and wire it to the controller. Most manufacturers provide an input on their board for this very purpose. The sensor continuously monitors the temperature of the air being delivered to it from the rooftop unit, and establishes the required mode of operation of the damper. Now, in order for the sensor to consistently determine the rooftop unit mode of operation, the air passing over the sensor must be continuous. So in addition to the requirement that the zone damper have a minimum position for ventilation purposes, this minimum position also serves to ensure that the inlet temperature sensor always has airflow across it, and can always properly determine the operating mode of the rooftop unit.

The past few paragraphs dealt with the "store-bought" version of a stand-alone zone damper. The fabricated option exists here as well, perhaps at a cheaper overall cost. It is feasible to purchase or fabricate a zone damper, and equip it with an actuator and enough "smarts" to perform subzoning as described above, at the same time eliminating some of the costs associated with buying a full function RTZ zone damper. Zone type proportional space temperature controllers exist, and can very easily be employed to modulate a zone damper actuator. The zone damper controller goes away, for the controlling function resides in the zone temperature controller. So the zone controller is wired directly to the damper actuator, with no "middleman." An inlet temperature sensor or controller, of course, must still exist, and be wired into the zone controller/damper actuator circuit.

As with anything, stand-alone zone dampers have their potential for misapplication, and their use could lead to disappointment in terms of what is expected from them. Nevertheless, when applied properly and not oversold, the stand-alone zone damper, and its application of subzoning, is an attractive and inexpensive method of getting "a little bit more" out of a single zone constant volume heating and air conditioning unit.

Chapter 13

VAV and Fan Powered Boxes

The basic VAV box is not much different than a zone damper. A typical "cooling-only" VAV box consists of a damper, an actuator, a controller, and not much else. Construction of a VAV box is more robust than that of a zone damper. As you will remember from our discussions on rooftop units and air handlers serving VAV systems, the air handler primarily operates to maintain 55-degree air at 1.5" W.C. A VAV box must be able to withstand pressures of this magnitude and greater, and still retain its functionality. A VAV box will normally have a round connection on the inlet (medium pressure) side and a rectangular connection on its outlet (low pressure) side. Other notable differences between VAV boxes and zone dampers will be discussed herein.

This chapter deals with VAV systems, how they operate as a whole, and more specifically, how the "terminal units" operate and what they need in terms of controls requirements. Before we dive into the sections on the various types of VAV and fan powered boxes, we first need to lay down some groundwork on the fundamentals. To that end, we list out the possible discrete components that can make up a VAV (or fan powered) box:

- Primary Air Damper
- Electric Reheat Coil
- Hot Water Reheat Coil
- Series Fan
- Parallel Fan

The **primary air damper** is functionally equivalent to the zone damper we introduced in the last chapter. This damper accepts air from the medium pressure loop (air handler side), and delivers it, in varying amounts, to the supply diffusers of the zone served.

VAV systems are primarily cooling systems. However, VAV boxes can be equipped with reheat coils, especially if the required VAV box minimums are substantial. If the temperature of the space served by a VAV box drops below setpoint, the heating coil will "reheat" the air being delivered into the space by the VAV box, in an effort to achieve and maintain setpoint. **Electric reheat coils** utilize electric resistance heating elements to heat the air passing through the coil. The electric reheat coil can typically have up to three stages of control. **Hot water reheat coils** are equipped with either a two-position or a proportional control valve.

Fan powered boxes incorporate the use of a fan as part of their operation. The fan operates to pull air from the plenum space. What is "plenum space?" Simply put, it's the area above the ceiling grid. In a typical VAV system, there is no return air duct, *per se*. Return air grilles are mounted in the ceiling grid, yet are not connected to anything. Air returning from the spaces passes through the return air grilles, into the plenum space, making its way back to the air handler's return air opening, which is open to the plenum. This type of return system is referred to as "plenum return." The air in the plenum is typically a bit warmer than room temperature, perhaps picking up some heat from the lights that lay in the ceiling grid. A fan in a fan powered box will pull air from the plenum as part of its temperature control process. How the fan is configured, with respect to the primary air damper, is how the fan powered box is described. A **series fan** is "in series" with the primary damper, and a **parallel fan** is… anyone? Okay, you get the point. These concepts will be more thoroughly explained in the upcoming sections on fan powered boxes.

Like zone dampers, VAV and fan powered boxes will have an onboard controller. It can be an "application specific" digital controller, as would be the case in a full-blown networked DDC system. However, an alternative is to have the VAV box equipped with an "analog electronic" controller. By looking at the two different types of controllers and comparing them, they don't appear to be that different from each other. The digital and analog controllers are both circuit boards, with a bunch of electronic components soldered onto them. The difference, of course, is that the digital controller has a chip, a microprocessor, that can convert and process analog information. The digital type controller is what's used with zone dampers that are part of an RTU zoning system. It is through the brain of the controller that many of the operating characteristics of the zone damper are defined and configured. A controller of this type has the ability to be networked, though can also operate "standalone," as long as it's properly configured.

If a VAV system is not going to be fully networked, there may not be much reason to go with digital controllers on the VAV and fan powered boxes. They are typically more costly than the analog controllers, and their brains must be "accessed," either via a laptop computer or via a "service tool" (portable human interface device), in order to properly configure them for the given application. Analog controllers are cheaper, mainly because they don't have the "horsepower" that their counterparts have, in terms of capability and flexibility. Yet they pack enough punch to effectively and economically control a VAV or fan powered box. Instead of setting operating parameters via access to an on-board microprocessor, these parameters are set through the controller's discrete on-board electronic components. For instance, setting the "personality" of the controller to match the type of terminal unit may be done with jumper wires, dip switches, or a combination of both. Setting minimum and maximum flow rates will likely be via on-board potentiometers, or something similar. The initial setup of an analog controller is comparatively simple, for the technician and air balancer do not need a requisite knowledge of DDC, nor will they need any special interface devices or computers. Typically all that is needed (in addition to a flow hood) is a screwdriver, and perhaps a voltmeter.

Each and every VAV (and fan powered) box in a system will have a zone sensor wired to it. A count of the zone sensors in a building served by a VAV system would thus provide you with the number of zones served by the system, and also with the number of terminal units, generally speaking. Sometimes a zone is created that is so large, it cannot be handled by a single VAV box. The designing engineer will show two boxes serving the same general area, with a single zone sensor location common to both boxes. In reality, the zone may end up having two sensors, mounted side-by-side, in this same location. This is controller dependent. Whereas the higher end VAV box controllers will allow for a single zone sensor to be tied to more than one VAV box, others may make no provision for this, and without getting overly complicated, will require you to install and wire a sensor for every box.

Please read on, as we take an in-depth look at the different types of VAV and fan powered boxes that are in popular use in this day and age. At the end of this chapter, the reader should have a well-rounded knowledge of these types of equipment, their operating principles, and their installation requirements. In addition, the reader shall be educated in the practical sense, thus gaining valuable tips and tidbits that allow for a more comprehensive overview of what's done and what should be done "in the real world."

Types of VAV and Fan Powered Boxes

Four types of terminal units fill the vast majority of VAV system applications. We've used that term (terminal unit) a number of times now, without formally defining its meaning. The term "terminal unit" is just another way of referring to VAV and/or fan powered boxes. The term is especially helpful when discussing a system consisting of both VAV and fan powered boxes. Anyway, the four most common types of terminal units are listed below:

- Cooling-only VAV Box
- VAV Box with Reheat
- Series Fan Powered Box with Heat
- Parallel Fan Powered Box with Heat

A **cooling-only VAV box** is just that. A primary air damper/actuator and a controller, and not much else (Figure 13-1). No heat, no fan. The primary damper modulates to allow varying amounts of cool air into the zone served. As the zone temperature drops down to setpoint, as determined by the zone sensor, the VAV box modulates toward its minimum position.

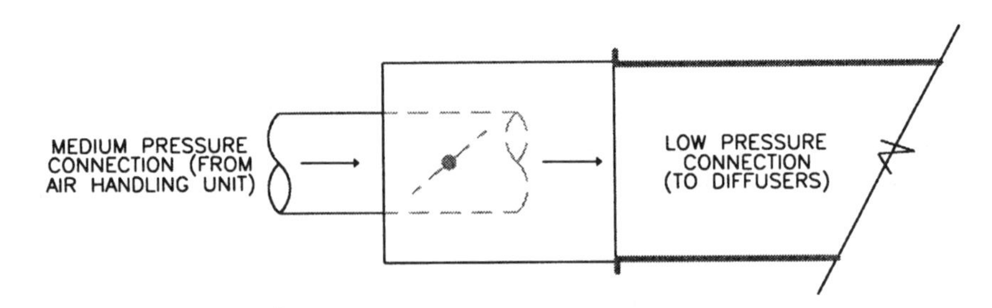

Figure 13-1. Cooling-only VAV box (controller and actuator not shown).

A **VAV box with reheat** has heating capabilities. The reheat coil is downstream of the primary air damper, so the air from the main loop passes through the damper, through the reheat coil, and then on to the zone. Operation of the primary damper is similar to that of a cooling-only VAV box. As the zone temperature falls, the damper closes toward its minimum position. If the zone temperature drops far enough below setpoint, the primary damper will typically open to a "heating" position, which is greater than the cooling minimum, and reheat will be engaged.

A **series fan powered box with heat** (Figure 13-2a) is comprised of a primary air damper, a heating coil, and a fan. As the name suggests, the fan is in series with the primary air damper, and also circulates air from the plenum. The fan runs continuously during occupied modes. The plenum air and the main "system air" passing through the primary air damper are "blended," passed through the fan, passed through the heating coil, and delivered to the zone. With a series fan powered box, the air delivered from the box is of "constant volume." As the zone temperature drops toward setpoint and the primary air damper closes down, the fan simply pulls more air from the plenum. Also with a series box, the air delivered from the box is of "variable temperature"; as the primary air damper closes down, the mixture of system air and plenum air rises in temperature. Operation of the heating coil in this type of box is similar to a VAV box with no fan; when the zone temperature drops far enough below setpoint, heating is engaged.

A **parallel fan powered box with heat** (Figure 13-2b) is made up of the same components as a series box, except that the fan is in parallel with the primary air damper. Configured as such, the fan does not circulate air from the primary air damper, and only serves to pull air from the plenum. Fan operation is intermittent. The fan is considered the first stage of heat. Upon a call for heat, the primary air damper has already assumed its minimum position, and the fan starts and draws air from the plenum. This air is blended with the minimal system air, passed through the heating coil, and delivered to the zone. With a parallel fan powered box, the air delivered from the box is of "variable volume." As the zone temperature drops toward setpoint and the primary air damper closes down (with the fan off), the quantity of air delivered into the zone decreases. When the fan starts, the air quantity delivered "jumps" to a higher value. Of course the air delivered into the zone before fan operation, as the primary air damper closes down, is of "constant temperature." In essence, the operation of a parallel box is identical to that of a

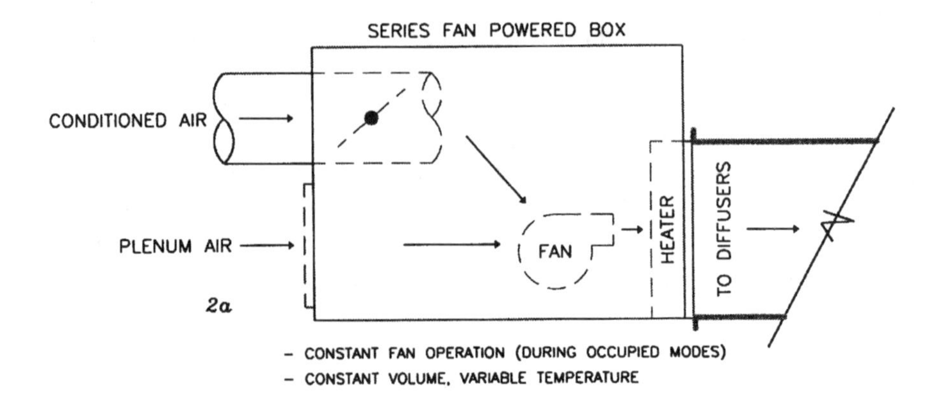

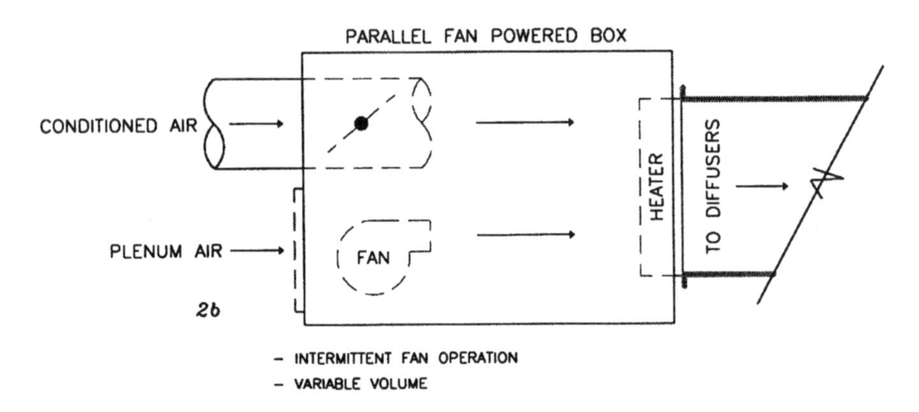

Figure 13-2a & 13-2b. Series and parallel fan powered boxes with heat (controllers and actuators not shown).

cooling-only VAV box, at least up to the call for heating. Heating coil operation again is similar to that of the other types of boxes we've talked about. When the zone temperature drops far enough below setpoint, the fan starts. If the zone temperature continues to decrease, then heating is engaged.

What are the functional advantages and disadvantages associated with the various types of VAV and fan powered boxes? In a given application, what drives the decision to go with one type of terminal unit as opposed to another? These are questions that are a bit beyond the intended scope of this writing. We can at least lay down a few tips. First, in a system having VAV boxes and no fan powered boxes, you may find the VAV boxes serving the perimeter zones equipped with reheat coils, because of the greater heat losses at the perimeters. Or you may find electric or hot water baseboard heaters zoned and interlocked with the operation of cooling-only VAV boxes on the perimeter. In a system having VAV and fan powered boxes, you will normally find the VAV boxes serving interior spaces, and the fan powered boxes serving the perimeter. This again has to do with the greater heat losses at the perimeters, and might have a little to do with the fact that fan powered boxes have the capability to operate and provide heat during unoccupied modes, when the main air handler is shut down.

When comparing series and parallel fan powered boxes, we notice that the series box fan runs continuously (during occupied modes), and the parallel box fan runs intermittently. While this translates to more energy usage at the box for the series box, first cost and energy usage associated with the main air handler is comparatively less than with a system of parallel boxes. The reason is that the main air handler's supply fan need not be sized to overcome as much "static loss," since the series box fans are helping the cause. So the supply fan is smaller in physical size, and consumes less energy. Another benefit offered with series boxes is that the air delivered from the box is of constant volume. This may be a requirement in a given application, to have a constant level of ventilation of the spaces served, regardless of what the temperature needs of the spaces are.

The following sections will take a look at VAV and fan powered boxes, from a mechanical/controls contractor's viewpoint, and establish basic installation and operational requirements for each type of terminal unit. Each section will build upon its preceding section, by introducing new concepts and principles that will help to build a well-developed understanding of all types of terminal units, their installation requirements, and their operating modes.

Cooling-only VAV boxes

We begin this section by elaborating on what has already been said about the construction of a VAV box. The very first line of this chapter stated that VAV boxes are not all that different than zone dampers (see Figure 13-3 for a typical application). Each has a damper, an actuator, and a controller. We did note that VAV box construction is in fact more

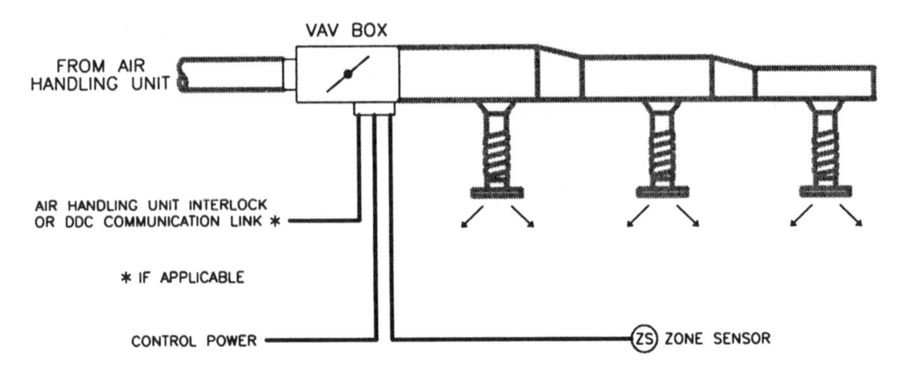

Figure 13-3. Cooling-only VAV box serving a single zone.

robust than zone damper construction. Aside from that, are there any other differences between a basic, cooling-only VAV box and a zone damper? The answer is "yes," if the box is "pressure independent."

Pressure Independence

To understand what is meant by that term, let us first define the term "pressure dependent." A zone damper operating as part of a rooftop unit zoning system is an example of a pressure dependent device. The position of the damper is strictly a function of the temperature sensed by the zone sensor. In other words, the further above zone setpoint, the more the damper is open. The amount of air delivered into the space, at any given damper position, is "dependent" upon the system pressure at that moment in time. Fluctuations in pressure affect the cfm quantity delivered.

With a VAV system, we have stated that the main air handler operates to maintain a constant supply air static pressure of around 1.5" W.C. As VAV box dampers open and close, this pressure is subject to fluctuation. A pressure dependent VAV box is at the mercy of these fluctuations. If a particular zone needs cooling, and it positions its VAV box primary air damper as a function of the deviation in temperature from zone setpoint, then the volume of cool air delivered into the space is primarily a function of the pressure in the system. If the pressure drops for a short period of time, due to whatever circumstances, then the volume of air delivered to the space drops off as well.

A pressure independent VAV box is equipped with a "velocity sensor," which can (indirectly) measure in feet per minute (FPM), the "speed" of the air passing through the primary air damper. With the square foot area of the damper known, the quantity of air in cfm passing through the damper can be calculated, by multiplying the two values. The VAV box controller itself performs this calculation, so the controller "knows" the amount of air passing through it at any point in time. This in itself is a very valuable piece of information.

Since the controller actually knows how much air is passing through it, it can feasibly control this value. In fact, that's exactly what it does. Instead of damper position being the controlled variable, it is the air quantity that is the controlled variable. In other words, as the zone temperature rises above setpoint, the "cfm setpoint" of air passed through the primary air damper is "reset" upward. The result is that fluctuations in system static pressure have little effect on the performance of the terminal unit. If a particular zone is calling for cooling, the VAV box primary air damper will be positioned by the controller to maintain the appropriate cfm quantity. If system static pressure drops, the primary air damper will open more to maintain the required cfm quantity. Bottom line is, the cfm delivered from the VAV box to the zone is independent of system static pressure. Hence the term "pressure independent!"

The difference in cost between pressure dependent and pressure independent terminal units is minimal, at least taking into account the advantage that the pressure independent box has over the pressure dependent box. For the dollar per zone paid for a VAV system, the added cost for pressure independence is often a "non-factor." Most engineers will default spec pressure independence, so the whole topic of pressure dependency is a moot point anyway. Still, at least now you know!

Power Requirements

Let's talk about power requirements for a cooling-only VAV box. Let's see, we have a damper actuator, and we have a controller. Chances are good that both of these items will require low voltage power. In fact, virtually all VAV box actuators and controllers work off of 24 volts AC. There are a couple of different ways of handling the power requirements for a system of cooling-only VAV boxes.

The first is to have each and every terminal unit equipped with a 120-volt by 24-volt transformer. This can be furnished and mounted by the manufacturer, usually as an option. Each VAV box then would require 120 volts wired to it. In conduit.

The second method is to size a single transformer for all of the VAV boxes in the system. The term used in the sizing of transformers is volt-amps, or "VA." This term describes the available power that a transformer can safely provide. To understand the magnitude of this term, consider that a simple general purpose control transformer is usually 40 or 50 VA. Line-by-low voltage transformers can range from 40 VA, up to 2 or 3 kVa (thousand VA). To size a transformer for a system of VAV boxes, the VA consumption of each box needs to be known; this is the combined power requirement of the actuator and the controller, and is usually in the range of 10-20 VA (max). By multiplying this value by the number of VAV boxes on the system, you come up with the total VA requirement for the system. Multiply this value by a "safety factor" of 1.25, and you have your transformer size. If you envision any future expansion of the system, account for this now by sizing your transformer for future boxes.

Consider a system with nine cooling-only VAV boxes. To size a single transformer for the entire system, multiply 20 VA (per box) by 9, and multiply the result by 1.25. The required VA for this system is 225. You might not be able to find a transformer with this exact VA, so just go to the next largest size transformer.

On larger systems with dozens of VAV boxes, it might be more practical to provide not one, but several transformers, one for each "group" of VAV boxes. This is a favorable alternative for several reasons. First, it conserves on circuit and wire size. A transformer sized for all terminals on a system would typically wire out to each box in a "daisy chain" fashion. In other words, from box to box. The circuit breaker, and the initial wire size, would have to be large enough to handle the current draw of every box on the system. By having groups of VAV boxes each served by a separate transformer, circuits remain in the practical range, and wire sizes generally stay small. While this may not necessarily translate to savings in labor and material costs, it does allow for a more "conventional" installation.

Another advantage to having several transformers serving groups of VAV boxes is that the whole system is not dependent upon a single transformer. On a single transformer system, if the transformer blows, the whole system is down, until that transformer gets replaced. On a multiple transformer system, if a transformer fails, then only the VAV boxes served by that particular transformer are affected, while the rest of the VAV boxes continue to go about their business.

A great advantage to running 24 volts to all the VAV boxes in a system, in lieu of 120 volts, is that, depending upon the municipality (among other factors), you could get away with running the 24-volt power loop(s) in "free air." This means that instead of having to install conduit and pull wire through the conduit to each box, the wire can be "fancy free" up there in the plenum space. The time and material savings could be substantial, especially with larger systems. Care must be exercised, however. The wire used to run the power loops must be plenum rated, meaning that it is manufactured specifically for use in a plenum space. Certain guidelines should govern the installation of the wire. It should be run neatly from box to box, and should be tied or clipped down and supported from structure, instead of just being thrown up there to ultimately end up lying on the top surface of the ceiling panels. Also, check your local authority and the codes that they follow, for regulations governing the installation of "conduit-free" low voltage wiring. There is a definite limit to how many boxes you can power with a single transformer, and still "stay legal." This limit is based on the fact that even low voltage wiring can carry some heavy current. Code may dictate that any 24-volt loop be "power limited," and thus would specify a maximum VA rating for the power source (transformer).

Operational Considerations

Switching gears from installation practices to operational concerns, let's talk about what the cooling-only VAV box does, or should do, during unoccupied modes. An unoccupied mode is defined as the main air handler being shut down by the system's time-of-day scheduling function. When the unit shuts down for the night, it remains off until the scheduling function transitions the unit into its occupied mode. If the unit is equipped with heat, then it will come on to provide heat during the unoccupied mode if and when the general space temperature drops below the "night setback" setpoint. A unit equipped with heat can also provide a "morning warm-up" cycle upon transition to the occupied mode, if required.

How does the VAV box operate in these different modes of air handler operation? The first thought is, if the air handler is shut down, who cares what the box does! But if the air handler is equipped with heat, and that heat is used for night setback and morning warm-up cycles, then we would want the primary air dampers in these cooling-only VAV boxes to go fully open upon entering either one of these modes. There are a couple of ways to accomplish this. One method is to "communicate" this information from the air handler. In other words, have the air handler inform all of the VAV boxes that it is in a heating mode, so that the boxes go fully open. For the networked system utilizing digital controllers, this information can be communicated over the communication link connecting all of the controllers. For the non-networked system, this consists of running a pair of wires from the air handler to each and every VAV box in the system, and installing a relay at each box. The task here is to have the local relays energize when the air handler enters a heating mode, and have the contacts of the relays disrupt the VAV box control circuit in such a way so as to drive the primary air damper fully open. This is in essence a 24-volt control loop, a "hardwire interlock," sourced from the air handler, driving 24-volt relays, local to each box. Sounds like a lot of work, and it is, when you consider the installation time and material costs associated with the relays. Running the control loop itself isn't that big of a deal, if you are already required to run a 24-volt power loop. Hey, what's an extra pair of wires?

There are other methods of accomplishing this task. Some VAV box controllers have a point to terminate a temperature sensor. The sensor gets installed at the inlet of the VAV box, and gets wired to the controller. Operation is such that if the sensor senses inlet air temperatures of the "heating variety," the primary air damper drives fully open. Call it automatic changeover. Call it VAV box heating mode operation. Whatever you want to call it, it's a pretty clean way to accomplish the task of getting the box fully open for air handler heating. There are concerns with this method, that the designer must be aware of. First, what is the temperature of the air delivered by the air handler in a heating mode? What is the temperature requirement at the terminal unit for it to go fully open? Is this an adjustable parameter, or is it fixed? If it is fixed, can the air handler provide heating air in excess of this value? What happens at the VAV boxes during the time when the air handler first turns on, until the time the air heats up to the changeover point of the VAV boxes?

All good questions. Okay, let's start with the first one. The temperature of the air delivered by the air handler in a night setback or morning warm-up mode is a function of the return air temperature and of the air handler's heating capacity. In a night setback cycle, for instance, the cycle is initiated when the general space temperature drops below the night setback setpoint, typically 60 degrees. So the unit starts up, with its outside air damper remaining shut, and proceeds to recirculate 60-degree air from the space, heating it up (in time) to a suitable heating temperature. If the available delta T of the heating section is known, then we can calculate the heating discharge air temperature. With gas or hot water heat, this will generally be pretty substantial, and the unit should have no problem discharging air at a temperature that is suitable for heating, say 80-100 degrees.

So what about the boxes? How hot does the air at its inlet have to be, in order for the primary air damper of the VAV box to drive fully open? Well, it has to be in excess of room temperature, you would think. How much in excess? Typically anywhere from 75 to 85 degrees is a good changeover point. Is this an adjustable parameter? Well that depends on the type of controller. If it is, great. If not, the fixed value of course must be less that what the air handler is capable of providing.

When the air handler enters a heating cycle, the supply fan turns on and air handler heating is engaged. The air circulated to the VAV boxes is not immediately hot. So what does a VAV box do until the air temperature at its inlet rises above its changeover setpoint? To answer this question, we have to back up a step and think about what the box is doing before the heating cycle is initiated. When a pressure independent VAV box is given a minimum cfm value other than zero, the box strives to maintain this value, regardless of what the main air handler is doing. During the unoccupied mode, there is no air blowing to and through the VAV boxes. The VAV box controller will do everything in its power to get its minimum cfm value. This translates to the controller driving the primary air damper fully open, of course without succeeding in achieving its task. So the primary air damper drives fully open upon loss of airflow, and stays there, until airflow is re-established.

Now as soon as the air handler initiates a heating cycle and airflow is established, the primary air damper in the VAV box will begin to close down. It will be likely, since the air handler is entering a heating cycle based upon space temperature conditions, that the VAV box will not be calling for any cooling. Therefore, the controller will try to position the primary air damper for the minimum cfm value. Now, once the air entering the VAV box has warmed up, the VAV box will again drive fully open.

To recap the above scenario: before the heating cycle is initiated, the VAV box primary air dampers are fully open. Once the cycle is initiated and airflow is established, the VAV boxes will begin to close down to their minimum positions, until the air entering the boxes is warm enough, at which point they all drive back open. So what's the big deal? Nothing, really. It's just interesting to note what happens in this scenario. This could be a concern if the air handler defaults to full fan capacity during unoccupied heating cycles. If the VAV boxes close down far enough before the air provided to them is above their drive open setpoint, enough pressure can be built up in the duct system to trip the high limit static pressure switch, thereby shutting down the air handler. With this method of VAV box changeover, it is important that air quantity control be imposed (upon the supply fan) in the unoccupied heating modes as well as the occupied cooling modes.

With networked DDC, all VAV boxes will be equipped with application specific digital controllers, and will be networked together. The main system controller will be able to communicate directly with all of the VAV boxes in the system, and command them to drive open at the same time the air handler is commanded to engage in a heating cycle. If a particular project is not specified as DDC, the control system designer may want to have the VAV boxes provided or fitted with DDC controllers, and have them networked back to a central controller, for the sake of communicating rooftop unit operating mode information to the VAV boxes. This may be a cost-effective alternative, especially on larger jobs. The controllers themselves may cost a few dollars more, but the results may be worth the added cost. Labor associated with running the network is negligible, if you are already running the power loop. And there are no relays or sensors to install. Of course, the designer would have to get approval for this. There may be a very good reason that the project isn't specified as DDC, one of which could be personal preference. So don't give DDC if DDC isn't specified (without approval), even if it's cost-effective and dramatically simplifies things.

To summarize this section, we can point out three major topics that this section dealt with. The first is the topic of pressure independence, what is meant by it, and what the VAV box needs, in terms of controls, in order to accomplish it. Second, we discussed power requirements for VAV boxes, and some different methods of providing it. Last, we talked in great detail about the importance of the VAV boxes having knowledge of what the main air handler is doing at all times, giving several methods of tackling this extremely critical requirement. We now move on to the next section, and build upon what we have learned here in this section. Hang on! With what you now know about cooling-only VAV boxes, the following section should be "a walk in the park!"

VAV Boxes with Electric Reheat

A VAV box with reheat has the ability to provide both heating and cooling. The normal method of operation is as follows: as the zone temperature served by the VAV box falls toward setpoint, the primary air damper closes toward its minimum position. The damper reaches minimum position at or around setpoint. If the temperature in the zone continues to drop, the primary air damper may or may not jump to an intermediate "heating position," and heating is engaged. The further the drop from setpoint, the more heat is produced. While this doesn't sound like a very energy efficient method of providing comfort control, it is one that is extremely precise, performance-wise. The precision comfort control gained with such a method is worth the sacrifice of having to reheat the cool air in order to provide warmth into the zone in need of it.

Power Requirements

With electric heating coils, the first thing we'll talk about is power requirements for the heater itself. Though the heater may be small enough to utilize single-phase power, most often the heater will be specified to be three-phase. And for good reason. Electric heat is very inefficient. With three-phase power being more efficient than single-phase power, it only makes sense to at least try to offset the inefficiency of the electric heater by operating it with three-phase power.

When a VAV box is equipped with electric heat, power for the heater must be provided to the box, in the form of single-phase or three-phase. Control power must also be provided to the box, for the damper actuator and the controller, in the form of 24 volts AC. With these types of terminal units, there are two options for providing the 24-volt control power. One is to run the 24-volt power loop(s) as described in the previous section. The other is to equip each VAV box with a control transformer. The power for the electric reheat coil must be run to each and every box. This power can be transformed down to 24 volts, at each box, with a factory furnished or field installed control transformer. In a VAV system consisting of nothing but VAV boxes with electric reheat coils, the 24-volt power loop can be virtually eliminated, by going this route. When offered as an option by the manufacturer, the transformer will be completely factory wired, thus eliminating control power labor altogether. Of course you're paying more for each box, but depending upon the difference, this added cost may be completely offset by the savings in labor cost, and will at best be the least problematic route to take.

When you have a system consisting of both cooling-only VAV boxes and VAV boxes with electric reheat coils, it may be simpler and more cost-effective to run the 24-volt power loop. The reasoning is that you need to provide the loop at least for the cooling-only VAV boxes. Why not just run it for all of the terminal units? In terms of consistency, this makes good sense. In terms of cost, it may or may not be cheaper. This needs to be evaluated on a job-to-job basis. The ratio of cooling-only boxes to boxes with reheat could have a lot to do with the decision here. If you only have a handful of terminal units with electric heat, and dozens of cooling-only units, it's an easy decision to go ahead and run the 24-volt loop out to all of the boxes. On the other hand, if the situation were reversed, it might make more sense to equip those reheat boxes with transformers, and run the power loop just out to those cooling-only boxes.

When a hardwire interlock or communication link needs to be run (for reasons discussed last section), then it is likely most economical to go ahead and run a 24-volt power loop, instead of equipping the boxes with transformers. The reasoning is that, since you're expending the labor to run a "daisy chain" anyway, how much more time and labor is involved in running an extra pair of wires? Not much. And the cost of the wire and the main transformer(s) will more than offset the cost of having each VAV box equipped with an individual transformer, in terms of savings.

Operational Considerations

Time to turn the corner here and talk about how the VAV box with electric reheat operates. We touched upon this in the first paragraph. Expanding on that, we can say that a VAV box with electric reheat will engage stages of heat as the zone temperature falls further and further below setpoint. As stated earlier, a VAV box with reheat may or may not have the operational characteristic of the primary air damper jumping up to a heating position, upon engagement of reheat. For a VAV box with electric reheat, the amount of air passing through the primary air damper during a heating cycle is of critical concern. The importance of this is twofold. First, the electric heater will be equipped with an air proving switch. The manufacturer will rate the heater to operate with a minimal amount of air passing through it, and require that the heater does not operate below this value. An air proving switch or pressure switch will be factory installed and set for this minimum flow rate. If the airflow through the heater is not above this value, the air proving switch will not allow operation of the heater. Eliminating the air proving switch voids the warranty and jeopardizes the safe operation of the heater.

The second reason that it is important to have a proper amount of airflow through the heater really stems from the same reason that the air proving switch is there for: safe and proper heater operation. If the airflow through the active heater is too low, the delta T through the coil may be quite high. In other words, the slower we pass air through an active heating coil, the hotter it comes out. The temperature of the air leaving the coil may get hot enough to trip the heater's high limit thermal cutout switch. If this is a manual reset device, the heater will from that point on be disabled, until the device is reset.

To get around this, the VAV box controller may have the ability to open the primary air damper to an intermediate "heating" position, somewhat higher than the cooling minimum position. This ensures that the electric heater gets the required airflow through it, for safe and proper operation. If the controller does not have this capability, it may be required to set the cooling minimum high enough to allow for the proper operation of the heater, upon a call for heating.

VAV box electric reheat coils will generally be staged equipment. A typical VAV box controller can handle up to three stages of electric reheat. The output stages of the controller are wired directly to the electric reheat coil control circuit. Depending upon the size of the electric heater and the application, the heater will typically have at least two stages, if not three. Only the smallest of electric heaters should be equipped with only one stage, where, because of its physical and electrical size, it is not (economically) feasible to break the heater down to more than one stage. For comfort considerations, electric reheat coils should be multistage; the VAV box controller has the capability to control multiple stages, why not take advantage of that? The cost of adding a stage to an electric heater is likely minimal, considering the payoff in temperature control and customer satisfaction.

What does the VAV box with electric heat do in the unoccupied mode? Upon a loss of airflow in the system, the primary air damper behaves the same as if the box was a cooling-only box. Upon a loss of airflow, the electric heater's air proving switch breaks, and disables the heater. That part is easy. What happens in, say, a night setback cycle, when the air handler comes to life to provide heat for the general space served? The answer to that question is not so simple, and a few different scenarios must be considered. To that end, we contemplate the following combinations:

- No heat in the main air handling unit, no heat at the terminal units
- Heat in the main air handling unit, no heat at the terminal units
- No heat in the main air handling unit, heat at the terminal units
- Heat in the main air handling unit, heat at the terminal units

The first and second scenarios were pretty much covered in the last section. For the first scenario, if there is no heat in the main air handler and no heat at the terminal units, there are no night setback and morning warm-up modes. The unit stays off for the entire unoccupied mode, and proceeds directly into cooling mode upon transition from unoccupied mode to occupied mode. It probably should be noted here that with such a system, it is not necessary that the terminal units know what mode the main air handler is in, thus eliminating any need for an air handling unit—terminal unit interlock.

For the second scenario, we discussed the need for the primary air dampers to go fully open upon an unoccupied heating cycle. This is so that the air handler can do its thing and heat up the space in a quick manner.

When an air handler has no heat, a night setback cycle can still be implemented, if there are enough terminal units out there equipped with reheat. If the general space temperature drops below the night setback setpoint, the air handler supply fan turns on, and the air handler recirculates air from the spaces. Once airflow is established at the boxes, the electric reheat coils are enabled for operation. The zone temperatures will likely be substantially below zone temperature setpoints on a call for night setback operation, so the electric reheat coils stage up to their full heating capacities. When the general space served by the air handler comes up in temperature, the night setback cycle is terminated, and the reheat coils are once again disabled. For VAV boxes equipped with heaters, it is not necessarily required that the primary air damper go fully open, though it can be implemented (via a hardwire interlock). Upon entering a night setback cycle, VAV boxes with electric reheat will automatically go into a heating mode, with primary air dampers jumping to their heating positions. Even if the VAV box controllers don't have that capability, the primary air dampers will still open to their requisite minimum positions, which must be set high enough to allow the electric heaters to operate. By not forcing the VAV box primary air dampers to go fully open upon entering a night setback (or morning warm up) cycle, the air handling unit's supply fan will operate at somewhat less than full capacity, and it may take the space longer to come up to temperature, compared to if the dampers were forced open. Yet for the savings in first cost and for the sake of simplicity, this is not a bad way to go.

In case you didn't recognize it, the above paragraph described the third scenario. Now to tackle that fourth scenario! When a VAV system is made up of terminal units all equipped with reheat, or even a combination of cooling-only boxes and boxes with reheat, it is customary to lock out the operation of the reheat coils, if the main air handler has heat. The reason being that, if the air handler has heat, it is highly likely that it's there for unoccupied heating! And sized appropriately! There is no need to employ the heat at the terminal units, during a night setback cycle. Furthermore, if the electric reheat coils were allowed to operate in this scenario, their high limits might be reached, thus completely disabling their operation anyway, perhaps requiring a manual reset.

When you have a system made up completely of VAV boxes with reheat, you might ask why the air handler would even be equipped with heat. If night setback and morning warm-up cycles can be accomplished by using the heat of the terminal devices, why spend the extra money on the air handler by ordering it with heat? It's a good question. If your project has some cooling-only VAV boxes, it seems to be more desirable to perform night setback and morning warm-up with the air handler's heater. However, if your project consists solely of terminal units with reheat, you could feasibly eliminate the heat from the main air handler. Then again, that heater might be there for another reason. Perhaps the system is designed to bring in large amounts of outside air, or maintain a constant volume of outside air. Then heat in the air handling unit is mandatory. This being the case, it's good practice to lock out terminal unit electric reheat, if the heating section of the air handler is able to be employed for these unoccupied heating cycles. Or impose high limit control on the air handler's heater.

As an example of this scenario, picture an air handling unit heating coil having a temperature rise of 40 degrees. During a night setback or morning warm-up heating cycle, figure that the temperature of the air entering the heater is around 60 degrees. So the leaving air temperature could approach 100 degrees. This air is delivered to the terminal units, whose reheat coils have delta Ts of 20 degrees (on average). Therefore, the temperature of the air leaving some of these coils could be 120 degrees or greater! If the high limit thermal cutout of the heaters is set at 120 degrees, then you're going to run into problems. To avert these troubles, lock out the reheat at the terminal units, or impose a high limit discharge air temperature setpoint of no greater than 90 degrees, back at the air handler's heating section.

That about covers it for VAV boxes with electric reheat. Now that wasn't so bad now, was it? One more point worth mentioning here, before we jump to the next section, is regarding electric reheat with respect to zone dampers. In the chapter on RTU Zoning Systems, we didn't really explore the possibility of adding heaters to zone dampers. However, this can absolutely be implemented, and often is, especially with trouble zones or zones in constant need of heat regardless of what the other zones are in need of. When a zone damper is equipped with a reheat coil, its zone has the ability to satisfy its needs even if the rooftop unit is not in the zone's desired mode of operation. The zone will cast its vote for heat, and if the temperature in the zone continues to fall, its reheat coil will be staged as necessary.

VAV Boxes with Hot Water Reheat

Rather than spending a whole lot of time rehashing what we've already talked about last section, in terms of terminal unit reheat, we will spend at least the first part of this section in a comparative mode. In short, we will compare VAV boxes with hot water reheat to those with electric reheat, and discuss the differences, in both installation requirements and operational characteristics.

Power Requirements

When a VAV box is equipped with a hot water reheat coil, it must also be equipped with a control valve. Whether equipped with a two-position or a proportional control valve, the valve actuator will require control power. Since we already require 24-volt AC control power for the VAV box actuator and the controller, we make things easy and provide a control valve rated for 24-volt power.

Power requirements for VAV boxes with hot water reheat are the same as for cooling-only VAV boxes. The methods discussed in the section on cooling-only boxes, for providing control power to each and every box, are the same for VAV boxes with hot water reheat. The only difference is that, for each VAV box, you have the added power draw of the control valve, and you must account for that when sizing the 24-volt transformer(s). Typically, a control valve used in a VAV reheat application will be small in size, ranging from 1/2" to 1", and the VA consumption of the valve actuator will be 10 or less. If you want to be on the safe side, add 10 VA per terminal unit for transformer sizing.

Operational Considerations

For main air handler heating cycles, i.e., night setback and morning warm-up cycles, it is important that the primary air damper of the terminal unit go fully open, for all of the same reasons discussed in the previous sections. Since there are no thermal cutout devices to affect the operation of the reheat coil, it is not *typically* necessary to lock out heat at the terminal units when the main air handler is performing a heating cycle. Just go ahead and let the hot water valves do their thing! Hey, the hotter the air dumped into the zones during an unoccupied heating cycle, the faster the general space comes up in temperature, and hence the quicker the cycle, right? There may be a legitimate concern that the temperature of the "twice heated" air could be hot enough to cause problems in the individual zones. If such a concern exists, then go ahead and lock out terminal unit reheat during air handler heating cycles, just as you would with electric reheat. Or impose high limit control. Practically speaking, though, you really shouldn't have to do this.

The most important topic to discuss in this section is the topic of two-position versus proportional control for the hot water reheat coil. As we have continually stated throughout this book, "proportional control is the way to go!", as is the case for most VAV reheat applications as well. There may be an argument hidden somewhere in here though, favoring two-position control. To understand why that may be, we need to know a little bit about how proportional reheat is accomplished with a VAV box.

First, to proportionally control a hot water valve in a VAV reheat application, the VAV box controller must have the ability. This normally is not a concern; most controllers worth their weight will have the ability to modulate a control valve. However, this must be addressed, and at least verified. If the controller that you are using on your VAV boxes does not have the ability to operate a proportional or floating type control valve, you might be forced to go two-position, or pay a premium for the added option of proportional control.

Remember that as a VAV zone drops in temperature toward setpoint, the primary air damper in the VAV box closes toward its minimum position. With hot water reheat, the damper in the VAV box need not jump to an auxiliary heating position; there is no air proving switch, associated with the operation of the heating coil, to be worried about. Still, if the VAV box controller has the ability to do this, it may be implemented. The cfm value associated with the heating position of the primary air damper is usually scheduled on the mechanical drawings, as specified by the mechanical designer.

At any rate, whether the primary air damper jumps to a heating position or not, the air passing through the primary air damper will typically be substantially less than the maximum air quantity scheduled to be delivered into the zone at full cooling. For instance, a typical VAV zone may require a cooling minimum position of 20 percent of the maximum deliverable air, with an intermediate heating position of 50 percent.

What is the point of all of this, anyway? Well, in trying to make an argument for going with two-position control and saving a few bucks, it is important to understand that, upon a call for heating, the amount of air that the reheat coil handles is quite a bit less than the maximum deliverable cfm for the particular zone. Let's consider for a moment, a constant volume reheat system, in which an air handler delivers a constant amount of air to a bunch of hot water reheat coils, each of which in turn operates to provide individual zone comfort control, by reheating the air delivered into the zone. Considering proportional control at the reheat coils, as a zone temperature drops below setpoint, the hot water valve is modulated to reheat the air delivered into the zone. So when the zone temperature is above setpoint, the zone gets 55-degree air dumped into it, at a constant volume. As the zone temperature falls from setpoint, the temperature of the air dumped into the zone is proportionally reheated, the amount of which is a function of the deviation in zone temperature from setpoint. The amount of air delivered into the space, however, remains constant. Two-position control of the hot water valve doesn't have much of a chance here of providing acceptable temperature control. Were it be, as soon as the temperature drops below setpoint, the valve opens fully, and the air delivered into the space jumps from 55 to 80 or more. At a constant rate of delivery. An occupant stands a good chance of feeling the change from cooling to heating, and vice versa, as the zone temperature fluctuates below and above setpoint.

The above paragraph demonstrates why you would never want to go with two-position control in a constant volume reheat application. So why would you try to get away with it in a VAV reheat application? The argument for two-position control is that we're "dialing down" the amount of air delivered into the space as we approach setpoint from above. When we finally drop far enough below setpoint and call for heat, we may jump to an intermediate heating position, but that position is still only a fraction of the maximum deliverable amount of air. With two-position control, it's true that we may be raising the temperature of the air delivered into the space from 55 to 80 or so. But we are doing so at a much lower rate of delivery, at least compared to the maximum. If you're buying all of this, you might buy the notion that the occupant in the zone will have less of a chance of feeling a change from cooling to heating, as compared with the constant volume reheat example described above. Customer satisfaction is the goal here. How far you have to go to meet that goal, and how much the customer is willing to spend, is what we're talking about here. Of course, if the application is critical in the least bit, it's best to go the extra yard and provide proportional control.

One other argument in favor of two-position control has to do with the amount of water passing through a typical VAV box reheat coil at full flow. The smaller the gpm value, the smaller the required Cv value, and hence the smaller the valve. For VAV boxes with reheat that serve smaller zones, the required flow through the hot water coil may be on the order of only a couple gpm or less. For gpm values this small, the control valve body will be 1/2". Depending upon the style of valve body, and the manufacturer's offering of differing port sizes and Cv values in this size valve body, you might be hard pressed to find a control valve that will match your application. For instance, a 1/2" valve body with a Cv value of 2, handling a flow rate of 1 gpm, will have a pressure drop across it of only 1/4 psi when fully open. Remember the 3-5 psi rule on sizing control valves? This is out of that rule, by quite a bit. What we have here is an oversized valve. How it will control is unpredictable. In essence, control could actually approach two-position. If the control valve is oversized, then when the valve is only slightly open, the flow through that valve approaches the full gpm value. This means that the effective operating range of the control valve is only a small percentage of its full stroke. In this case, the benefits offered by proportional control may be lost, yet the price tag for it remains.

Despite all that has been said over the past few paragraphs, most (if not all) engineers will specify proportional control for VAV hot water reheat applications, whether the system will benefit from it or not. It's good practice. Still, the price for modulation as opposed to two-position control is something to consider. If there were no price difference, then there would be no argument; go with proportional control always! Since a price difference does exist, the above material was presented.

Fan Powered Boxes with Electric or Hot Water Heat

What is a fan powered VAV box and why is it used (Figure 13-4)? We've touched upon the definition and description of a fan powered box in the beginning of this chapter. We didn't really discuss why such a terminal unit is utilized. Let's attempt to do that right now. As stated before, in a conventional VAV air handling system, there is no return air duct.

Let us begin our discussion on fan powered boxes by again describing the physical construction and operational attributes of series and parallel boxes. As we stated earlier, both configurations consist of the same basic components, but that's probably where the similarities end. Air returning from the spaces passes through ceiling grilles and into the plenum space above the ceiling. From there it makes its way back to the

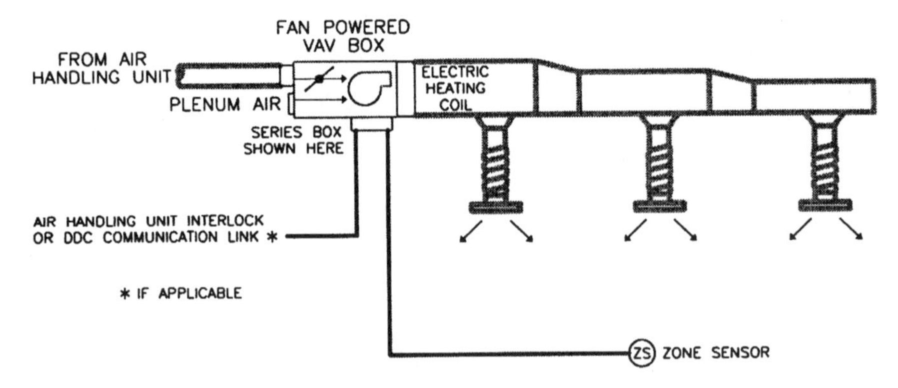

Figure 13-4. Fan powered box with electric heat, serving a single zone.

air handling unit. This air has picked up heat from the space, and on its way back to the air handler, picks up more heat from the lighting. Fan powered boxes make use of this "free heat," by recirculating plenum air back to the spaces, and heating it as well, if required. The inherent inefficiency of reheating cold air, as is the case with terminal units without fans, is minimized, or even virtually eliminated, with the use of fan powered boxes.

Series vs. Parallel Boxes

Let's take a quick look once again at the two types of fan powered boxes prevalent in today's VAV applications, and of the benefits each style has to offer. We'll start with the series box. Since the fan draws air from both the plenum and the primary air damper, the temperature of the air delivered into the zone from a series fan powered box is variant, and dependent upon the position of the primary air damper. Since the fan runs continuously during the occupied modes, the air delivered into the zone is of constant volume, dependent upon the operating point of the fan. So the two discerning characteristics of series fan powered boxes are that they operate as constant volume, variable temperature, terminal units. Applications requiring a constant volume of air at all instances are good candidates for series boxes. Spaces such as lobbies and atriums, hallways, bathrooms, and conference and meeting rooms, are well served by series type boxes. Series boxes are also chosen for their "sound qualities." Not that the sound they produce is entertaining or anything! But since the fans run continuously during occupied modes, the sound level in the zones served remains constant, which is often less annoying and thus more desirable than a fan that runs intermittently, turning on and off, on and off...

Now for the parallel box. Since the fan draws air only from the plenum, and only runs upon a call for heat, the air delivered into the zone from a parallel fan powered box is of constant temperature and variable volume, up to the "call for heating." Of course, when the fan turns on upon a call for heat by the zone, the air quantity delivered into the zone increases. During a heating cycle of a parallel fan powered box, the air delivered into the space is of constant volume, as dictated by fan speed. The temperature of the air delivered now becomes variable in nature, a function of the temperature of the plenum air, the temperature of the primary air, the mixture of each, and the amount of heat added. Parallel boxes are commonly applied to perimeter zones, where loads tend to fluctuate during occupied hours, more so than interior zones, which may be served by cooling-only VAV boxes. Where intermittent fan operation and variable airflow is acceptable, parallel boxes can provide a less expensive alternative to series boxes, in terms of terminal unit operating costs.

Power Requirements

Regarding power requirements for a fan powered box, it's simple. You need power for the fan. The manufacturer will provide a control transformer to take the voltage required for the fan down to 24 volts AC. This will virtually always be standard factory equipment, mounted and wired. So control power is taken care of for a fan powered box. All you need to worry about is power for the fan (and the heating coil, if electric).

Motors for fan powered boxes are typically small in electrical size, and are therefore normally supplied as single-phase. For fan powered boxes with hot water heat, the power for the boxes will be what is required for the fan, and should be specified to match what the facility has by way of single-phase power. For fan powered boxes with electric heat, the power for the boxes will most likely be driven by what is required for the heater. Electric heating coils for fan powered boxes will often be three-phase, this determination driven by efficiency issues, as discussed in the section on cooling-only boxes with electric reheat. The single-phase power required for the fan will be derived from the three-phase power required to serve the electric heater.

Operational Considerations

Operating characteristics of each type of fan powered box have been touched upon thus far, at least in terms of how they behave in the occupied modes. Reiterating what is meant by occupied mode, it is when the VAV air handler, via its scheduling function, operates to blow cold air. So what happens with fan powered boxes upon transition into the unoccupied mode, when primary air ceases to be delivered?

We know that the series box fan runs continuously during occupied modes, so the box must be required to know what mode of operation the main air handler is in, either via a hardwire interlock, or by some other means. The parallel box fan, however, runs intermittently, regardless of which mode the air handler is in. That being the case, is there a requirement for an interlock of some sort, between the main air handler and the parallel box? If so, what purpose would it serve?

You can argue that the series box requires knowledge of air handler mode, via an interlock or some other means, because the fan must operate continuously during occupied modes. With the same argument, you can say that the parallel box does not require knowledge of air handler mode, because fan operation of the parallel box is not dependent upon it. Actually, both types of fan powered boxes require knowledge of air handler mode, if night setback operation (or night shutdown) is desired. Take for instance, a parallel box having no knowledge of air handler mode. Upon transition of the main air handler to the unoccupied mode, the parallel box would continue to operate via its individual zone setpoint. For zone cooling, nothing would happen. The primary air damper would stroke fully open, but with no primary air behind it, there is no effect on the zone. There is nothing wrong with that; it's the same thing we would expect with a non-fan powered (pressure independent) box. The issue is with the flip side: heating. The parallel fan powered box will continue to operate to maintain setpoint, and that goes for heating as well. So when the zone temperature drops far enough below setpoint, the box's fan will start, and the heat will operate to maintain zone setpoint. Occupied zone setpoint. Yet this is the unoccupied mode that we're talking about.

What can be done about that? Well, the first thing is that we have to communicate air handler status to the fan powered box. We can do this in one of several ways. First is a hardwire interlock, whether it's a 24-volt control loop between the unit and all of the boxes, or, in the case of boxes with digital controllers, a communication link between the controllers and "the main brain" in charge of air handler and VAV box operation. Another method, one that may be more costly componentwise yet less labor intensive, is to monitor airflow to each fan powered box, with an airflow or pressure switch. The fan powered box manufacturer may even offer this as a factory option, pre-installed and pre-wired. If not, required labor would consist of installing the switch, wiring it into the control circuit of the fan powered box, and verifying its proper operation. This for each and every fan powered box in the system. Still, it may be less costly in the end, when you compare it with the labor involved in running a daisy chain.

It is important to note the distinction between the unoccupied mode of the entire system, and whether or not the main air handler is blowing air. During occupied modes, the air handler supply fan runs continuously, and the air handler operates to supply cold air. During unoccupied modes, the air handler supply fan is shut down, unless it enters an unoccupied heating cycle, a night setback cycle. For a fan powered box, it is more important to know of air handler operation, rather than simply knowing of which mode the entire system is in. If knowledge is given to the fan powered box controller, it should be knowledge of whether or not the air handler's supply fan is running, and not simply knowledge of the time-of-day mode of the system (occupied or unoccupied).

So now that the fan powered box knows whether or not the main air handling unit is blowing air, what does it do based upon that knowledge? The fan can simply be disabled. Let the primary air damper do its thing, with the fan completely prohibited from operating. What you essentially have then, when the air handler is shut down, is a VAV box with reheat! The primary air damper will go full open upon a loss of airflow to it, provided that a cooling minimum has been set. We already know that from our prior discussions on pressure independence and VAV box operation in unoccupied modes. We also know that, once airflow is re-established, the primary air damper will likely begin to close down. Suppose the air handling unit enters a night setback heating cycle. What do we want to happen? Well, the fan powered box will recognize the cycle, be it through a hardwire interlock, via digital communication, or by way of a local airflow switch. So the fan of the fan powered box will be re-enabled. What about that primary air damper, closing down as we speak? Do we want to force that bad boy fully open upon an air handler heating cycle? Good question! Hold on, let's try to find the answer!

So during the unoccupied mode, the fan powered box sits there, fan disabled, primary air damper fully open, waiting for something to happen. Then suddenly, the air handler enters a heating cycle. The fan is re-enabled, and depending upon the zone temperature (and the type of box), the fan may go ahead and turn on. Let's make the general assumption that if the air handler is entering a heating cycle, all VAV and fan powered zones served by the air handler are below their individual zone temperature setpoints (a good assumption). Therefore, the fans of all fan powered boxes will start, the heaters will become active, and likely will go to full capacity.

This begs the question: "If the fan powered boxes go to full heat upon a main air handler heating cycle, why do we have to force that primary air damper fully open?" Honestly, for a fan powered box, you don't. While it may be important for the primary air damper of a non-fan powered box to go fully open upon air handler heating, for reasons discussed in prior sections, it is not as important for this to take place with a fan powered box. Think about it. As long as that fan (in the fan powered box) is re-enabled, the fan powered box will act as a little self-contained heater, circulating air from the plenum, heating it up, and dumping it into the zone. Why bother with the primary air damper? For a system with a mixture of non-fan powered and fan powered boxes, you may want to force open the primary air dampers of the non-fan powered boxes, and just leave the dampers of the fan powered boxes alone. The only reason that you might want to mess with the primary air damper of a fan powered box, is if there is a morning warm-up cycle. Upon transition to the occupied mode, the air handler enters a heating mode. The purpose of this mode is to get the general space up to room temperature, as quickly as possible, before going into a normal cooling mode. If the primary air dampers of the fan powered boxes opened fully upon entering this cycle, in addition to the non-fan powered boxes in the system, the morning warm-up cycle can theoretically be shortened, for the air delivered into the fan powered zones would be warmer in temperature. That's "twice heating" the air though, once with the main air handler, and once with the fan powered box's heating coil. Depending upon the application, and the type of terminal unit heat, this can lead to problems. It is simpler to just leave the primary air damper of a fan powered box alone, so that it goes to its minimum position upon an air handler heating cycle.

Wouldn't it be nice if the fan powered box could "set back" to operate via a reduced zone temperature setpoint during air handling unit shutdown modes? Imagine it. The air handler shuts down for the night, only to come on, if at all, when the general space temperature drops below 60 degrees, as measured by some centrally located space sensor. In a system with fan powered boxes serving the perimeter zones, we can all but alleviate the air handler from coming on for a night setback heating cycle, in all but the most extremely cold outdoor air conditions. Consider that, upon shutdown of the air handler's supply fan, the zone setpoints of the perimeter fan powered boxes could be reset to a lower value, say 60 degrees. These terminal units in essence become unitary heaters, cycling to maintain their local unoccupied zone setpoints. The air handler's night setback temperature sensor, located somewhere in the interior, would never feel temperatures below 60, as long as the fan powered boxes have the capacity to maintain the zones at their individual night setback setpoints. This is good practice. To alleviate a big old air handler from coming on to perform night setback heating is a good thing to do, and can be accomplished as described, as long as the fan powered box had this capability. This is primarily a function of the box controller, and may or may not be able to be performed, depending upon the brand, style, etc. If the controller is capable, then by all means, do it!

What about the primary air damper in this scenario? Again, we know that the damper goes fully open upon loss of main air. If the fan of a fan powered box turns on for a "zone specific" night setback cycle, should the damper be forced shut? Better yet, why not just shut the darn damper for all times that the main air handler isn't running? Good idea, if your controller has the ability to do it. Otherwise, it may not be worth the hassle. To understand what happens in this scenario, we actually have to consider both types of fan powered boxes individually. For the series box, the fan is in series with the primary air damper, and if not forced closed, the fan can feasibly *pull* from both the primary air damper and from the plenum. However, the path of least resistance for the air pulled by the fan is the plenum opening of the box, by a long shot! It is very unlikely that the fan will be able to pull any air from the primary side. And even if it did, it would be inconsequential. For the parallel box, the fan is in parallel with the primary air damper. If the damper is not forced shut, the fan can feasibly *push* air through both the heating coil and through the primary air damper. Again, the path of least resistance will dictate that the majority of the air will be pushed through the heating coil, through the branch ductwork, to the diffusers, and into the space.

If there is a concern with any of this, then when the main air handler stops blowing air, why not just automatically close the primary air dampers of all fan powered boxes on the system? The integral nature of the pressure independent control loop of the damper will try to open the damper fully, upon loss of main air. This function would have to be overridden. When the fan powered box receives knowledge that the air handler has stopped blowing air, however that's done, the controller must be able to override the control loop of the damper, and force the damper shut. Again, this is dependent upon the controller, and its ability to do this. Also note that this would only have merit if the fan powered box itself has the capability to "set back." If the box simply shuts down during periods of main air handler shutdown, then there is no reason to force the damper shut in these periods.

As we are seeing here, much of what we can do and can't do with a fan powered box is dependent upon the controller. The more flexible the controller, the more can be done with it. Some controllers are limited in what they can provide. Others are quite flexible, this being especially true with digital type controllers. The purpose of this section is to explore many of the different possible scenarios, point out what's important, and what's maybe not so important, so that you can formulate your own opinions on how you feel a fan powered box should operate, in all scenarios. To that end, we can summarize what might be one individual's (the writer's?) "ideal" sequence of control for a fan powered box. So here goes...

During the occupied modes, the (series or parallel) fan powered box operates in the manner that is typical for the type of box. When the air handler shuts down for the night, the fan powered box goes into a "setback" mode. Its fan shuts down, and the primary air damper closes completely. If the temperature of the zone served by the fan powered box drops below the setback setpoint of 62 degrees, then the fan energizes and the heating coil is engaged, to pull air from the plenum, heat it up, and dump it into the space. Once the space temperature rises back above this setback setpoint, the fan and heat are disengaged. If the main air handler comes on for a general night setback cycle, the box will know of this, and simply think that the air handler has entered the occupied mode. The primary air damper will be enabled, and will likely go to minimum position. The fan and heat will operate via occupied setpoint, and likely go to full capacity. The air delivered into the zone during an air handler night setback cycle will be a mixture of minimal (warm) primary air and plenum air, heated by the heating coil. Once the main air handler's night setback cycle is satisfied, and the air handler shuts down, the box reverts back to its setback mode. Upon morning warm-up cycle, the box behaves in the same way as it does for a night setback cycle. Once morning warm-up is terminated, and the air handler enters its normal occupied mode of operation, the fan powered box returns to its normal mode of operation as well.

To wrap up this whole topic of fan powered box operation in unoccupied modes, it is interesting to note that the differences between the operation of series and parallel fan powered boxes are mainly found in occupied modes. These two types of boxes don't really behave any differently from each other during unoccupied modes, when the main air handler is shut down. Hey, take away that primary air, and these two types of boxes are virtually identical! Just fans and heaters!

Speaking of heaters, can we finally talk about them? Yes, and we'll keep it short! Okay, two types of heat available for a fan powered box, just as for a non-fan powered box: staged electric heat and hot water heat. With electric heat, depending upon the "point count" on the controller, the number of heating stages may be limited to only two. This because the fan normally takes up an output point on the controller. Remember that, for a parallel box, fan operation is actually the first stage of heat. At any rate, the controller, which is relatively small and limited in the number of inputs and outputs, may not be able to accommodate three stages of heat in a fan powered application, seeing that the fan has gobbled up one of the available outputs. No problem. Two stages of electric heat will typically work just fine in this application.

As far as hot water heat goes, we basically have the same arguments as we had before, regarding two-position versus proportional control. In a nutshell, the argument in favor of two-position control is even stronger with fan powered boxes than it is with non-fan powered boxes. If you think about what is happening with each type of fan powered box, as it approaches a call for heating, you can make the justification of using two-position control on the hot water valve.

First, look at the series box. The fan is running continuously, and the temperature in the zone served is falling toward setpoint. The primary air damper is closing off, and the air being delivered into the space is smoothly rising. At or around setpoint, the temperature of the air delivered into the space will be close to that of the plenum, say 75 degrees. The heating coil of a fan powered box will typically be sized at a delta T of only 10 or 15 degrees, if that, since it is not burdened with the task of reheating 55-degree air. So with a series box, if at this point the temperature falls below setpoint, the two-position hot water control valve opens, and the temperature of the air delivered into the space rises from 75 to 85 degrees or so. This difference in temperature delivered into the space may neither be noticeable to nor unpleasant for the occupants, and may therefore be quite acceptable.

Now for the parallel box. The fan is off, and the zone temperature is falling toward setpoint. The primary air damper is closing off, and the air being delivered into the space, at 55 degrees, is decreasing in volume. At or around setpoint, the fan turns on, as the first stage of heat, and the volume of air delivered into the zone jumps to a new level. The temperature of that air increases to around that of the plenum, again 75 degrees or thereabouts. Upon a further drop in zone temperature, the two-position hot water control valve opens, and the air delivered into the space rises to 85-90 degrees. If you can justify the difference in air volume and air temperature delivered to the zone resulting from a call for first stage heating (fan operation), then you might be able to justify the difference in air temperature resulting from a call for second stage heating (hot water valve opening).

Again, all of this is good "food for thought," but when you have an engineer specifying proportional control for everything you can imagine, it's a hard sell, to push two-position control (no slight against engineers, the author is one!). The argument will normally be that the controller equipped with the box will come with the capabilities already built into it, whether its specified or not. So why not use it? With that being the case, then what it really comes down to is the difference in price between two-position valve actuators and proportional valve actuators.

It's time to put an end to this section on fan powered boxes. Hopefully this has served as a primer on the topic. The subject matter in this section is complicated in parts. That's to be expected, for fan powered boxes are complicated devices, in terms of construction, operation, and control. There's a lot to know about these critters. A lot of different possible scenarios can surface, and it's important to at least know of these scenarios, how they affect the overall operation, and what can be done, or better yet, what should be done, to prevent unacceptable conditions resulting from such scenarios. The more that is known about the ins and outs of fan powered box operation and control, the better the chances of avoiding problems down the road. Don't let all of this scare you too much. Manufacturers of controllers have pretty much figured out how to control VAV and fan powered boxes a long time ago. Their product is time-tested, and their control techniques are widely accepted by the industry. Still, its nice to know what the controllers are actually doing, to better understand box operation as a whole, and to avoid doing something that perhaps shouldn't be done.

Constant Volume Terminal Units

Oftentimes design engineers will specify the use of a VAV box to serve an area, set up to deliver a constant amount of air, regardless of what the temperature needs of the zone are. This in essence is a VAV box set up so that its minimum and maximum cfm settings are equal. Most always, a box specified to be set up as such (if supplied with 55-degree air) will have a reheat coil.

Why would anyone want to take a VAV box and set it up to operate as a constant volume terminal unit? That there is a darn good question! To answer this, consider first that certain spaces require constant airflow. Spaces such as lobbies, hallways, and conference rooms, may need continuous, constant ventilation. We understand now that a series fan powered box can fill this need. We can also use a VAV box with reheat to do this, at a lower first cost, and without the noise that would be generated by the fan in a fan powered box. Of course operating costs would be greater; the cost of reheating air would eclipse the cost of operating a fan. Nevertheless, this is often enough specified to warrant coverage in this chapter. By now you're probably pretty worn out from all of the previous discussions on VAV and fan powered boxes, so we'll keep this section short. I promise!

A VAV box with reheat, set up for constant volume operation, is done so to provide a constant cfm rate into the zone served at all times. The zone sensor wired to the controller does not have any say-so in what the primary air damper does. This damper operates, in a pressure independent fashion, strictly to maintain a constant cfm value. The only reason the zone sensor is needed is for the operation of the reheat. If there were no reheat, there would be no need for a zone temperature sensor. Of course, why would you install a constant volume terminal unit without heat; wouldn't you continuously overcool the space, with a fixed amount of 55-degree air?

So we've concluded that the constant volume terminal unit typically requires some form of reheat, and a zone temperature sensor as well. When the temperature in the zone is above setpoint, 55-degree air, at a constant volume, is delivered into the space. When the zone temperature drops below setpoint, the reheat is engaged. Sounds kind of like a typical constant volume reheat application; why even have the VAV box to begin with? Why not just connect that reheat coil right to the main duct system?

Well, there are two reasons why you don't want to do that. The first is that the main duct system is a medium pressure system. Pressures in this duct are typically maintained at an inch and a half to two inches of static pressure. That's pretty substantial. The branch ductwork and diffusers downstream of a VAV box are not rated for these pressures, and the accompanying air velocities. The VAV box serves as a medium pressure to low pressure "regulator." If the VAV box weren't present, the air would be "Whistling Dixie" through the ductwork and diffusers!

The second reason is maybe less obvious, yet perhaps more fundamental. The purpose of a constant volume terminal unit is to provide a constant cfm rate of air into a space requiring it. With a VAV air handling system, the static pressures in the duct tend to fluctuate as things happen at the terminal units. These fluctuations are overcome by the pressure independent controls of such VAV boxes. Thus, a pressure independent VAV box utilized as a constant volume terminal unit will hold its cfm value, regardless of the fluctuations in static pressure at its inlet.

We now close the door on VAV and fan powered boxes, and move on to an in-depth look at heating coils, when they aren't an integral component of a terminal unit. Before we do, though, in closing let's lay down a few generic Sequences of Operation for all of the VAV and fan powered boxes we've talked about. The following section is not mandatory reading, does not lay down any new prerequisite material for the forthcoming chapters, and therefore can be happily skipped over by the reader (hooray!).

Typical Sequences of Operation

This section is presented to give the reader a brief, concise sequence of control for each of the types of terminal units that have been discussed in this chapter. The sequences are not meant to reflect any particular manufacturer's methods and modes of operation. Rather, they are meant to serve the purpose of providing a "generic" description of operation of the various types of terminal units. These descriptions can be "paraphrased" to match a particular brand of terminal unit, for use in a control systems designer's overall sequence of operation.

Cooling-only VAV Boxes

Occupied Mode

An individual zone sensor (one per zone/box) transmits zone temperature and zone setpoint information to the VAV box controller. The controller in turn modulates the primary air damper in order to deliver the appropriate amount of cool air from the medium pressure duct into the zone. As the zone temperature falls toward setpoint, the damper is modulated closed toward its minimum po-

sition. When the zone temperature reaches setpoint, the damper reaches its minimum position.

Unoccupied and Morning Warm-up Modes

The primary air damper is driven fully open during the unoccupied mode, to allow for main air handler night setback and morning warm-up heating cycles.

VAV Boxes with Electric Reheat

Occupied Mode

An individual zone sensor (one per zone/box) transmits zone temperature and zone setpoint information to the VAV box controller. The controller in turn modulates the primary air damper in order to deliver the appropriate amount of cool air from the medium pressure duct into the zone. As the zone temperature falls toward setpoint, the damper is modulated closed toward its minimum position. When the zone temperature reaches setpoint, the damper reaches its minimum position. If the temperature in the zone continues to fall, then the primary air damper opens to an intermediate "heating" position, and electric heat is staged. The further the temperature falls from zone setpoint, the more stages of heat are engaged.

Unoccupied and Morning Warm-up Modes

The primary air damper is driven fully open during the unoccupied mode, to allow for main air handler night setback and morning warm-up heating cycles. VAV box electric heat is disabled.

VAV Boxes with Hot Water Reheat

Occupied Mode

An individual zone sensor (one per zone/box) transmits zone temperature and zone setpoint information to the VAV box controller. The controller in turn modulates the primary air damper in order to deliver the appropriate amount of cool air from the medium pressure duct into the zone. As the zone temperature falls toward setpoint, the damper is modulated closed toward its minimum position. When the zone temperature reaches setpoint, the damper reaches its minimum position. If the temperature in the zone continues to fall, then the primary air damper opens to an intermediate "heating" position, hot water heat is engaged, and the hot water valve is modulated. The further the temperature falls from zone setpoint, the more hot water is allowed to flow through the coil.

Unoccupied and Morning Warm-up Modes

The primary air damper is driven fully open during the unoccupied mode, to allow for main air handler night setback and morning warm-up heating cycles. VAV box hot water heat is disabled.

Series Fan Powered Boxes with Heat

Occupied Mode

An individual zone sensor (one per zone/box) transmits zone temperature and zone setpoint information to the fan powered box controller. The controller in turn modulates the primary air damper in order to allow the appropriate amount of cool air from the medium pressure duct into the box. The air is then blended with warm air circulated from the plenum by the fan, which is in continuous operation during the occupied mode. The resulting air delivered into the zone is of constant volume (as dictated by fan speed) and variable temperature. As the zone temperature falls toward setpoint, the damper is modulated closed toward its minimum position, allowing less and less cool air to be blended with the plenum air. When the zone temperature reaches setpoint, the damper reaches its minimum position. If the temperature in the zone continues to fall, then terminal unit [electric/hot water] heat is engaged. The further the temperature falls from zone setpoint, the more heat is engaged.

Unoccupied and Morning Warm-up Modes

The primary air damper is driven [fully open/fully closed] during main air handling unit shutdown modes. The fan and heat cycle to maintain a reduced zone temperature setpoint. During air handler night setback and morning warm-up heating cycles, the primary air damper assumes its [fully open/minimum] position, the fan runs, and terminal unit heat is controlled to occupied mode zone setpoint.

Parallel Fan Powered Boxes with Heat

Occupied Mode

An individual zone sensor (one per zone/box) transmits zone temperature and zone setpoint information to the fan powered box controller. The controller in turn modulates the primary air damper in order to deliver the appropriate amount of cool air from the medium pressure duct into the zone. The air delivered into the zone is of variable volume and constant temperature. As the zone temperature falls toward setpoint, the damper is modulated closed toward its minimum position. When the zone temperature reaches setpoint, the damper reaches its minimum position. If the temperature in the zone falls below setpoint, the fan energizes to circulate warm air from the plenum. The plenum air is blended with the minimal cool air from the primary air damper, and delivered to the zone. If the temperature in the zone continues to fall, then terminal unit [electric/hot water] heat is engaged. The further the temperature falls from zone setpoint, the more heat is engaged.

Unoccupied and Morning Warm-up Modes

The primary air damper is driven [fully open/fully closed] during main air handling unit shutdown modes. The fan and heat cycle to maintain a reduced zone temperature setpoint. During air handler night setback and morning warm-up heating cycles, the primary air damper assumes its [fully open/minimum] position, and terminal unit fan and heat are controlled to occupied mode zone setpoint.

Constant Volume Terminal Units with Reheat

Occupied Mode

An individual zone sensor (one per zone/box) transmits zone temperature and zone setpoint information to the terminal unit controller. The controller modulates the primary air damper in order to deliver a constant amount of air from the medium pressure duct into the zone, independent of the temperature needs of the zone. If the temperature in the zone falls below setpoint, then terminal unit [electric/hot water] heat is engaged. The further the temperature falls from zone setpoint, the more heat is engaged.

Unoccupied and Morning Warm-up Modes

The primary air damper is driven fully open during the unoccupied mode, to allow for main air handler night setback and morning warm-up heating cycles. Terminal unit heat is disabled.

Chapter 14

Reheat Coils

The word "reheat" is standard terminology in HVAC, though is often overused and misapplied. In the pure sense of the word, a reheat coil is one whose job is to heat up air that has already been conditioned by some cooling process. We spoke about reheat systems, in which cool air from a central air handling unit is delivered to reheat coils serving individual zones, and zone controllers operate the reheats as required to maintain zone temperature setpoints. We also spoke about reheat coils in their role in dehumidification sequences, in which the air is cooled down in order to wring out moisture, only to be heated back up in order to maintain desired comfort level.

Finally, we spoke about reheat in VAV applications. You may have noticed that the author elected not to use the term reheat when referring to fan powered boxes. Technically, the air heated with a fan powered box's heating coil is primarily from the plenum space and not from the primary air system, so we veered from using the term reheat, and simply used the term "heat." Whether or not it is correct to use the term reheat in this type of scenario is really just a matter of opinion. As long as the concept at hand is understood, then there should be no confusion derived from the use, or misuse, of the term.

We've discussed heaters and heating coils, as applied as part of an air handler. In most cases, these coils will be referred to not as reheat coils, but simply heating coils, or perhaps even "preheat coils," depending upon the location of the heating coil within the air handler, and the function that it serves.

Whether used correctly or not, the terms heating coil and reheat coil refer to the same physical type of equipment. The terms simply relate to the application that they are serving in. We will title this chapter as it is, and while we will base most of our discussions on reheat applications, the subject matter is surely not restricted to this, in terms of application guidelines, rules of thumb, and operational considerations.

So for the sake of this chapter, we define the term "reheat coil" as a coil installed in a branch duct, applied as zoning equipment, served by some upstream air handling unit that has cooling capabilities. We will also restrict our discussions to constant volume applications, in which the air delivered to and through the reheat coil is of a constant value.

What's the "typical" entering air temperature of a heating coil? Well, that depends upon the application. If part of an air handling unit, the temperature of the air entering the coil can vary greatly, even down to 0 degrees or less, in the case of a make-up air unit! In a system consisting of nothing but reheat coils all downstream of the main air handler, the temperature of the air that the air handler is providing, and therefore the temperature of the air entering all of the reheat coils, is ordinarily around 55 degrees or so, at all (occupied) times. A reheat coil can be applied as a subzone, served by an upstream single zone constant volume air handler or rooftop unit. In such an application, the air handler operates to maintain some general space temperature as per an appropriately located thermostat or temperature controller. The subzone served by the reheat coil will have a thermostat or temperature controller as well, to operate the heater. At any point in time, the air handler may be in a heating mode, cooling mode, or strictly a ventilation mode. If the air handler is in a cooling or ventilation mode, the subzone has the ability to maintain comfort level, by heating the ventilation air, or reheating the cool air entering the coil.

In this type of application, how is the coil sized? Is it sized to raise the temperature of "cooling air" up to a suitable "heating" temperature? Or is it sized to simply temper the cooling air? For example, if a reheat coil is sized to simply temper the air entering it when the upstream air handler is in a cooling mode, then it may be sized for a delta T of 15-20 degrees. So that when the air handler is blowing cold air, the reheat coil can raise the temperature of the air up to room temperature. In this scenario, the heater is sized so that the space won't be subcooled during air handler cooling cycles. Also in this scenario, the reheat coil would be able to provide "heating air" into the space whenever the air handler is simply in a ventilation mode.

For reheat coils as part of a full blown reheat system, the Delta Ts will be more along the lines of 20-30 degrees or more. The coils in general must be sized to raise the entering air to a suitable heating temperature.

Who determines how a reheat coil is sized? The design engineer will most likely have run a load on the space served by the particular system. As such, he will have the task of determining reheat requirements, whether he is determining if a certain single zone system needs a subzone, or whether he is determining the heating requirements of each and every zone in a reheat system. In a typical reheat system, all zones will not be equal. Hence, the scheduled heaters will almost always differ in size, and also in achievable delta Ts for the zones they are serving. For instance, you might gather from the reheat coil equipment schedule that the delta Ts of some coils are only 20 degrees or so, and others are more like 30 degrees. The sizing of reheat capacities is a direct result of the load run by the mechanical designer. The difference in coil sizes and available delta Ts should not lead one to "second guess" the designer's selections. Yet knowing the delta Ts of the reheat zones is important information for the control systems designer, for reasons to be discussed further on. At the very least, these values can provide some insight into the system design; the controls designer can evaluate the systems, and make sure that the selections are "in line" with the application at hand.

Electric Reheat Coils

We've covered some of the principles of electric heating back in the chapters on make-up air units and built up air handling units. We'll reiterate for the sake of this section. Refer to Figure 14-1 for the typical application of the electric reheat coil.

Electric heating coils (duct heaters) are constructed of electric re-

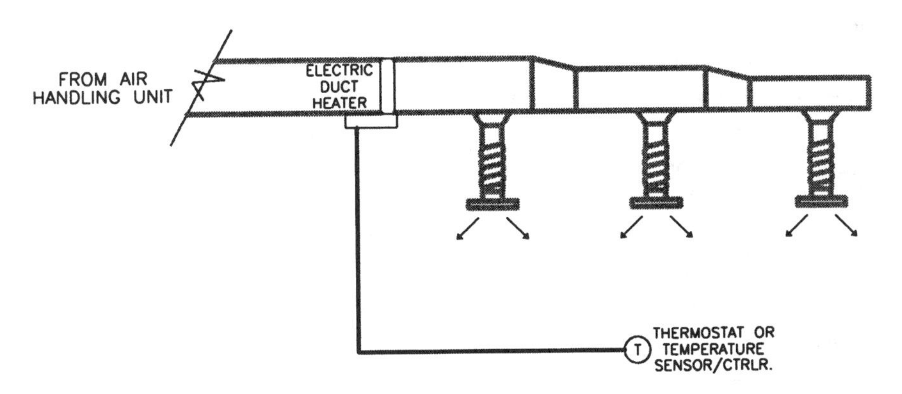

Figure 14-1. Electric reheat coil installed in a branch duct, serving a single zone.

sistance heating elements, which are placed in the air stream. When an element is given power, it heats up and glows. The heat from the element is transferred to the air passing over it. Electric heaters are typically staged equipment, with each stage consisting of one (or more) heating elements. A stage of heat is accomplished by giving power to the heating element(s) that make up the stage, by means of a contactor. When the contactor is energized, power is allowed to the element(s), and the air passing over is heated up accordingly. A simple two-stage electric duct heater is made up of two electric resistance heating elements, two contactors, and a control transformer and terminal strip. The terminal strip serves as the point of interface for external control of the heater. A simple two-stage heating thermostat can be wired to the terminal strip, to operate the heater. Other components found in a typical electric duct heater are a control circuit fuse, an air proving (pressure) switch, and a high limit thermal cutout switch. The air proving switch prevents heater operation unless there is sufficient airflow through the heating coil. The high limit cutout switch prevents the heater from getting dangerously hot.

Required Number of Stages

Determining the number of stages of electric heat requires looking at the total delta T of the particular heating coil, as well as considering the degree of control required. For precise space temperature control, the delta T per stage should be no more than 5 degrees. For mediocre control, the delta T per stage can be in the range of 6-12 degrees, and for coarse control, 13 degrees and up. The rule of thumb is, 10 degrees per stage, for mediocre control.

In sizing and selecting an electric reheat coil for an application, the mechanical designer will run a load on the zone, and come up with the Btuh required, as well as the cfm. By plugging these values into the "airside" equation for Btuh, which is ***Btuh* = 1.08 × *cfm* × *delta T***, the total required delta T can be found. Consider for example that a certain zone requires 30,000 Btuh at 1,000 cfm. The required delta T, using the above formula, is found to be around 27.8 degrees. For mediocre control, a three-stage heater should be selected, at the corresponding Btuh capacity, thus offering a delta T per stage of around 9 degrees.

Electric heater capacities are not normally categorized in units of Btuh. As pointed out in the chapter on built up air handlers, the term used to describe electric heater capacity is kilowatts (thousands of watts).

The formula for converting Btuh to kW is ***kW = Btuh/3410***. Considering the above example, in which a heater with a capacity of 30,000 Btuh is required for a specific zone, the kW of the required heater is 8.79. The mechanical designer now must make a decision, because electric heaters are offered in discreet sizes, and the designer will not be able to secure a heater with a capacity of exactly 8.79 kW. Depending upon the manufacturer he is specifying, the designer may be able to choose between an 8 and a 10 kW heater. Being conservative, he will likely select the 10 kW heater, and specify it to have three stages of control.

Since the designer selected a heater that does not precisely match his original requirement, he must "rework" the equations, to find his delta T per stage, using 10 kW. Solving for Btuh, he comes up with a heater capacity of 34,100. Solving for delta T, he comes up with 31.6. Following the "mediocre control" rule of thumb, his original decision on three stages of control still holds, for 31.6 divided by 3 is 10.5 degrees per stage.

It is a common misconception that the larger the electric duct heater, the more stages of control are needed. From what we have just gotten through discussing, we see that the number of stages needed is solely a function of the total delta T and the desired delta T per stage. A 20 kW duct heater handling 3,150 cfm of air yields the same delta T as a 3 kW heater handling 475 cfm of air: 20 degrees. Thus, given the same application, both of these heaters can be supplied with simple two-stage control.

The whole origin of the "mediocre control" rule of thumb stems from the notion that the larger the delta T per stage, the more noticeable the difference in supply air temperatures is when staging takes place. With a 10-degree delta T per stage, the occupant may be likely to feel the difference when a stage of control is added or dropped out. The larger the difference is, the more likely he/she is to feel the difference. This is normally an undesirable and often an unacceptable condition for HVAC comfort control. For more critical applications, the delta T per stage should be lessened, by increasing the number of stages.

The Case against Single-stage Reheat Control

You would never want to use single-stage control in a constant volume reheat coil application. Bold statement! What is the reasoning that backs it? Well, let's consider the application. We have a constant volume reheat air handling system, serving a bunch of zones with a bunch of electric reheat coils. Look at one particular zone, a row of offices, served by a single-stage duct heater with a total delta T of 25 degrees, and controlled by a single-stage thermostat located in one of the offices. On a particularly mild day, the thermostat calls for heat, and the heater is engaged, to raise 55-degree air up to 80 degrees. The zone comes up in temperature, and the heater turns off. The occupants will feel the difference in supply air temperatures upon the engagement and disengagement of the electric heat. That's a given. Whether or not it's acceptable to the occupants depends upon their individual sensitivities, comfort zones, and tolerances. On a mild day, the change in supply air temperature from 55 degrees to 80 degrees, and vice versa, may not be that unacceptable. Consider the same scenario on a chilly day. The occupants have entered the building, and are trying to shake off the chill from being outdoors. The thermostat is cycling the heater to maintain zone temperature setpoint. When the air delivered into the offices is 80 degrees, there are probably no gripes. However, when that zone temperature comes up, and the thermostat cycles off the heater, that air delivered into the offices drops down to 55 degrees rather quickly. Not only is the difference in supply air temperatures noticeable, it's downright offensive! Especially if you're an occupant and your desk is right under the diffuser in your office!

Consider again the above scenario, only now with two stages of control at the zone level. On the mild day, the heating load in the zone may be such that the two-stage thermostat is cycling on first stage. In other words, the thermostat calls for heat, first stage heating is engaged, and the supply air temperature jumps from 55 to 68 degrees or thereabouts. The zone comes up in temperature, and the thermostat disengages the first stage of heat. The zone temperature again drops, and the thermostat engages first stage heating. And so on. The change in supply air temperature, from 55 degrees to 68 degrees and vice versa, is likely not too noticeable, and definitely not very offensive. Now on that chilly day, the heating load in the zone may be such that the thermostat is cycling on second stage, with the first stage kind of "locked on" because of the load. In short, the zone temperature falls to the point in which the second stage kicks in, the zone temperature rises, the second stage cycles off, the zone temperature begins to fall again, the second stage turns on, etc. What the occupants are feeling is supply air temperatures varying between 68 and 80 degrees; when the second stage is off, the supply air temperature is around 68, and when the second stage kicks in, the supply air temperature increases to about 80 degrees. The difference in supply air temperatures when staging takes place is not all that bad, especially when you consider that when that second stage cycles off, the supply air temperature is pretty close to room temperature! The degree of comfort control is a little bit better than when you have only one stage, wouldn't you agree?

The above example falls upon the outer limits of the mediocre control rule. With a total delta T of 25 degrees, the delta T per stage, with two stages of control, is 12 or 13 degrees. With a total delta T any higher than 25 degrees, you might want to consider going to more stages of control. Yet by going to more stages, you potentially sacrifice the simplicity of thermostatic control, as is explained under the following heading.

Step Controllers and SCRs

Special consideration must be taken when selecting duct heaters with three or more stages. One and two-stage thermostats are common and inexpensive. Three-stage heating thermostats are rare (and often kinda ugly!), and four-stage thermostats are virtually nonexistent. Therefore, when selecting a duct heater with three or more stages, a decision must be made, prior to purchasing, on the control method. Duct heater manufacturers offer different control packages, such as step controllers and SCR controllers, which are factory installed inside the duct heater controls cabinet. These packages must be specified up front, as any control package is often an option, and it is the purchaser's responsibility to ask for these options.

Step controllers are multistage temperature controllers or "sequencers," requiring a control signal from a space temperature sensor/controller. The step controller monitors this signal and determines the appropriate number of stages of heat to energize. SCR (Silicon Controlled Rectifier) controllers are proportional type electronic temperature controllers. An alternative to the step controller, the SCR controller, though relatively more expensive, offers precise temperature control, for those applications that demand it. Like the step controller, the SCR controller requires a signal from a space temperature sensor/controller. Unlike the step controller, the SCR controller actually varies the average power output of the electric heater, as a whole, in proportion to the deviation in space temperature from setpoint. Both step controllers and SCR controllers are available as factory installed control options for electric duct heaters.

For a multistage application (say three or more stages), it is feasible to equip the multistage duct heater with a separately purchased control system. In this case, the heater is ordered without a step controller, and the stages of heat are controlled by some other means. The alternatives have to be weighed carefully. Although it may be less expensive component-wise to do it this way, the labor and design time may outweigh the advantages. In most cases, ordering the duct heater with the factory installed control package is the least problematic alternative.

Okay, consider this. The mechanical designer got it right. He specified his duct heater selection to have the appropriate number of stages for the given application. Specifically, he specified four stages of control. Unfortunately, the purchaser of the heater did not request a factory installed step controller. Hence, the heater is ordered with four stages, but with no control. The contactors are simply wired to a customer terminal strip, of which needs to be interfaced properly, in order to make it work. Four-stage thermostats are unavailable, so what do you do? Like we mentioned above, it is quite feasible to equip the heater with an "after the fact" control system. It just may not be the least problematic route taken. Now the control systems designer must select and purchase a temperature control system for the heater, draw up a wiring diagram, and provide direction to the installer. Any multistage temperature controller will do. Of course we're typically talking about an electronic temperature controller, with control power requirements. If the electric duct heater is not furnished with a factory control package, its control transformer may not be sized to have the added capacity required for the electronic temperature controller. That being the case, a separate control transformer must be provided and installed. Other issues are those relating to installation and location requirements of the controller (this stuff might not fit in the heater's controls cabinet), setup time, etc.

A reason to purposefully specify and order a multistage duct heater without a factory control package is if the heater is to be controlled via a Building Automation System (BAS). In this scenario, a digital controller with binary outputs could directly control the stages of heat. Of course, the controller would have to have enough binary output points available to accommodate all of the stages of electric heat. This may not be a very economical use of binary points, depending upon the application and the system. It may be more cost-effective, in the long run, to buy the heater with a multistage step controller, utilize an analog input on the digital controller as a space temperature input, and utilize an analog output on the digital controller to feed the step controller. Though in this case the digital controller isn't performing the actual staging of electric heat, it at least allows for the heating setpoint to be established (and processed) via the BAS.

A final word regarding SCR control. While it is possible to provide an "after the fact" multistage controller for a duct heater not equipped with a factory installed step controller, this is typically not the case with SCR control. In other words, an SCR controller cannot readily be installed by a contractor on a duct heater having no controls. The reasons, many of which are too complex to go into here, have to do with installation requirements and performance criteria. For instance, an SCR controller gives off a lot of heat when in operation. When factory installed, a "heat sink" is typically provided to dissipate the heat. This device is a finned piece of metal that must be thermally secured to the SCR controller, with the proper and liberal amount of heat conducting grease. For all practical purposes, "don't try this at home!" In other words, leave SCR control up to the experts: the manufacturers of electric duct heaters.

Hopefully this section has brought forth some guidelines and some useful rules of thumb in applying and controlling electric reheat coils. As stated before, many of these guidelines and rules are applicable to electric heating coils in all applications, and not just in reheat applications. Adhering to the suggestions laid out in this section should, in most cases, provide for good temperature control and customer satisfaction, and at best, keep you out of trouble!

Hot Water and Steam Reheat Coils

A hot water or steam coil used in a reheat application will be equipped with a control valve (Figure 14-2). For hot water applications, the valve body may be two-way or three-way, depending upon the application. For steam applications, the valve body will always be two-way. The actuator may or may not need to be spring return; this type of application does not typically require spring return operation upon removal of power from the valve actuator. If specified, it's mainly a matter of preference.

The valve actuator can theoretically be a two-position device, though really should be a proportional device, for the same reasons that we wouldn't want to go with only one stage of heat with an electric

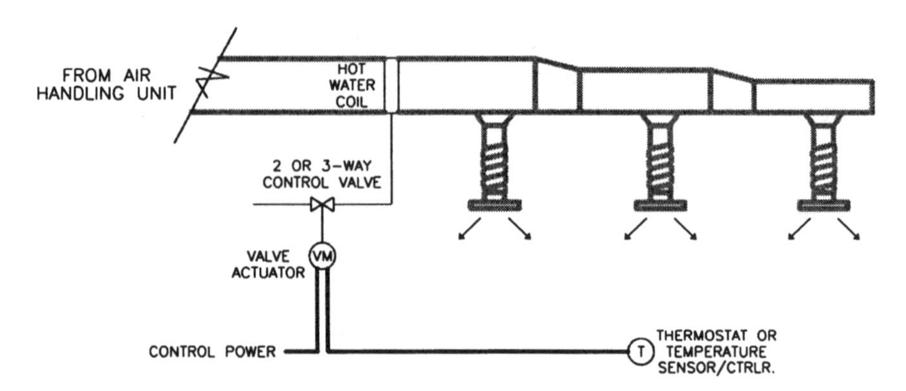

Figure 14-2. Hot water reheat coil installed in a branch duct, serving a single zone.

reheat coil. Instead of rehashing the reasons, suffice it to say that proportional control is the way to go with a hot water or steam reheat coil.

For the sake of giving equal coverage to all scenarios, let us at least take a look at the choices we have with hot water reheat control. The same concepts apply to steam, so we will generalize by talking in terms of hot water, and make the correlation to steam by simply stating that all that is said about hot water reheat control is true for steam as well. There are some subtle differences, of which we can cover after we exhaust the subject of hot water reheat control.

Hot Water Reheat

First, with hot water reheat, we don't have as many choices of control at our disposal as we do with electric reheat control. Specifically, we don't have the choice of staged control. It's two-position (single-stage) or proportional control, and no in-betweens. There are no multiple stage hot water control valve actuators out there, at least not at the time of this writing. So we're looking at either two-positioning that control valve, or modulating it.

Two-position reheat control is cheap and simple. That probably being a good reason why it's sometimes misapplied. The valve actuator is a non-proportional device, and the controller is nothing more than a single-stage heating thermostat. What could be simpler? On a call for heat, open the valve, and when the call for heat is satisfied, close it! Simple, yes. Suitable for zone temperature control, probably not!

Proportional reheat control is more costly than two-position control, but the benefits reaped from going this route should in most cases outweigh the cost differential between the two methods of control. The valve actuator is a proportional device, and the controller is not a simple thermostat, but a proportional space temperature controller. Unlike the simple thermostat, the controller is most likely an active electronic device, and therefore requires control power. Yet the valve requires power as well, so this is kind of a "moot point" in terms of cost. Whether two-position or proportional control is chosen, control power is required for the valve, normally the 24-volt AC variety. In a typical reheat system, several zones are served by several reheat coils. To provide control power to each control valve, we have two choices. One is to utilize a single 24-volt control transformer and run a control power daisy chain to each and every valve. Care must be exercised with this method, with regard to proportional control. Technical precautions need to be considered, to determine if this is even allowable for a given manufacturer's product. The more conventional method to provide power for the control valves is to equip each valve with its own control transformer. Anyway, with either method, the space temperature controller will derive its required power from the valve's control power source, and therefore will not require any additional parts or labor to "power it up."

So really, the only difference in cost between two-position and proportional control is in terms of the components; specifically, the valve actuator and the space temperature controller. With proportional zone valves becoming more and more popular, and proportional space temperature controllers becoming smarter and less expensive, the price gap between two-position control and proportional control is getting smaller, and the attractiveness of proportional control is getting larger. There is no doubt that proportional control will become the standard in reheat applications, if it hasn't already.

Steam Reheat

As far as steam goes, there really isn't too much to add here in terms of control. Certain control valve manufacturers may require you to purchase a linkage extension kit, that allows the valve actuator to be a greater distance from the valve body. This is so that the actuator electronics are somewhat protected from being destroyed by the high level of heat generated by the steam passing through the valve. Other manufacturers may simply require that the steam control valve be mounted such that the actuator is on a 45-degree angle (or greater) to the vertical. This is probably a good idea for all steam control valve applications, though the valve/actuator operation and installation manual should be checked out, to ensure that the valve can properly and safely function in such an orientation.

The only other issue with steam reheat control has to do with capacity. In the chapter on Common Control Schemes, we discussed what is meant by 1/3, 2/3 steam control. That section basically said that steam is not the easiest thing to control, and you're better off using two control valves in lieu of one, for larger capacities. For zone control, however, seldom will the capacity be (for a single zone) so large that there is a need for two valve control of the steam. For an explanation of two valve control of steam, and the rules of thumb governing its use, please refer to the appropriate section of the Common Control Schemes chapter.

Chapter 15

Exhaust Fans and Systems

Exhaust fans are a fundamental component of most HVAC systems. Exhaust requirements of a typical office building range from bathroom, kitchen, and conference room exhaust to equipment and telecomm room exhaust. Industrial exhaust applications encompass fume hood exhaust systems and general warehouse exhaust systems. Applications that call for the movement of air from an interior space to the outdoors are those requiring the removal of odors, fumes, smoke, heat, contaminants, or just "stagnant air" from the spaces served. Ventilation requirements for most commercial and industrial facilities mandate the continuous introduction of outside air via the main fan systems that are conditioning the building. As such, in addition to all of the above-mentioned needs for exhaust systems, additional exhaust may be required simply to balance out the amount of outside air being brought in.

The mechanical designer often has his hands full with trying to 1) account for all of the required exhaust for a building, 2) determine how to handle all of the various exhaust requirements, 3) strike a balance between supply and exhaust requirements in the building, and 4) select exhaust fans and systems in accordance with his design. Needless to say, the issue of exhaust in a typical HVAC system is of substantial concern, and must be considered not only fan by fan, but also on a building-wide level. In general, as a whole, the entire ventilation system (the "V" in HVAC) must operate to maintain the building at a fairly constant, fairly neutral building static pressure. The mechanical designer must be mindful of this, and must address this issue up front, with at least a minimal amount of foresight into how (and when) the various exhaust systems will operate, what effects they have on the overall balance of the building pressure, and what can be done to counteract the effects in order to maintain an acceptable pressure in the building.

Hopefully the last two paragraphs demonstrated the importance of exhaust and ventilation in buildings requiring HVAC systems. While this chapter is not meant to explore the design concepts behind ventilating systems, it should at least be known that it is an important and crucial part of the overall mechanical design process of an HVAC system. It's not as simple as saying, "Hey, this room needs exhaust, put a fan in there!", without being aware of the effect that this exhaust system has on the overall building-wide ventilation requirements. With all of this said, we'll move on and leave all of that fun stuff to the mechanical systems designer!

Exhaust fans take on various shapes and styles. **Roof mounted exhaust fans** are, yep, usually up there on the roof. They are ducted from the fan, through an opening in the roof, down to the spaces that they serve to exhaust. Roof fans are the popular choice for most general exhaust applications (see Figure 15-1). **Sidewall exhausters** poke through the side of a wall. Depending on the setup, they may or may not be ducted. This is the type of fan that you would commonly find in a warehouse, there to provide general warehouse ventilation. **Ceiling fans** are ducted fans that reside in the ceiling space and typically serve bathrooms and other such spaces. The fan itself sits in the ceiling, and is ducted from there to the outdoors. **In-line fans** are fans that are ducted to and from, the fan being "in line" with the duct system. An alternative to roof and sidewall fans, in-line fans have their place in HVAC applications.

Before we talk about the different methods of controlling exhaust fans, it might behoove us to cover a few of the electrical aspects of exhaust fans, as they do impact the methods of controlling them. Exhaust fan motors (or any motors, for that matter) will be either single-phase or three-phase, this basically being a function of the required size of the motor. When we use the term size, we refer not to the physical size of the motor, but to its strength, or "horsepower" (HP). Small motors with less than 1 HP are commonly referred to as fractional horsepower motors. There is a practiced cutoff point between single-phase motors and three-phase motors, that typically occurs in that fractional horsepower range. Specifically, the amp draw of a motor gets to a point to where the use of a single-phase motor is simply not economical, and a three-phase motor must be used. For instance, a 3/4 HP fan motor, at 120 volts single-phase, draws close to 15 amps of current. The same size motor at 460 volts three-phase draws only a couple of amps. The reduction in current trans-

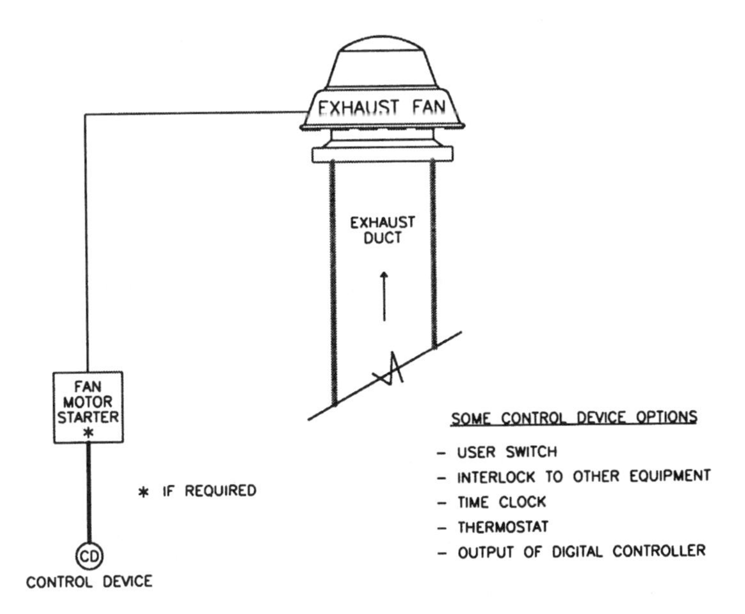

Figure 15-1. Roof mounted exhaust fan, controlled in some manner.

lates to smaller wire sizes and operating costs. Simply put, three-phase motors are more efficient, and more economical, than single-phase motors.

So where is that "cutoff point?" Generally speaking, any motor smaller than 1/2 HP will be specified as single-phase. At 1/2 HP, a 120-volt single-phase motor will draw around 10 amps, which isn't necessarily unacceptable, yet economics dictate the use of three-phase power for a fan of this size. There is some overlap, in terms of what motor manufacturers offer, of single-phase and three-phase motors. Motors that are 1/2, 3/4, and even 1 and 2 HP, can be found as single-phase. The amp draw of these motors is excessive, yet there may be applications out there where three-phase power just isn't available, and single-phase power is all you have.

For the purpose of the rest of this discussion, we will take the liberty of generalization and call any motor less than 1/2 HP single-phase, and any motor 1/2 HP or greater three-phase. This actually is a pretty good rule of thumb, and common practice in terms of motor selection.

Another generalization that we will make, that echoes a statement made back in the section on motor starters, is that three-phase motors require starters, whereas single-phase motors don't. For more on this, please refer back to that particular section of Chapter 5.

What does any of this have to do with controlling an exhaust fan? The purpose of the last few paragraphs was to give the reader an idea of how motor size relates to the method, or means, of controlling the motor. With what has been said, we basically drew a correlation between motor size and whether it needs a starter or not. By generally stating that small motors are single-phase, and that single-phase motors don't need starters, we generally conclude that small motors (less than 1/2 HP) don't need starters, and larger motors do. This is important to understand, when considering a project with several or dozens of exhaust fans. By having a "feel" for the sizes of the exhaust fans and of their starter requirements, a direction is established, in terms of how they can (and will) be "practically" controlled.

To control an exhaust fan, we need to intercept its power supply, for what we typically mean by "control" of an exhaust fan is "two-position" control of its state (having it energized or de-energized). With small (single-phase) fans, the motor is fed by a circuit breaker panel, located somewhere in the building, and not necessarily anywhere near the fan itself. We can feasibly intercept that power feed anywhere between the fan and the panel, for the purpose of control. We directly intercept the power supply to the motor with our control device, whether it be a switch, temperature controller, contactor, or what have you.

With larger (three-phase) fans, we cannot intercept the actual power supply to the fan with our control device. The fan is controlled by a starter; in order to control the fan by some remote means, we have to break into the control circuit of the starter with our control device. So our control device operates the starter, turning it on and off, and in turn the starter allows or disallows power to the fan. This requires that our control device be physically wired back to the starter, wherever that may be.

What we see here is that, with larger fans, it doesn't matter where the fan is, for purposes of remote control. The starter is what's important here, more so the location of it. This is of potential impact on the estimating of a project, and on the installation thereafter. Consider the application of an exhaust fan serving an equipment room, controlled by a thermostat in the space. The fan motor is three-phase, and the starter for it is located back at some main "motor control center" (MCC) 100 feet away. Even though the thermostat controlling the operation of the fan is virtually in the same room as the fan, say within 15 feet of it, the thermostat needs to be wired all the way back to the motor starter. The seemingly simple task of wiring the thermostat to the fan is in all actuality a labor intensive task, especially if there is a mandated requirement that the control wiring be run "in conduit."

Okay, all of that to relay a simple concept, of which we now summarize. A difference exists between the methods of control of small exhaust fans and large exhaust fans. Whereas the small fans, those of which are single-phase, do not have starters, the larger three-phase fans require starters, the location of which is often arbitrary with respect to the actual fan. For fans without starters, we directly control the power supply to the fan motor with our control device, and can do this at any point between the origin of power (circuit breaker panel) and the fan itself. For fans requiring starters, we cannot intercept the power supply to the motor, and must wire our control device to the starter, wherever that may be located, even if it seems illogical to do so. So now that we have some insight as to what we need to do with our control device, we can finally talk about some of the different methods of controlling exhaust fans in typical HVAC applications.

Manually Controlled Exhaust

The simplest method of controlling an exhaust fan is just manually turning it on and off. The controller is a switch, operated by the user, at the discretion of the user. Not really too much to say here, so instead of dragging out a simple topic and making it appear to be more complicated than it is, we'll just stop right there.

Applications requiring manual operation of exhaust fans include laboratory and kitchen fume hoods, janitor's closets, small bathrooms (sometimes interlocked with the light switch), conference rooms, and any other type of space that doesn't require automatic or interlocked control of the exhaust fan serving it.

Interlocked Exhaust

Applications of which require the operation of an exhaust fan to be a function of the status of some other piece of equipment or system, are the topic of this section. "Interlocking" an exhaust fan to other equipment is a common practice. Consider a piece of equipment that starts and stops randomly, and also produces fumes and/or excessive heat when in operation. An emergency generator, for instance, fits this description. The generator may sit idle for most of its life, but upon a detection of loss of main electrical power to the facility, the generator comes to life to produce at least some of the electricity that the facility needs to maintain its operations. The generator at the very least will give off heat, enough to require automatic ventilation upon initiation of the generator. Depending upon the type, the generator may also give off fumes that are undesirable or even just downright dangerous! Hence, a ventilation system is required to be interlocked to the operation of this piece of equipment.

Notice the term "ventilation" in the above paragraph. The system required to be interlocked doesn't necessarily consist of just an exhaust fan. An exhaust fan is required to pull air out of the space and to the outdoors. Where that air is pulled from may be of some concern. In an application such as this, a motorized outside air intake damper may be interlocked as well. Upon a call for exhaust fan operation, as signified by the generator "coming to life," the fan will start, and at the same time the outside air damper will open. Thus air is pulled into the space through the intake damper, through the space, and out of the space by the exhaust fan. Hence the term "ventilation!"

For this type of application, it is important to know at least a little bit about how the equipment that the exhaust fan is interlocked to operates. More specifically, how it starts and stops, and how this is performed electrically. To interlock the operation of an exhaust fan to another piece of equipment requires that we be able to determine when the unit "turns on." The equipment may have, as a point of external connection at its customer terminal strip, a set of contacts that close when the unit starts. We can wire this set of contacts to directly control the exhaust fan. If the equipment does not have this factory furnished feature, then it's up to us to figure out how to make the interlock. If we were to install a relay, and somehow wire the coil to the piece of equipment so that when the equipment starts, the relay energizes, then we can use the relay's normally open contacts to control the exhaust fan. This is basically how it's done, and is relatively straightforward, yet still requires an understanding of how the equipment operates electrically.

Chiller rooms are another application for an interlocked ventilation system, yet the point of interlock is not a piece of equipment, but a refrigerant monitor. This electronic device monitors refrigerant levels in the air of the chiller room, and upon detection of an excessive level, perhaps due to a leak in one of the refrigeration lines, issues an alarm and presses the ventilation system (exhaust fan and intake damper) into operation. Carbon monoxide monitors perform a similar function, and are commonly found in garages and parking structures. Upon detection of excessive CO in the garage, perhaps caused by a substantial amount of traffic in the garage at a given time, the CO monitor turns on the ventilation equipment to flush the space of toxic fumes. Smoke purge systems work the same way. Upon detection of smoke (by a supervisory fire alarm/smoke detection system) in a space such as a warehouse, sidewall exhaust fans are energized, in an effort to quickly purge the space of smoke.

One final word about the term "interlock." It is sometimes said that an exhaust fan is interlocked to a make-up air unit. Recalling our discussion on make-up air units, we remember that make-up air units exist because exhaust fans exist. A better utilization of the term is to say "make-up air unit is interlocked to exhaust fan." It is more "technically correct." It's kind of like "Which came first, the chicken or the egg?" For interlocking scenarios, equipment that exists, because of the precedented existence of another piece of equipment, is the "interlocked" equipment. An exhaust fan exists in a generator room because the generator exists. The exhaust fan is interlocked to the generator. Likewise, a make-up air unit exists in a warehouse because an exhaust fan exists. The make-up air unit is interlocked to the exhaust fan.

Time-of-Day Controlled Exhaust

Exhaust systems that are in continuous operation during occupied periods and are shut down otherwise are the topic of this section. Not much to cover here, control-wise. Time-of-day control of equipment is performed by a time clock, whether it be a stand-alone device, or a scheduled output from a DDC system. If an exhaust fan is to be controlled by a stand-alone time clock, and the fan has a starter, the simplest thing to do is to locate the time clock in the immediate vicinity of the starter, for ease of installation, wiring, and control. If controlled from a DDC system output, then the output must be wired back to the starter.

Exhaust fans and systems that generally operate via time-of-day scheduling include bathroom exhaust systems, lunchrooms, and exhaust systems that are part of a general building ventilation system, such as those that operate in concert with a building's main fan and air conditioning systems.

Temperature and Pressure Controlled Exhaust

Exhaust fans can be automatically operated by temperature and pressure controllers. Temperature controlled exhaust is considered first.

An exhaust fan controlled by a temperature controller is always a "cooling" application. The space that the exhaust fan serves to exhaust becomes too warm, and the exhaust fan turns on. This is primarily "forced ventilation," serving the purpose of maintaining a space at an acceptable "high limit" temperature level. There is no cooling, *per se*, yet the heat generated in the space is removed by the exhaust/ventilation system. This is definitely not comfort control; we're usually talking about setpoints here of 80 degrees or so! Temperatures of which are unacceptable for human comfort, but are acceptable for the heat-generating equipment that occupies these spaces. Electrical/switchgear rooms, boiler rooms, and elevator machine rooms all fall under this category.

The temperature controller performing the "call for cooling" is normally a "line voltage stat." This is especially true when the fan is single-phase. Line voltage thermostats are typically capable of handling up to 8 amps, which is on the level of 1/3 HP at 120 volts. So for a small single-phase fan, the line voltage thermostat directly controls the fan.

If the fan is three-phase and requires a starter, then the thermostat is wired to the starter. The thermostat chosen for this scenario is also most often a line voltage thermostat, since the starter's control circuit is often 120 volts. Line voltage thermostats are typically constructed as single pole double throw, "snap-acting" controllers, capable of being applied to either a heating or a cooling application.

For single-phase applications, often a switch is wired "in parallel" with the thermostat, and mounted adjacent to it. This allows the user to "override" the thermostat and manually turn on the exhaust fan. For three-phase applications, this is normally handled at the starter, with the starter's HAND-OFF-AUTO switch.

Pressure control applications employing exhaust fans are "standard procedure" these days for many air handling and ventilation systems. We've actually already considered some of these types of applications when we spoke of rooftop units and air handling units. For such applications, we are normally not only engaging and disengaging fan operation, but are also controlling deliverable air quantity of the fan, via IGVs or a VFD. Pressure controlled exhaust applications typically consist of monitoring space static pressure conditions, and controlling exhaust fan air quantity in an effort to maintain a suitable space static pressure level. These kinds of applications are often just a part of one larger ventilation system, as exhaust requirements of this caliber are usually dictated by another component of the ventilating system, and its arbitrary and varying modes of operation. As such, pressure control of an exhaust fan will be enabled for operation if and when its operation is warranted. This determination could be made perhaps by an "end switch" on an outside air intake damper actuator set at 20 percent of its stroke. Upon reaching the 20 percent point, the damper actuator's end switch closes, thereby allowing exhaust fan operation. Upon engagement, the fan operates as required to maintain desired space static pressure setpoint.

Another method of performing space static pressure control with an exhaust fan is to use a constant volume fan (no air quantity control) in conjunction with a proportionally controlled damper. The fan sits behind the damper, in a configuration that allows for the fan to pull air from the space served, and push it through the damper and out the building. Static pressure control is engaged perhaps in the same manner as described above; by an end switch on some outside air intake damper. Once engaged, the exhaust fan runs, and the damper is modulated to maintain space static pressure setpoint. For periods of substantial exhaust requirements, the damper is modulated near fully open. Yet for periods of lesser exhaust requirements, the exhaust fan is blowing against a partially to substantially closed damper. While this may be acceptable operation, and is definitely a cheaper alternative to a VFD in terms of first cost, efficiency is sacrificed in terms of operating costs. A fan blowing against a closed damper as a method of maintaining space static pressure doesn't necessarily sound like it would be very cost-effective, especially during those periods of light exhaust requirements. And is this even acceptable operation for the fan motor? Is it decreasing the expected life span of the motor? Maybe. Nevertheless, it's still a popular method of control, and will continue to be for as long as the first cost of this method remains less than the alternative.

Chapter 16

Pumps and Pumping Systems

The pump is the "heart" of the waterside element of any HVAC system! As the fan is the fundamental component of any airside system, so too is the pump for the waterside. As such, we will see in the upcoming sections just how much importance is placed upon pumping systems, and how much reliability and redundancy is often built into such systems. A failed pump means bad news for a typical HVAC system! This is something that will particularly be explored in the section on Two and Three Pump Systems.

The types of pumps and pumping systems that we will be taking a look at in this chapter will be primarily limited to those of which are used in heating and cooling applications. Hence, we are basically talking hot and chilled water pumps, though we will at least touch upon condenser water pumping systems.

For our purposes, we will forego discussions about mechanical construction of a pump, and make the assumption that the reader knows the basic principles behind pump construction, application, and operation. Furthermore, we will not explore mechanical design considerations and concerns, except as they might apply to the proper control of pumps. What this chapter will do, is describe pump control as it applies to several different common HVAC applications. The first section will list out some basic single pump applications, and explain typical control schemes. The next section looks at two and three pump systems, why they exist, and what can be done to adequately control them. The third section talks about pumping systems that utilize variable frequency drives, and the fourth section explores primary/secondary systems, touches upon their purpose and their need, and discusses the requirements for proper control. Finally, the last section touches on system bypass control methods, when and why they are needed, and the pros and cons of implementing them.

As far as electrical requirements go, pumps are really no different than fans. As with fan motors, pump motors are either single-phase or three-phase, depending on their size and application. The smaller, single-phase pumps will not require starters, whereas the larger three-phase pumps will. The one thing nice about pump control applications is the fact that pumps requiring starters usually will have these located in close proximity. Unlike exhaust fans, whose location is typically arbitrary and often overhead (on the roof!), pumps for the vast majority of applications tend to reside in a room suited for mechanical equipment, such as a chiller or boiler room. Therefore, the requisite starters for these pumps are simply installed local to the pumps, perhaps on a nearby wall. This makes the "assumed" control point of these pumps a relative certainty; the point of control for the pump(s) in question will almost always be right at the pump! Even if the motor driven equipment of a facility is served by a central motor control center (MCC), chances are good that this piece of electrical equipment will be in a nearby room, or even the same room, as the pumping equipment.

Single Pump Systems

With so much reliance placed on pumping systems, there aren't too many applications out there that rely upon the use of only a single pump (Figure 16-1). Redundancy is built into most of the more critical pumping systems, by having two (or more) pumps piped in parallel, serving the same system. Nevertheless, there are applications in which a single pump will do, as we see in this section.

One of the most common applications for a single pump is "coil circulation." This is the practice of installing a circulating pump right at the hot (or chilled) water coil of an air handling unit. The pump serves as a means of "freeze protection" for the coil, and is typically controlled from an outside air temperature controller, set at 50 degrees or so. Whenever the outside air temperature is below setpoint, the pump runs to continuously circulate water through the coil. This type of setup usually involves the use of a three-way control valve, configured so that when there is no call for heating, the pump simply recirculates water through the coil. As a call for heating is established, the three-way valve begins to modulate open to allow the pump to pull some water from the main hot water loop, and mix it with recirculated water. The whole premise to this setup is that

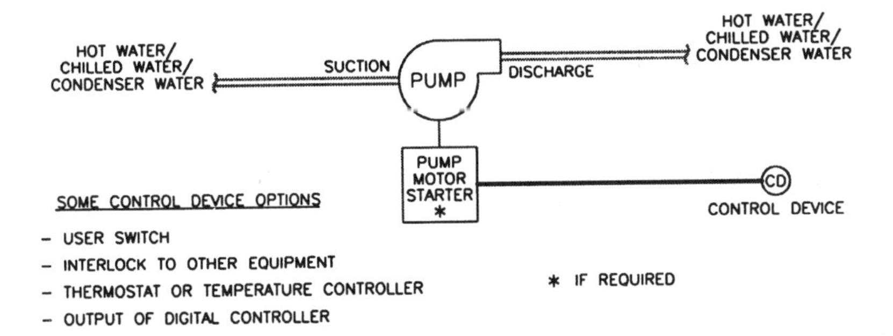

Figure 16-1. Single pump, controlled in some manner.

water that's continuously flowing doesn't stand much of a chance of freezing. The only scenario of potential concern here with this setup is when there is a call for heating when the outside air temperature is above the pump's outside air controller setpoint. The valve would modulate open and, providing the main system pump is running and circulating hot water through the mains, would attempt to allow water to flow through the coil. With the coil circulating pump off, the system pressure may or may not be large enough to overcome the in-line obstruction of the circulating pump that isn't running. A simple solution to this is to have an "end switch" on the valve actuator make and override the pump into operation upon the valve beginning to stroke open to the coil.

In this application, often the coil circulating pump will be small enough to where a starter will not be required. Hence, the point of control can be right at the pump, by directly interrupting the power feed to the pump motor. For larger coils, the pump may very well be large enough to be three-phase and require a starter. If so, the starter may or may not be mounted locally at the pump. Remember that for motors requiring starters, the "point of control" is the starter. This being the case, that outside air temperature controller (and the control valve's end switch as well, if applicable) would have to be wired to the starter, wherever that may be.

Another application requiring the utilization of single pumps would be that of a "zoned" system. Consider a hot water system consisting of a single boiler and a half dozen zones of baseboard radiation. Instead of a single system pump and thermostatic control of baseboard zone valves, the system under consideration accomplishes its zoned temperature control with zone pumps. Each zone of baseboard has a dedicated pump. The pumps are simply controlled by two-position temperature controllers (thermostats) in their respective zones. So a zone in need of heating would energize its zone pump to flow water through its section of baseboard, and would terminate the operation of its pump when the call for heating by the zone temperature controller is satisfied. Even though there are several pumps in this system, we still consider it here, for no redundancy is built behind any of the individual zones.

This is another application where the pumps will tend to be relatively small. Since zones in general are kept small (the more zones and the smaller they are, the better the individual comfort control!), it follows that the gpm requirement of any particular zone will be small as well, as will the pump required to provide it. Therefore, the pumps will most likely be able to be controlled directly by the zone temperature controllers, provided they are of the "line voltage" variety.

A third application calling for single pump operation would be that of a boiler circulating pump. A boiler manufacturer may require that his boiler have continuous flow through it in order for it to be operational. As such, he may recommend (or even mandate) that a small pump be installed in such a configuration. The controlling entity of this pump is simply the boiler itself; the pump is interlocked to the boiler's operation so that whenever the boiler is "enabled" (though not necessarily firing), the pump runs.

As these applications are relatively simple in concept, they are controlled by simple means. Moreover, the criticality of their application does not merit redundancy, which would take the form of an additional pump. The following section discusses those applications that do merit redundancy, and explores the various means and methods of operating the pumps in these type of systems.

Two and Three Pump Systems

Pumping systems that require redundancy are those of which a whole HVAC system is dependent upon the operation of the pumping system. These would be main hot and chilled water pumping/piping systems, and even condenser water pumping/piping systems, for water cooled systems. Most often the engineer of an HVAC project entailing major waterside systems will specify not one main pump, but two, for any given pumping/piping system. The pumps are piped in parallel, and either run both at the same time, or only one at a time.

Two Pump Systems

If two pumps are each sized for half the gpm rate of the total required, then they are meant to run at the same time. For the control systems designer, this is a piece of cake! Sure, there may be a requirement that they both run depending upon some other condition, perhaps via an outside air temperature controller, or an interlock to some other entity. But that's fairly simple to do as we saw in the last section. The added dimension here is that instead of one motor, we have two, and therefore two separate points of control, whether they be starters or the power feeds to the motors. For control of pumps via an outside air temperature controller, for instance, either the controller itself would have to be a two-pole device, or it would have to operate a two-pole contactor, with each pole controlling either pump.

For automatic operation of two pumps that are each sized for half the flow rate, control is simple. For manual operation of two pumps each sized for half the gpm, control is extremely simple. Just turn em' both on and let 'em run!

The inherent redundancy of two pump systems, in which each is sized for half duty, is demonstrated upon the failure of one of the two pumps. If the mechanical designer played his "pump curves" right, the pump that remains in operation actually makes up for some of the capacity lost by the failure of the other pump. In other words, flow does not drop off to 50 percent, but to a point somewhere greater than that. The remaining pump, in essence, takes up some of the slack. In lighter load periods, this is normally adequate to maintain desired comfort levels. The reduced flow of the hot or chilled water still packs enough Btus to satisfy the heating or cooling needs of the spaces served by the pumping system.

Even during periods of heavy load, the system can still provide at least partial capacity. Of course, how long people will stand for partial environmental comfort control is of issue here, and if one of those pumps breaks down in the dead of winter (heating) or in the "dog days" of summer (cooling), better get it fixed pronto! On the other end of the spectrum, if one of the pumps breaks down during milder periods, will the failure even be recognized?

While this method of redundancy has its strong points, it does have its drawbacks as well, as illustrated in the last paragraph. It is cheap and extremely simple to set up and control, and "backup" is performed automatically with no additional controls required. However, as we saw above, there may be some unacceptability to it, subject to the end user's needs and his willingness to pay more for a better method of redundancy. Furthermore, it is imperative that the mechanical system designer selects the pumps appropriately. Specifically, he must not only select pumps so that they operate efficiently together, but also so that either pump can operate alone in the same piping system, without riding too high off its curve into the zone of inefficiency. Or worse yet, into a state of "cavitation." Without taking the time to explain this phenomenon, suffice it to say that this is not a good thing for the pump, and can actually cause physical damage to the mechanical inner-workings of the pump.

A second, more popular (albeit more costly) method of system redundancy is to again have two pumps, but each sized for the full system gpm. Only one pump runs at a given time, with the other serving as a backup in case of a failure of the primary pump. Of course the first cost of this method of redundancy is more than the previously discussed alternative; the pumps are larger and more costly. Operating costs pan out to be the same, though, and the overall insurance of having flow when it's needed is greater. Rest assured, if that primary pump fails, you have one just sitting there waiting to take over!

How these pumps are sequenced and alternated normally depends upon the requirements of the specific project. Sometimes nothing more than a primary pump selector switch is required, as may be the case if the facility in question is suitably staffed to monitor and maintain the mechanical systems and equipment. In other cases, the addition of "automatic backup" is required. In this case, the primary pump is selected manually, and a differential pressure switch or flow switch monitors primary pump operation. Upon failure of the primary pump, the backup pump automatically starts (refer to Figure 16-2).

Another requirement may be "automatic alternation" of the primary pump. In addition to the automatic backup function, this feature would allow the pumps to be alternated automatically, instead of manually via a selector switch. Primary pump alternation can occur daily, semi-weekly, or weekly. This feature ensures even run times on both pumps.

Even though we have already covered the concept of primary/backup in a previous chapter (see Common Control Schemes for a quick overview), how it relates to pumping systems is of the utmost importance these days, and is often misunderstood (and underestimated!). The

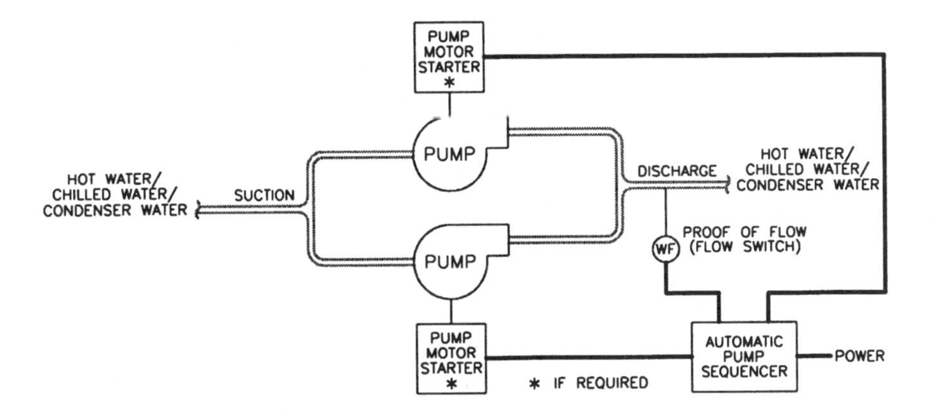

Figure 16-2. Two pumps, controlled by an automatic pump sequencer. Each pump is full-sized, and only one pump runs at a time, with the other serving as a backup in case of primary pump failure. Proof of flow is required for automatic backup operation.

decision on what to provide, in terms of automatic control features, is often driven by the engineer's specifications, or by the customer's needs or desires. Naturally, the more that's provided, the more costly it is. And if what is provided falls short of expectations, somebody ends up paying for it in the end.

In order to gain an understanding and an appreciation of what is meant by the term "primary/backup" as it applies to pumping systems, it is important to establish definitions of the various terminologies used:

Primary/Backup Pumping System—A pumping system (hot, chilled, or condenser water) in which two pumps are installed and piped in parallel, each pump sized for the full system gpm, with the intention of having only one pump run at any given time, while the other serves as a backup in the event of failure.

Primary Pump—The pump chosen, for any given period of time, to be the operating pump.

Backup (Standby) Pump—The pump chosen, for any given period of time, to be the idle pump.

Lead/Lag—A term normally confused with the term "primary/backup." Lead/lag implies "staging," where stage one is the lead, and stage two is the lag. The term is better suited to equipment fitting this description, such as: two boilers, each sized at half capacity, or two compressors that are part of the same chiller or condensing unit.

Manual Backup—Failure of the primary pump is not detected automatically. Upon failure, the backup pump must be placed into operation manually.

Automatic Backup—Failure of the primary pump is detected automatically. Upon failure, the backup pump starts automatically. Failure is most commonly detected by a differential pressure switch installed across the common suction and discharge headers of both pumps. Other acceptable methods of failure detection include water flow switches and current sensing switches.

Alternation (Rotation)—A procedure by which the primary pump is taken out of service (thus becoming the backup pump), and the backup pump is placed into service (thus becoming the primary pump). This can be performed on a daily or semi-weekly basis, and is normally done so as to ensure even wear on both pumps.

Manual Alternation—Selection of the primary pump is a manual procedure, accomplished by a selector switch. The maintenance personnel must manually flip the switch periodically, if even wear on both pumps is desired.

Automatic Alternation—Selection of the primary pump is done automatically. The most common method of alternation is time-based, in which the pumps are automatically alternated on a daily or semi-weekly basis, via the scheduling function of a time clock.

With the terms defined, let us now investigate the various methods of implementing primary/backup pump control. Specifically, we will explore three different combinations of backup and alternation. Each method builds upon the previous method, with each being more complex and more costly to implement, though at the same time offering more in terms of "insurance."

The fist method to consider is **Manual Backup/Manual Alternation**. There are virtually no controls associated with this method, except perhaps a pump selector switch wired between the two pump starters. A switch is not even required if the pump starters are equipped with pilot devices (switches). However, care should be exercised so as not to have both starters energized at the same time.

The second method is **Automatic Backup/Manual Alternation**. With this method, you get the automatic switchover upon failure, yet you still have to perform alternation manually. For the automatic backup function, a means of detecting failure is required (differential pressure or flow switch). This controller can't on its own switch control from the failed primary pump to the backup pump. Something needs to be in between the controller and the pump starters. A "pump sequencer," if you will. A controls designer worth his salt will be able to design a custom-built pump sequencing control panel using ladder logic and off-the-shelf components. Alternately, the heart of the pump sequencer can be a programmable (digital) controller. Either way, the components will sit inside of an enclosure, and a couple of pilot devices will don the cover of the enclosure.

The pump sequencer serves as the central point of control; the pump starters and the failure detecting switch are wired back to the control panel (refer again to Figure 16-2). In addition, a control input may be provided for an external "call for pumping." This may be wired out to an outside air temperature controller that calls for pump operation when the temperature drops below some point (as might be the case for a hot water system). A two-position selector switch determines which pump is the primary pump. During pumping operation, a green pilot light (one per pump) will indicate which pump is running. Upon failure of the primary pump, the sequencer automatically brings on the backup pump, and a red "Failure" light is lit.

The above describes what is minimally required of such a pump sequencer. Of course, more features can be added, however along with the added cost and complexity. A comparably priced "store-bought" sequencer can also be used, with primarily the same features and functions, and may be an attractive alternative to "building your own," application permitting.

The third and final method of handling primary/backup pumping systems that we will consider is **Automatic Backup/Automatic Alternation**. With the addition of a scheduling function, you get time-based alternation! For the custom-built pump sequencer, this means the addition of a time clock or, if microprocessor-based, additional programming.

To perform this with ladder logic is cumbersome at best, and requires a little more work than just mounting a time clock inside the enclosure. Additional design time, added panel building labor, and added component cost may warrant the use of a programmable controller for this type of control. The store-bought option for this exists as well, and is a very attractive option for applications that are completely "stand-alone." Of course, these manufactured products are naturally limited in what they can provide, in terms of features and functions, and if something is required of your pump sequencer control scheme that is "above and beyond" what the store-bought sequencer has to offer, you might find yourself back designing a custom-built controller, one that incorporates all of your pump sequencing needs, and any other related needs.

Following is a concise sequence of operation for a pump sequencer offering both auto backup and auto alternation. We will base the sequence on the use of a differential pressure switch as the means of proving flow and detecting failure.

Normal Operation

1) *Control input (external call for pumping operation) closes, signaling the panel to start the primary pump.*
2) *Primary pump, as selected by a scheduling function, starts and runs continuously. Primary pump alternation occurs weekly.*
3) *Green indicator light respective to this pump is lit.*
4) *After a short time, differential pressure switch installed across the common suction and discharge of both pumps makes, thus proving pump operation. Built-in time delay allows for enough time for this switch to make, once a command for pumping has been issued, before transferring to "failure mode."*
5) *Primary pump continues to run until the control input opens, or until the scheduling function selects the other pump to be the primary pump.*

Failure Mode

1) *External command for pumping exists, and primary pump is running continuously.*
2) *Primary pump fails. Differential pressure switch breaks, indicating a failure.*
3) *After a short time delay, system enters "failure mode":*
   —Primary pump is electrically disabled.
   —Backup pump automatically starts.

—Green indicator light respective to this pump is lit.
—Red failure light is lit.

4) *For as long as the system is in this mode, the backup pump will run whenever the control input is closed.*
5) *The failure indicator light will remain lit for as long as the system is in this mode.*
6) *At this point, the failed pump should be checked out and repaired as necessary. Once this pump has been repaired, the reset push-button can be pressed to restore the system to normal operation.*

It is interesting to note the requirement of a "built-in time delay" in the above sequence. The purpose of the time delay is actually two-fold. First, it allows time for flow to be established upon a call for pumping operation. Second, it prevents "nuisance trips." This means that there is a short time period instituted between the "loss of flow" and the backup pump coming on line. The reason is that there could be events, or "glitches," that cause momentary loss of flow, that aren't indicative of a failure of the primary pump. For these events, we don't want to falsely transfer to the failure mode.

As promised, we now briefly touch upon condenser water pumping systems. For hot and chilled water systems, the above requirement for a time delay should be of little or no consequence to the operation of the entire system. For condenser water systems, however, this time delay is of crucial importance. If the primary pump fails, the time between failure and initiation of the backup pump must be quite short. The reason is that, for water cooled equipment, if they are calling for cooling and not getting their condenser water flow, they tend to "trip out" on their refrigeration pressure limit switches. This is not a good thing; if the time delay between failure and backup is too long, all of the water cooled units calling for cooling at that time will potentially shut down via their "manual reset" pressure switches, thus requiring maintenance personnel or service technicians to go around and reset all of these units.

For a custom-built pump sequencer, an adjustable time delay can be utilized, and the commissioning technician can set the delay appropriately, to strike a balance between "loss of flow nuisance trips" and "water cooled equipment pressure switch trips." For the store-bought version, the buyer must be wary of the application, and must verify that the sequencer either has an adjustable or selectable time delay period, or if not, that the "fixed" time delay period is short enough to be properly applied to condenser water systems.

Well that about does it for two pump systems. The length and content of this section demonstrate the importance of building redundancy into pumping systems. In fact, the next section expounds on that notion, by introducing yet another pump into the mix!

Three Pump Systems

Another means of building redundancy into pumping systems is to have not two, but three pumps serving the same system. Each pump is typically sized for half the system gpm, so that two pumps run at any given time, with the third serving as the backup. Additional reliability is gained by having a third pump. Upon failure of either of the two primary pumps, the backup pump starts, with no loss of flow capacity. Now, if (Heaven forbid!) a second pump failed, one pump still lives, and at least partial flow is still provided.

To apply the three methods of control that we talked about for two pump systems, we see some differences, and run across some added complexities. For the **Manual Backup/Manual Alternation** method, a three-position selector switch may be utilized to select the backup pump. The switch would require a "supporting cast" in terms of controls; a couple of relays and some ladder logic to perform selection. These of which would be housed in a custom-built control panel.

For the **Automatic Backup/Manual Alternation** method, the means of detecting failure is more complex. In fact, each pump must be monitored individually, via a flow switch or a current sensing switch. A single controller won't cut it! With two pump systems, we could get away with using one controller, installed common to both pumps. If the primary pump failed, then flow would be lost, and the pressure or flow switch would indicate this. Yet if we tried to do this with a three pump system, upon failure of either of the two primary pumps, main flow will not be completely lost, and the controller would not signify the failure. No, we need to monitor each pump individually, in order to know of a primary pump failure, meaning that we need not one but three controllers.

The store-bought option doesn't necessarily exist for this application, as the application is not quite as common and popular as the two pump sequencer applications. The control systems designer will either need to come up with some heavy duty ladder logic, or will need to use a programmable controller. Ladder logic for such an application runs high in terms of complexity and in terms of component count and required labor. Yet it is feasible, and the possession of ladder logic skill level by the control system designer, coupled with the absence of DDC capabilities, makes this a viable option.

For the **Automatic Backup/Automatic Alternation** method, a programmable controller is most definitely required. The added complexity of the scheduling function, and how it needs to alternate not two, but three pumps, all but prohibits the use of ladder logic. Not to say that it can't be done, but at this point we have indeed crossed the line of needless complexity, and a programmable controller is better suited for this application.

VARIABLE FREQUENCY DRIVE PUMPING SYSTEMS

The use of variable frequency drives is energy efficient, and perhaps nowhere is that more evident than when applied to pumping systems. With water being an incompressible fluid, and with a piping system being required to have absolutely no leaks in it, we can see how this statement can be made. With air systems, efficiency is lost if there are leaks in the ductwork, of which there will likely be. For a fan controlled by a VFD, the speed that the fan is operating at is a function of the pressure that is trying to be maintained in the ductwork. For a ductwork system with leaks, the VFD needs to operate the fan at a higher speed in order to maintain pressure, as compared to a system with no leaks. So efficiency is lost, as air seeps through the leaks in the ductwork system.

With a piping system, there are no leaks (there better not be!). So when a VFD is controlling a pump in order to maintain proper pressure in a piping system, no efficiency is lost due to leakage. So energy efficiency is one reason for applying variable frequency drives to pumping systems. Another reason has do to with control. Giving a pump proportional speed control opens up some new possibilities in terms of control schemes, as we will see in this section.

Pump/Variable Frequency Drive Combinations

Let's consider some pump/VFD combos. The first is easy, one VFD controlling one pump. For two pump systems, however, there are a few different possibilities here. Consider these three:

- Two pumps, each sized for half the total system gpm, one VFD (Figure 16-3)

- Two pumps, each sized for the full system gpm, one VFD
- Two pumps, each sized for the full system gpm, two VFDs

The first combination begs the question "Why not two VFDs?" The first step to answering that is to explain that a single variable frequency drive can operate two motors at once, provided that it is sized accordingly. So in this type of application, where both pumps will run at the same time, they can more easily be controlled by a single VFD. To further clarify, by way of example, picture if you will, that each pump is controlled by its own VFD. The pumps are to operate together, perhaps to maintain system pressure. As such, a pressure controller would need to monitor system pressure and transmit pressure and setpoint information to not one but two VFDs. In turn, both VFDs would have to be configured identically, so that they operate their respective pumps in the same manner. This is simply not done; there is too much to go wrong here. Besides, simplicity and first cost likely dictate the use of one VFD to begin with; the use of two VFDs doesn't even come to mind up front.

The second and third combinations deal with two full size pumps.

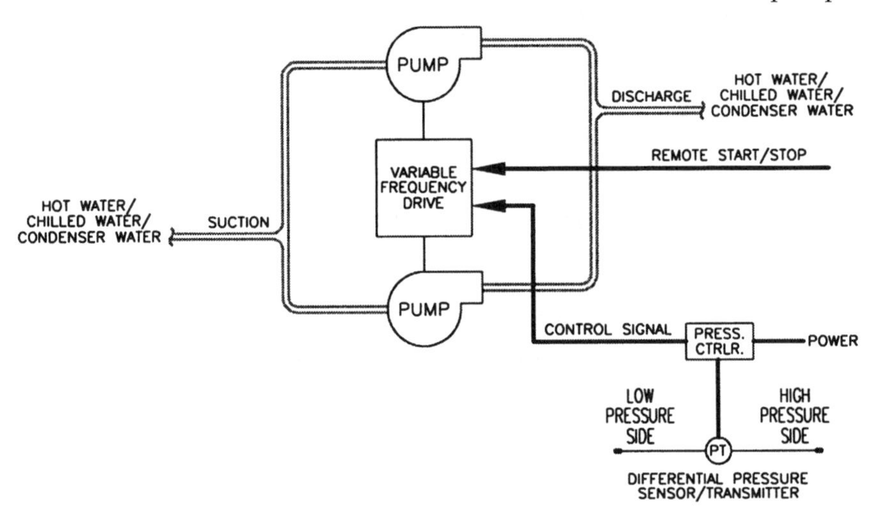

Figure 16-3. Two half-sized pumps controlled by a single variable frequency drive. A pressure controller monitors differential pressure across the supply and return mains. A setpoint is established, and a control signal is generated. The VFD accepts the control signal, and regulates the speed of the two pumps in preferential accordance with setpoint.

While it is feasible that one VFD can operate two full size pumps at the same time, it would warrant the use of a VFD sized for the full load amp draw of both motors. The VFD would control the speed of both motors, say, to maintain system pressure. Theoretically, if the system is at full load, the VFD would operate the motors at approximately half-speed. If one of the two pumps failed, the VFD would "ramp up" the remaining pump, in order to maintain pressure setpoint. On paper this seems to be a viable option, yet for some reason engineers are apprehensive about the idea of having two full size pumps in operation at the same time. Supposing the VFD fails, and the maintenance personnel turns the VFD switch to "bypass," not only one but two full size pumps would operate at full capacity, perhaps during a period of light system load. That could potentially cause enough problems to be apprehensive about it up front.

Another possibility of the second combination is to have the VFD sized for only one pump. So only one pump is in operation at a time, with the VFD physically connected to and controlling only that pump. Upon a failure, the three-phase power leads from the VFD to the primary pump must be transferred to the backup pump. This can be done by a small, manually operated transfer panel, consisting of a selector switch and a couple of contactors. In essence, a manual backup/manual alternation panel. To implement automatic backup and/or automatic alternation with this setup is prohibitive; to switch VFD power from one pump to the other pump, without ceasing the operation of the VFD, will most definitely incur a fault at the VFD, one that would shut down the VFD, and one that would have to be manually cleared.

So now we see why we have that third combination as an option. Two pumps, each sized for the full system gpm, each with their own VFD. At any one time, only one pump and one VFD are in operation, with the other pump and VFD "resting comfortably." With this type of setup, any of our previously discussed methods of backup and alternation can be implemented. Instead of wiring out to pump starters from the pump sequencer, we wire out to the VFDs, which contain starters, and have "start/stop" inputs. The control signal can be sent to both VFDs at once, or can be directed, through the pump sequencing panel, to the VFD in operation.

Control of the VFD

A variable frequency drive will accept a signal from a proportional controller, and operate motor speed as a function of the signal received.

The most common application is pressure control, though there are temperature control applications out there as well. Let's stick with pressure control for our purposes, and call temperature control utilizing variable pumping a unique, "lesser seen" application.

To lay the groundwork for the discussion on VFD pressure control, consider a steam-to-hot water heat exchanger, producing hot water for a building full of fan coil units. For the sake of simplicity, pretend that we have only one system pump (in the wake of all that's been said about the importance of redundancy??). Okay, it's not realistic, but for the purpose of our example, it'll do.

The fan coil units all have hot water coils and control valves. Whether proportional or two-position control is employed, the hot water control valves must all be two-way valves. We require the pressure in the piping system to be affected by the opening and closing of the control valves. Three-way valves would defeat the purpose of having a VFD controlling the pump.

A pressure controller must be furnished, the transmitter of which must be installed across the supply and return mains of the piping system. The high pressure port of the transmitter is teed into the supply main, and the low pressure port is teed into the return main. Location of the transmitter is critical. As a rule, the transmitter should be located at the end of the furthest piping run. This ensures both optimum energy savings and adequate flow for all loads in such a system.

The general rule of thumb laid out in the last paragraph certainly deserves some background. Perhaps we elaborate on this using our example at hand. So consider, if you will, a VFD pumping system whose loads consist of (fan coil unit) hot water coils all fitted with two-way control valves. It is determined that the furthermost coil requires 5 psi in order to get its scheduled flow rate (in gpm). If the pressure transmitter was located right off the pumping station, for instance, we would have to set the pressure controller to maintain a pressure that would ensure that the furthermost coil would always be guaranteed to get its scheduled gpm. In other words, it would need to be set for the worst case scenario. The worst case scenario, defined, is when all valves on the system are fully open to their respective coils. For the sake of illustrating with real numbers, picture that, under worst case conditions, the pressure at the transmitter, right off the discharge of the system pump, needs to be maintained at 10 psi, in order for the furthermost coil to get its 5 psi, and therefore its scheduled gpm.

So what am I getting at here? Hang on, we'll get there! So to recap, if all the control valves on the system are open, then 10 psi at the pump discharge would be required to maintain 5 psi at the extremity. The more flow, the more friction in the pipes to overcome. However, as valves close off, this actually helps the extremity (the furthermost coil), for the less required flow, the less friction there is to overcome. If 10 psi is continued to be maintained at the pump discharge as valves close off, the pressure at the extremity will potentially increase, to a value that is more than is required to satisfy the load's scheduled flow rate. Potential energy savings are lost; the pressure at the pump discharge does not need to be maintained at 10 psi, for virtually any operating condition other than worst case. It only needs to be maintained at a value that would correlate to a resulting pressure at the extremity of 5 psi.

If we locate the pressure transmitter at the furthermost coil, i.e., at the extremity, we can directly control to the needs of the extremity. We are assuming that if we control to the needs of the extremity, all prior loads on the system will be guaranteed their scheduled flow rates. Now, at worst case, all control valves on the system will be open, and we would control to maintain 5 psi at the extremity. This would result in a pressure at the pump discharge of 10 psi. At reduced demand, when valves are closing off, we would continue to control to maintain 5 psi at the extremity. However, the pressure at the pump discharge would be less than 10 psi. Hence the (optimum) energy savings!

The "system setpoint" is set via the pressure controller, and the output signal is sent to the VFD. As control valves open and close, the pressure in the system will tend to vary. For instance, as demand falls off, control valves close and the pressure tends to rise. The pressure controller signals the VFD to slow the pump down in order to maintain pressure setpoint. Likewise, as demand rises and control valves open, the pressure tends to fall, and the controller signals the VFD to increase pump speed.

So how is the system setpoint determined? Good question! An idea of what the setpoint will be must be had prior to purchasing the pressure controller, so that it can be bought with the proper range. For instance, a controller with a 0 to 10 psi range will not do if the system setpoint (that which is trying to be maintained at the furthermost coil) is to be 15 psi. For a typical HVAC piping system, required system "design" pressure is normally in the range of 10 to 20 psi. A controller should be selected in accordance with the pumps selected. The mechanical designer has already done the legwork on estimating system design pressure, and has selected pumps accordingly. If the pumps are selected to provide so and so gpm at, say, 40 "feet of head," this value can be divided by 2.31 (***psi = ft. hd./2.31***), yielding a result of 17.3 psi. A controller with a 0 to 20 psi range will suffice for the application, and since you are trying to maintain not system design pressure, but furthest load pressure, you may want to opt for the next smallest available range. A controller with an oversized range will tend to not have the accuracy or the resolution that a "right-sized" controller will, so be careful with your selection, and put some thought into it instead of arbitrarily selecting a controller sized for total system pressure.

Establishing a setpoint for the controller is something that has to be done in the field, upon startup of the system. The water balancer will need to open all control valves and run the pump at full speed, in order to set flows at all of the coils, as scheduled on the mechanical plans, and balance the system. Once the system is balanced, and all of the coils are getting their scheduled flow rates (with the control valves fully open), the pressure at the location of the transmitter can be measured. This pressure is the system setpoint!

The setting of the controller is typically "set and forget." Presuming that the commissioning technician goes through the exercise of determining the setpoint and setting the controller accordingly, the setpoint should not have to be readjusted again, except perhaps if equipment was added to or removed from the system. If the setpoint was not established by "scientific method" as described above, it may have been arbitrarily set at some value. If set too high, excessive system pressures will result. If set too low, loads may not be satisfied, resulting in service calls in the dead of winter.

VFD pumping systems are commonplace in this day and age, yet there is a lot of opportunity for misapplication. Hopefully this chapter nailed down some of the more important things to consider when applying variable frequency drives to pumping systems.

PRIMARY/SECONDARY PUMPING SYSTEMS

As a segue-way into this topic, consider a chilled water application in which variable speed pumping is desired. A pair of chillers produces the chilled water for the system, and they are piped in parallel, perhaps to be centrally controlled by some chiller sequencer. The sequencer measures chilled water supply temperature common to both chillers, via a remote sensor, and operates the chillers to maintain chilled water temperature setpoint. The system is driven by a pair of pumps, each sized at half capacity and piped in parallel, configured to pull water through the pair of chillers and out to the chilled water utilizing equipment.

This seems like a pretty straightforward application. With chilled water demand high, the pumps run near full speed (via the VFD), and the chillers operate at close to full capacity. When demand drops off, the pumps slow down, and the chillers de-stage accordingly. The problem with this design is that most (if not all) chiller manufacturers require a minimum flow rate through their product, to ensure proper and acceptable operation. When demand is high, this is not a concern. But when demand drops off, the chillers may see flow rates below their minimum requirements, which tends to lead to an array of potential mechanical and operational problems with the chillers.

What can be done, design-wise, to overcome this stumbling block? For variable flow applications employing chillers, another piping loop can be integrated. The "primary loop" consists of the chillers and constant speed pumps, while the "secondary loop" consists of variable speed pumps and the chilled water utilizing equipment. The loops are connected together by a section of pipe referred to as the "common" pipe, or "decoupler." The pumps in the primary loop operate at full flow, and the chillers in this loop stage and de-stage as required. The secondary loop pulls chilled water from the primary loop, at variable rates, as a function of chilled water demand. With demand low, very little chilled water is pulled from the primary loop, and primary flow is mostly recirculation through the chillers, via the common pipe. As such, chiller return water temperatures remain at or near chilled water setpoint (typically 45 degrees), and the chillers operate "unloaded." When chilled water demand increases, the variable speed pumps on the secondary side begin to ramp up in speed, pulling more water from the primary loop. Less water flows through the common pipe, and thus the water returned to the chillers is more so from the secondary loop. This water that is returned from the chilled water utilizing equipment is warmer than the chilled water setpoint, thereby causing the chillers to operate at increased capacities in order to maintain setpoint.

What you have here, as illustrated in Figure 16-4, is your basic primary/secondary pumping system! The primary loop, consisting of the chillers and low horsepower constant volume pumps, is where chilled water is produced, and is also referred to as the "production loop." The secondary loop, consisting of chilled water utilizing equipment (all with two-way valves!) and higher horsepower variable volume pumps, is where chilled water is distributed, and hence is also referred to as the "distribution loop." While there are other reasons to employ primary/secondary pumping systems than the one described here (ask your favorite engineer!), we will restrict ourselves to the application at hand for our discussions, and focus on operational and control aspects, rather than on the mechanical design concepts and principles behind primary/secondary pumping systems.

Okay, so let's start with the secondary loop, since we've basically already covered this in the previous section. Not much to say here, in terms of new material. The pumps operate at a speed necessary to maintain adequate system pressure. As demand falls and control valves close,

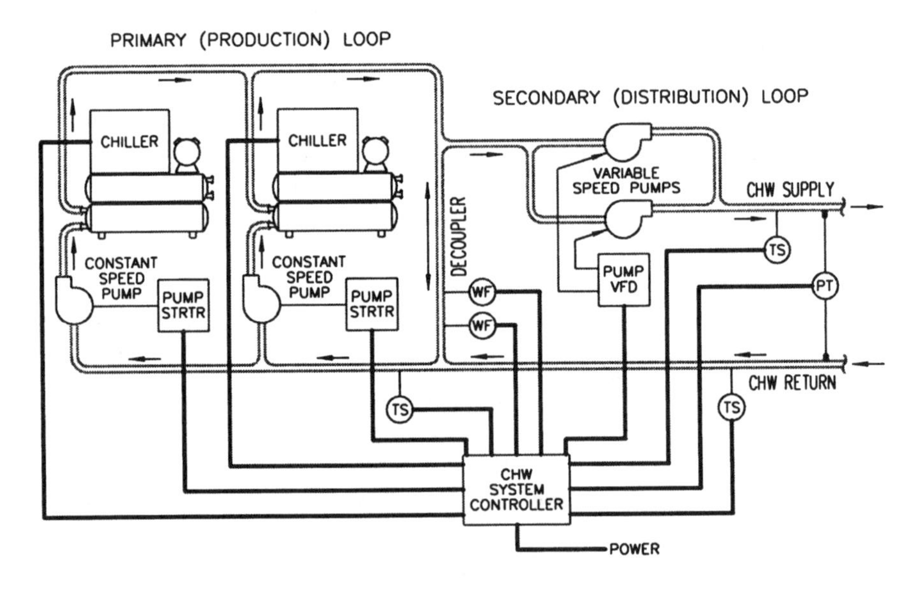

Figure 16-4. Primary/secondary chilled water pumping system. Central controller takes on all facets of system operation. Temperature sensors (TS) monitor temperatures at key points in the system, and report back to the controller. Differential pressure sensor/transmitter (PT) conveys demand information, and flow switches (WF) indicate flow direction in the decoupler.

the pumps slow down in response to the corresponding rise in system pressure. All of the application details and requirements discussed in the last section apply here as well

Now for the primary loop. The important thing here is how to control the pumps and chillers. Consider a primary loop consisting of two pumps and two chillers. A couple of different piping scenarios come to mind here. The first is that of two (half-sized) pumps in parallel, serving two chillers in parallel. The pump combination is piped in series with the chiller combination. The second, more efficient yet perhaps more complicated method, at least in terms of control, is to have a dedicated pump for each chiller, piped in series with it (as in Figure 16-4). The two pump/chiller combinations are piped in parallel.

With the first configuration, a simple method of control is to have the chillers operate via a stand-alone chiller controller. Both pumps run continuously, and both chillers are enabled for operation. As the temperature of the water returned back to the chillers rises, supply water temperature control will determine and dictate the appropriate staging of the chillers, in order to maintain chilled water temperature setpoint. This is elementary chiller plant control, using a central chiller controller to stage chiller capacity as required. The chapter on chillers will cover this topic much more thoroughly.

With the second configuration, control becomes a bit more complicated, to the point of warranting a programmable controller. With this method, chiller/pump combinations are brought on line only as demand dictates. For instance, when demand in the secondary loop is low, only one of the primary side pumps will run, and its associated chiller will be enabled for operation. The other pump remains off, with its respective chiller disabled, until secondary demand reaches a level that would require the additional production of chilled water. For this to happen, the "system controller" needs to know how to evaluate secondary demand. This in itself does not appear to be a difficult task, yet it is one that requires a lot of information to be disseminated and processed, before a decision can be made on whether to bring another chiller/pump on line, or take it off line. The control system designer may want to look at several temperature points within the piping system, and come up with a clever algorithm relating this information to primary production and secondary demand.

Another piece of useful information in this type of setup is the direction of flow through the common pipe. In the first method that we considered, the primary pumps run all the time. In this system, primary flow is virtually always greater than secondary flow, and the flow in the common pipe will always be in one direction. A flow switch installed in the common pipe will therefore not provide any insight to system operation, except to tell the controller that there is flow. However, with the second configuration that we discussed, flow in the common pipe can be in either direction. If only one chiller/pump is in operation on the primary side, and demand is low on the secondary side, secondary flow may be less than primary flow, and the direction of flow through the common pipe would reflect this. As demand increases, secondary flow surpasses primary flow, and the direction of flow in the common pipe is reversed. This information can be used in helping to decide when/whether to bring on an additional chiller/pump on the primary side. If secondary flow surpasses primary flow, and the direction of flow in the common pipe reverses, then it's probably a good time to bring on the next chiller/pump combo on the primary side. Of course, as soon as this is done, primary flow will again exceed secondary flow, and flow in the common pipe once again changes direction. While the information of flow direction in the common pipe will not be "all you need to know" to make the decision on when to bring chillers and pumps on and off line, it will, at the very least, continuously inform the controller as to which loop has greater flow. This in itself is a nice little piece of information, and can be implemented by installing two flow switches, in opposing orientations, in the common pipe.

Primary/secondary pumping systems are useful for applications that require constant flow on the production equipment side, and energy savings on the distribution side. If there were no requirement for variable volume pumping, there might be little need for a primary/secondary system as described herein. The principles behind the mechanical design of such systems run high in terms of complexity and criticality, and are out of the realm of what this writing is attempting to achieve. The chapter on chillers will at least explore the concept of chiller control, and will touch on chiller requirements and methods of operating and control.

System Bypass Control

The use of two-way valves in a typical pumping/piping system is cheaper than the use of three-way valves. Three-way valves are more expensive to purchase, and more labor intensive to install. On a project with dozens and dozens of control valves, a considerable chunk of change can be saved by going with two-ways in lieu of three ways. Yet if you are not planning on installing a variable frequency drive on your system pump(s), you are faced with the prospect of deadheading the pump if demand is low and enough valves are closed down.

There are a couple of ways to handle this. A relatively simple method is to have at least a few three-way valves, strategically located throughout the piping system. With all valves in the system closed off to their respective coils, there would be at least a minimal flow in the system, through the three-way valves. The mechanical system designer would be best suited to inform as to what that minimum flow requirement is. Selection of the system pumps is performed by this individual, and is done using "pump curves"; graphs that illustrate the operating characteristics of a pump, and how pressure relates to flow. By looking at the curve for a system pump, the designer should be able to decipher the minimum flow rate that the pump can safely operate at. With this information, three-way valves can be selected, the quantity of which would allow for the required amount of "system bypass."

Another method of system bypass entails the use of a single control valve and a differential pressure controller. The control valve has a two-way valve body, and is installed across the supply and return mains of the piping system. The pressure controller operates the proportional control valve, in order to maintain proper system pressure. With demand high and two-way control valves open, system pressure remains within reason, and the system pressure control valve remains closed. As demand decreases and control valves close down, the system pressure tends to build. The controller's pressure transmitter measures system pressure and transmits this information to the controller, which in turn compares it with setpoint, and modulates the pressure valve open accordingly. Figure 16-5 demonstrates how this would be applied to a hot water piping/pumping system.

This pressure controller is of the very same variety that we spoke of back in the section on VFD pumping systems, when a VFD is used to maintain system pressure. Placement of the transmitter is the same, as well as the means of establishing and setting the setpoint. The only difference is that, instead of being wired to a VFD, the controller is wired to a control valve.

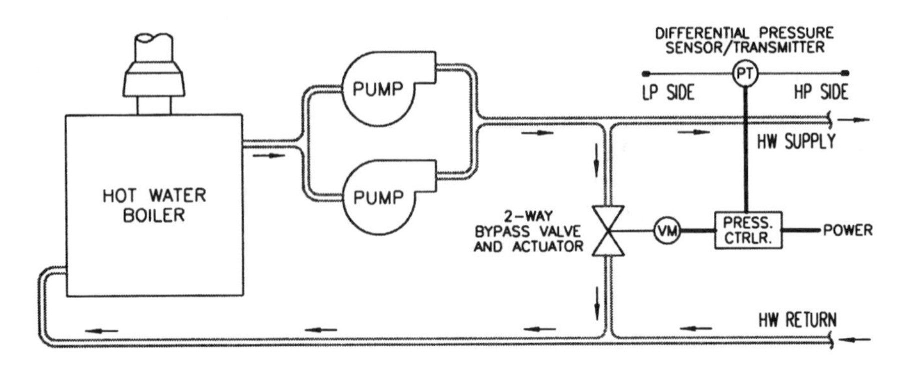

Figure 16-5. Pressure bypass valve serving a hot water system. As pressure in the system builds, the pressure controller modulates open the bypass valve.

Valve Sizing, Selection, and Location

The control valve in this application is of critical concern. How to size it, where to put it, should it be spring return, and if so, normally open or normally closed? These questions must be answered, and for that a little bit of insight into the operation of the system is required.

How should this valve be sized? It's a proportional control application. How about that 3-5 psi rule that we discussed back in the section on control valve sizing? That doesn't apply here, as there is no single coil that we are sizing the valve for. What about sizing it for system pressure? Sounds good. Size the valve for the pressure that we want to maintain in the system. Yet, do we have a good enough idea up front of what that will be? Or is it, at best, an educated guess that will hopefully be validated at the end of the project? What about diversity? Will the system load ever be such that every single two-way valve in the entire system is closed at the same time? Will there be diversity in the system, such that the spaces served throughout the facility will vary (with respect to each other) in required capacity throughout the course of an entire day? If so, should the pressure bypass valve still be sized for the full system gpm, or perhaps for somewhat less than the full gpm? What criteria should be used, in terms of valve sizing, that will ensure a properly selected valve, up front?

To size the valve, we need to know two things: pressure (psi) and flow (gpm). With these two values known, we can plug them into the formula relating these two variables, solve for Cv, and make our valve selection. As a rule, size the valve for system pressure and for half the system flow. Deviate from this rule if insight into the system directs you to do so. With the unknown aspects of the system and how it will operate when all is said and done at hand here, it's difficult to "dial in" up front on a perfect valve selection. An undersized pressure bypass valve will result in higher than normal system pressures when there is little or no demand. An oversized valve will result in poor control, the evidence of which would be fluctuating system pressures. With a little bit of insight, either predicament can be avoided, and a control valve can be selected that will perform adequately for the application.

Where should the valve be located? Does it matter? Differing opinions on placement of the pressure bypass valve suggest that location of the valve can vary. Practically speaking, it should go between the supply and return mains, at a point where the mains can handle the flow that the valve is sized for. It wouldn't make much sense to locate the bypass valve down toward the end of the mains, where the piping has been reduced to a size that can't properly handle the presumed maximum flow through the valve. For example, if you size a bypass valve to handle up to 100 gpm, you wouldn't want to install it across the supply and return mains at a point where the size of the mains can't handle 100 gpm. Makes sense. So closer to the pumps, right? Yeah, where design flow in the mains will be greater than (maximum) design flow through the valve. How about right at the pumps? Well, put a little distance between the pumping station and the bypass valve. In fact, the more distance you put between the pumps and the valve, the closer you get to the location of the pressure controller and transmitter, generally speaking. Which makes good sense, at least from a wiring and commissioning standpoint. Your limitation on how close you can get to the pressure controller/transmitter is again the size of the supply and return mains that the bypass valve will be teed into.

Should the valve actuator be of the spring return variety? In other words, upon a loss of power to the valve actuator, should the valve spring return to a "fail-safe" position? If so, what position should that be? Open or closed? There are no "default" answers to any of these questions. The control system designer must evaluate the situation, mainly on a job-to-job basis, and decide the proper course of selection. What would drive the need for spring return operation? If the valve is selected as spring return, normally open, then upon failure, the valve would open. From a safety standpoint, this may be desirable. Yet if there's a valve failure, it may go undetected unless or until system capacity suffers as a result of it. If the valve is selected to spring shut upon failure, system capacity won't suffer, but excessive pressures may ensue, causing the pump to deadhead.

The control valve and pressure controller will often end up being installed in a location that is not readily accessible, perhaps above the ceiling grid somewhere, in some all but random location. This is fine for the initial setup of the controller, for it's pretty much a "set and forget" control system. Yet if the location of this valve and controller isn't documented, it may be forgotten about altogether. Improper operation of the system as a whole, due to a failure of the control valve, may be difficult to detect at best, and may be misdiagnosed unless the service technician has knowledge of the valve and of its location. Consider a "forgotten" bypass valve in a hot water application, that has failed in the beginning of the heating season. If it has failed closed, excessive pressures would ensue during the early weeks of the heating season, when most of the two-way control valves would be shut. This may never even be detected, or perhaps only after the excessive pressures accelerated the development of another problem. If the bypass valve fails open, the problem would only be discovered upon reaching periods of heavy system demand. The equipment utilizing the hot water would not get their scheduled gpms, and the spaces served would not get satisfied. Cold zones lead to a service call, and the problem is diagnosed, hopefully correctly. Yet without knowledge of that bypass valve, it sure makes things more difficult.

Pressure operated system bypass valves do have their place in HVAC applications. For one, they are an extremely inexpensive method of system pressure control, at least in terms of the initial implementation of it, as compared to the alternative (VFDs). They are fairly simple to install and set up, and perform well if sized correctly and located appropriately. Disadvantages of handling system pressure control utilizing a bypass valve have generally been outlined herein, and are worth considering when applying such a control scheme. The listing of these "potential" disadvantages in the past few paragraphs is not meant to "dissuade" the reader from implementing this course of action. Rather, it is to serve the purpose of demonstrating the importance of proper selection, maintenance, and documentation of such a system.

Chapter 17

BOILERS

The basic gas-fired boiler is a packaged piece of equipment that utilizes natural gas to produce hot water or steam. Electric boilers exist as well; we won't concern ourselves with these types of boilers in this chapter. We will concern ourselves with the types of gas-fired hot water and steam boilers that are in popular use in today's heating applications.

While we won't get too much into the packaged controls of boilers, suffice it to say that the factory control systems of gas-fired boilers are pretty critical, to acceptable operation, and to safety. When you consider that a boiler utilizes natural gas to heat up water, you have to realize just how critical the control systems that govern the process are. From the ignition module to the safety controls, a boiler control system is designed with safety in mind. For the most part (with the exception of the operating controls), a boiler control system shall be left alone, the way it came out of the factory. Leave boiler controls to the pros. Tampering with the factory package will, at best, void the equipment warranty, and at worst, jeopardize the safety of the individuals in and around the facility being served by the boiler.

Whether hot water or steam, all boilers essentially have the same "utility" connections. First and perhaps most obvious is the gas connection. Second is the "make-up water" connection. This is simply a water line connected into the boiler system, to maintain the proper volume of water in the system. The third utility connection to a boiler is the electricity. Of course gas-fired boilers don't heat water with electricity, but it is still required, for the processes that govern the safe operation of the boiler. The final connections to a boiler, though not considered a "utility," are the mains, whether they be the hot water supply and return lines of a hot water boiler, or the steam supply and condensate return lines of a steam boiler.

Hot water and steam boilers are, for the most part, pretty similar in terms of operating characteristics. Both operate to heat water via natural gas. The real difference between the operation of hot water and steam boilers is the "variable" in which they operate to maintain. Whereas with hot water boilers the controlled variable is hot water temperature, with steam boilers the controlled variable is, let's see, uhmm… steam pressure! Seriously though, the difference between these two variables poses some unique operating characteristics of each type of boiler, some of which we will explore in the upcoming sections. But in the simplest of terms, all we are really doing any different with a steam boiler is heating the water to a greater extent, with no regard for temperature, bringing it up to a boil, and maintaining the pressure of the generated steam.

Before we break into discussions about boiler systems, we can still talk in general terms in regard to boiler operating and high limit controls. An **operating control** is simply the boiler's main on-board controller, whether it be a temperature controller or a pressure controller. Smaller boilers will usually have simple single-stage operating controls. Upon a call for operation, the boiler's operating control makes, and the boiler proceeds to run through its ignition cycle. When the operating control becomes satisfied and breaks, the boiler ceases to fire. And the cycle continues…

Larger boilers will most certainly have more than a single stage of control, and can even be equipped with proportional firing rate control. When a boiler is equipped with two stages of control, it is customary to refer to the stages as "low fire" and "high fire." The two stages of control are typically accomplished in the same manner as is done with the gas-fired heating section of a packaged rooftop unit: with a two-stage gas valve.

When a boiler is equipped with proportional firing rate controls, a proportional controller will exist (in addition to the operating control), that operates a modulating burner. Upon a call for boiler operation by the operating control, the boiler will fire, at a rate governed by the proportional firing rate controller. Often a switch and a potentiometer are provided on the cover of the burner cabinet, allowing the operator to switch over to a fixed firing rate, as established via the potentiometer.

The operating control and the proportional controller are to be set in conjunction with each other, as follows. The operating control shall be set so that it makes when the proportional controller has modulated up to the burner's ignition start point, or minimum firing rate. Huh? Okay, the burner of a typical boiler traditionally has some minimum firing rate, of which the manufacturer recommends that it should not be fired below. For our sake, let's call it 20 percent. So the operating control should be set to make once the proportional controller has modulated to its 20 percent point. Looking at this another way, assume that the hot water temperature is at a point higher than the setting of both controllers, and is on its way down. The boiler is idle, at least for now. At some point, the hot water temperature drops down into the throttling range of the proportional firing rate controller, and the controller begins to modulate. Yet the burner is disabled, because the operating control has not yet made. The hot water temperature continues to fall, and the proportional controller continues to modulate up through its band of control. Once this controller reaches the burner's minimum allowable firing rate, the burner may be allowed to operate. It is good practice to set the operating control at this point (or lower), so as to ensure that the proportional controller does not try to fire the burner at a rate below its (manufacturer recommended) minimum.

All boilers will be equipped with a high limit controller, normally of the manual reset variety. The high limit controller is adjusted a bit higher than the operating controller, and serves as a safety device, shutting down the operation of the boiler if the high limit setpoint is exceeded. Other safety devices may be furnished as well, such as high and low gas pressure controls. All of these safety devices will typically be wired in series, such that if any device trips, the boiler is disabled. This series connection of safety devices, which also includes the operating control, is commonly referred to as the **limit circuit**.

Other controls that are furnished as part of the package run the range from flame safeguard controls to the ignition module itself, which is typically an electronic device that oversees the ignition process, and ensures that ignition does indeed take place when commanded to. The factory furnished control systems of packaged boilers are wired, tested, and validated prior to shipment. For the most part, a boiler is factory equipped with everything that it needs to operate. So why this chapter? Good question. This book is all about "engineered systems." How do boiler systems fit into this category? By the second section of this chapter, the answer to that should become apparent. So read on, as we explore the different methods of boiler control when part of a mechanical system.

HOT WATER BOILERS—SINGLE BOILER SYSTEMS

In the simplest form, a hot water system consists of a boiler, a system pump (or two), and heating equipment that utilizes hot water. Not much to it. The pump runs continuously, circulating water from the boiler out to the heating equipment, and back to the boiler. The boiler operates, via its own on-board operating control, to maintain hot water temperature setpoint. What, if anything, can be added to this system by way of controls?

External Controls

A simple method of interlocking boiler operation to overall system operation, as illustrated in Figure 17-1, is to install a water flow switch in the supply main. The system pump(s) will be commanded to operate via some occurrence, whether it be via an outside air temperature controller, or perhaps by a manual switch. When placed into operation, flow is established, the flow switch makes, and the boiler is enabled for operation. If flow is interrupted, the flow switch breaks, and the boiler is disabled. For "single loop" hot water systems, the boiler really should only operate when there is flow in the mains. The system pump, once commanded to run, operates in a continuous manner. In other words, the pump doesn't "cycle." Once it has been determined that the hot water system needs to be placed into operation, the system pump will start and operate continuously, and the boiler will be enabled for operation, via the "flow switch interlock."

How about a time clock? With a typical HVAC system utilizing hot water, the facility will normally have occupied and unoccupied modes.

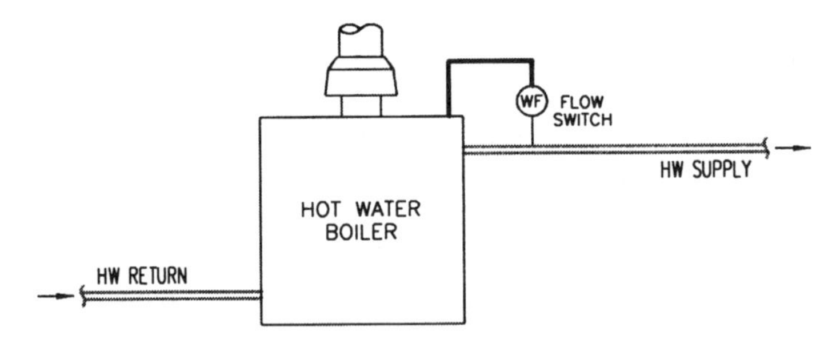

Figure 17-1. Single boiler serving a hot water system, operated via its integral, factory furnished control system. Flow switch is wired into the control system, to prevent boiler operation unless flow is present.

In an effort to conserve on energy usage, why not incorporate a "night setback" cycle for the boiler based on time-of-day, just as we do with the equipment utilizing the hot water? This is generally not a recommended practice (note: author's humble opinion), for more than one reason. First, the notion of saving energy by setting back the boiler's operating setpoint during unoccupied modes is, to an extent, a fallacy (again, author's opinion). If the heating equipment being served by the hot water system has night setback capabilities, then it follows that this equipment will be operating via reduced setpoints during the unoccupied modes. Hot water consumption is less, and the boiler works less to maintain hot water temperature setpoint. Why set back that setpoint? Leave the hot water temperature setpoint alone. The energy savings is realized by the fact that the equipment utilizing the hot water uses less of it during unoccupied modes, when their own operating setpoints are set back. The boiler simply "idles" during the unoccupied modes to maintain its operating setpoint, whether it's set back or not.

Another way of looking at it is as follows: If you don't set back the operating setpoint of the boiler, then upon entering the unoccupied mode, the boiler continues to operate via its setpoint, only it needs to work much less at doing it. If, on the other hand, we allowed the boiler's operating setpoint to be set back during unoccupied modes, then upon entering the unoccupied mode, the boiler would shut off. At least temporarily, until the hot water temperature fell to the night setback setpoint. So that's where the energy savings are, right? Okay, maybe so. So the hot water temperature falls to its night setback setpoint, and the boiler proceeds to cycle to maintain this reduced setpoint. What happens when the system transfers back to the occupied mode? Now the boiler's operating setpoint reverts back to the occupied mode setpoint, which is quite a bit higher than the unoccupied mode setpoint. The boiler goes to full fire, in an effort to bring the water temperature up to occupied mode setpoint. So much of that energy saved by allowing the hot water temperature to drift down and be maintained at an unoccupied mode setpoint is being used now to bring that water temperature back up. If we never allow the hot water temperature to be set back, then upon entering the occupied mode, we are basically at setpoint, and the boiler doesn't have to "recover," so to speak.

So what's better? To incorporate night setback operation into the hot water temperature control, or to just leave the boiler's operating setpoint alone? You can argue that there is perhaps a minimal savings realized by setting back the hot water setpoint and having the boiler operate via this reduced value. The truth of the matter is, doing this is simply not good for boilers. Starting up a boiler is rough, on the boiler. The temperature rise incurred upon startup causes the mechanical components of the boiler (and the materials of which they are composed of) to undergo a certain degree of stress. Frequent start-ups can shorten the life span of a typical boiler. Same goes with instituting night setback cycles on a boiler. Allowing the boiler to "cool down" (during night setback), and placing it back into full fire (upon transition to the occupied mode), is hard on a boiler. On top of that, many boilers are not even physically rated to operate at temperature setpoints of much less than 150 degrees.

Alternative Control Schemes

What about reset control? Can a hot water boiler be equipped with a temperature reset controller, so that the setpoint can be reset downward as the outside air temperature rises? First, recalling from our discussion on reset control from a prior chapter, implementation of hot water reset control based on outside air temperature is more of a control issue than one of energy savings. It is desirable, from a controllability standpoint, to reset the temperature of the water supplied to the equipment utilizing it, as a function of outside air temperature. Flowing 130-degree water to a heating appliance on a 60-degree day is a much better match to the heating load, as opposed to flowing 180-degree water to the appliance. The control device operating the heating appliance stands a much better chance of precisely maintaining setpoint, when the temperature of the water flowing to it is matched to the heating requirements of the space served by the appliance.

So reset control is good then, at least for purposes of control. Then it should be acceptable to apply reset control directly to a hot water boiler, right? Not necessarily. Caution must be exercised here. A paragraph ago we stated that many boilers aren't rated to operate at reduced setpoints. Typical hot water temperature setpoint is 180 degrees. A boiler's operating specifications must be evaluated by the designer. The boiler's minimum operating point must be verified, and this must be incorporated into the reset calculation. For instance, if a particular boiler's minimum operating point is 150 degrees, and reset control is applied to the boiler, the reset schedule would need to look something like this:

O.A Temperature:	–10	60
How Water Setpoint:	180	150

So as the outside air temperature rises from –10 degrees to 60 degrees, the hot water temperature setpoint is reset downward from 180 degrees to 150 degrees. Boiler reset control can be specified as a factory option. The boiler manufacturer can provide a reset controller as the operating control, in lieu of a standard temperature controller. Reset control can also be provided "after the fact," by simply replacing the boiler's on-board operating control with a reset controller. There is additional labor required for either option; the outside air sensor has to be installed and wired back to the reset controller, whether the controller is factory or field provided.

In the above example of reset control, we only get about thirty degrees of reset through the entire range of outside air temperatures. That because of the limitations on the boiler's operating point. Doesn't seem like a very worthwhile implementation of reset control. Another method of hot water temperature reset control is to utilize a three-way proportional control valve (Figure 17-2). The boiler itself operates to maintain a constant hot water temperature setpoint, via its on-board operating control. The three-way valve is a mixing valve configured such that the pump pulls water from the common port of the valve. Port "A" of the valve is piped to the supply side of the boiler, and port "B" is "teed-in" to the return. So the pump can pull varying amounts from both

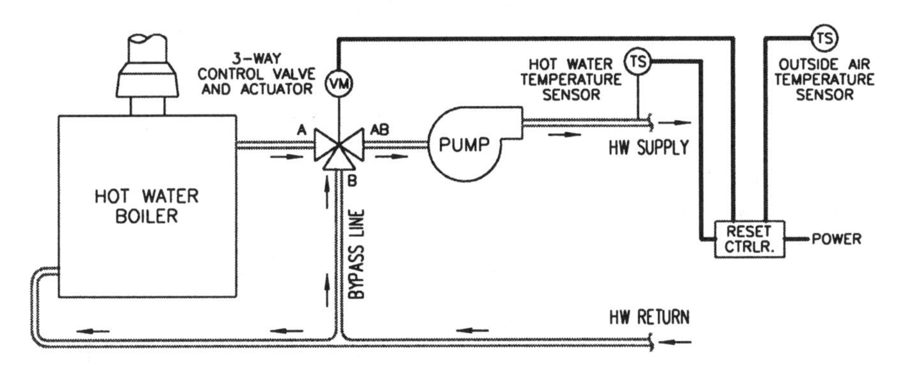

Figure 17-2. Single boiler serving a hot water system, operated via its integral, factory furnished control system. Reset control is implemented with a three-way control valve and an electronic reset temperature controller.

ports, and deliver the mixture to the heating equipment.

A reset controller, with its hot water temperature sensor located in the supply piping after the pump, proportionally controls the three-way valve. Reset control is accomplished as follows: when the outside air temperature is at its extreme (say –10 degrees), the system hot water temperature setpoint is 180 degrees. The boiler is operating on its own to maintain a boiler water temperature setpoint of 180 degrees, via its operating control. The system hot water temperature setpoint is maintained by the reset controller, by modulation of the control valve. If the system hot water temperature setpoint is 180 degrees, then the reset controller will position the valve to pull all of the water from port "A"; in other words, straight from the boiler. Now, as the outside air temperature rises, the system hot water temperature setpoint will decrease. The reset controller continues to operate the control valve in order to maintain system hot water setpoint, but now the setpoint is somewhat less than the boiler's setpoint of 180 degrees. Therefore, the controller must position the control valve to allow some "return water" to bypass the boiler and mix with the boiler supply water. As the system hot water temperature setpoint is reset downwards, the controller must position the valve to allow more and more water to bypass the boiler and mix with the 180-degree water provided by the boiler. On mild days, the system setpoint may even be such that the control valve is positioned to allow the water to fully bypass the boiler, perhaps repositioning every once in a while so as to "inject" a small amount of boiler water into the loop. In these scenarios, the boiler "idles," cycling on and off every so often in order to maintain the 180-degree setpoint as established at its operating control.

Performing reset control with a three-way control valve as described above is a commonly accepted practice. In doing so, it allows the boiler to retain its recommended minimum operating point, and also allows for a much wider band of reset. With this method, it's not unthinkable to implement reset down to 120 degrees, or even less. Some engineers may have a problem with the boiler being allowed to operate with little or no flow through it, which happens when the control valve is positioned to allow substantial bypass around the boiler. Some manufacturers may even require constant flow through their boiler. If this is the case, then often an additional pump will be specified, to run continuously and provide constant, minimal flow through the boiler.

Another common hot water application is "waterside zoning." As an example, consider the following system: one boiler and six zone pumps, with each pump serving a heating zone, perhaps baseboard radiation. Each pump is controlled by a thermostat located in the zone that it serves. Upon a drop in temperature below zone setpoint, the thermostat makes and energizes its respective zone pump, to flow hot water through its respective section of baseboard. When the temperature in the zone rises back above setpoint, the thermostat de-energizes the zone pump. In this manner, all six zones receive individual zone temperature control, from one central boiler. This type of control is more of a residential thing, though it is seen often enough in commercial applications.

For such a system, the boiler must continuously maintain hot water temperature setpoint, even when all zone pumps are off. An alternative method to controlling a boiler in such a system is to only control to setpoint if (at least) one of the zone pumps is running, and control to a "low limit" setpoint if all pumps are off. This particular method of control can be accomplished by a controller specifically designed for this type of application, or can be implemented via a digital controller configured to perform the control scheme. The controller will typically have a flow switch input, so it can know when any pump is running. The flow switch must be installed in the common supply header, before the zone pump takeoffs, in order for it to be able to determine flow via any single pump. If flow is present, the controller, which replaces the boiler's on-board operating control, will fire the boiler as required to maintain operating setpoint, say, 180 degrees. When flow is not present, the controller will operate the boiler to maintain a reduced "low limit" setpoint of around 150 degrees. This low limit setpoint really should not be any lower, for as we recall from a previous paragraph, it's not good for the boiler to be allowed to cool down too far, only to be pressed into full operation again, and so on. The apparent benefit to this type of control is that the boiler doesn't try to maintain a 180-degree hot water temperature when there is no flow through it, yet still maintains a high enough temperature during periods of no flow, so that recovery is not very harsh.

Concluding this section, we can say that, for the simplest of hot water systems, there is one boiler, and normally a pair of system pumps. Temperature control is performed by the boiler's on-board operating control, unless replaced by an external controller. A simple means of interlocking the operation of a boiler with the system pump(s) is to install a flow switch and wire it into the boiler's limit circuit, so that when flow is established, the boiler is allowed to operate. If the boiler manufacturer requires constant flow through his boiler, then this should be addressed, either with the control system design, or by adding a pump that continuously circulates water through the boiler. Firing a boiler with no flow through it isn't in itself a bad thing, as long as the manufacturer is okay with it, and is sometimes necessary, depending upon the type of application. More importantly for boiler operation and life span is the requirement that once a boiler is enabled to operate, it should stay enabled, at least for a considerable amount of time, and not be allowed to constantly cool down and heat up.

Got all of that? Good, because now we get into the real nitty-gritty on hot water system control. Specifically, when there is more than one boiler serving the same system. Sound simple? You will be surprised at how much there is to know on this subject, and how much can go wrong when controls are inappropriately applied, or "unapplied," as the case may be.

Hot Water Boilers— Systems with Multiple Boilers

When more than one boiler serves a common hot water system, it is often a good idea to control the boilers from one central controller, rather than allowing the boilers to operate from their on-board operating controls. Figure 17-3 illustrates a system of two boilers operated by a central boiler controller. Boiler "sequencers" can be purchased as a manufactured product, or can be implemented utilizing electronic or microprocessor-based controllers. The store-bought versions, albeit expensive, typically offer quite a few features, such as, giving the ability of either manually or automatically alternating the lead boiler, outside air reset control, single point outside air temperature lockout, and single point system shutdown. The engineered versions would have to be designed to incorporate any features required for the given application. A boiler sequencing system utilizing an electronic temperature controller as the basic "building block" of the design would have to be supported with other discreet components, such as relays and maybe a time clock, in order to provide the necessary features. A boiler sequencer implemented with a digital controller would have to accommodate the required input and output points, and be programmed to perform the desired sequence of operation.

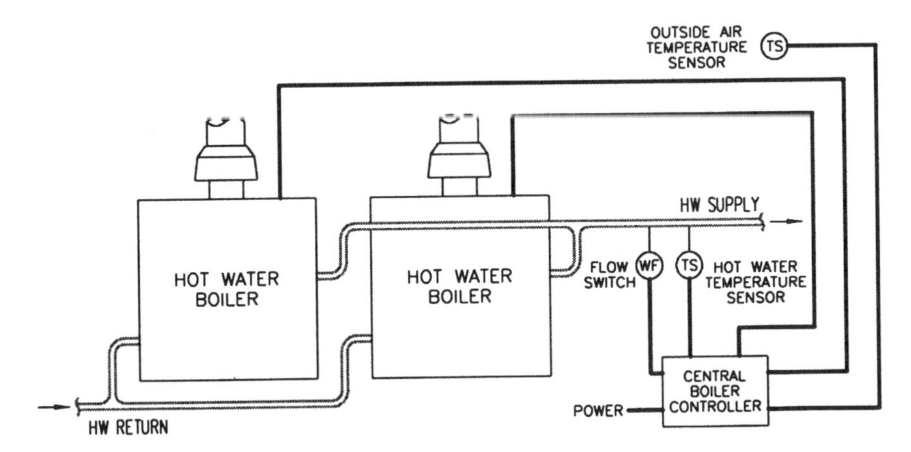

Figure 17-3. Two boilers serving a common hot water system, controlled (to some extent) by an outboard central boiler controller.

The Case for Central Boiler Control

A typical boiler sequencing controller will minimally have an input for a hot water temperature sensor (that gets installed in the supply piping common to both boilers), and outputs for boiler control. In addition, the sequencer may have an input for an outside air temperature sensor (for reset control, and/or for enable/disable based upon outside air temperature), and perhaps another input to land a flow switch.

Hot water temperature setpoint is established via the sequencing controller. For staged boilers (single-stage or low/high fire boilers), the boiler control portion of the sequencer can be a simple multistage temperature controller, that will fire boiler stages as a function of the deviation in hot water temperature from setpoint. In other words, the further the hot water temperature is below setpoint, the more stages are active. For boilers with modulating burners, the boiler controller should have modulating capability, so that the "firing rate" of the boilers can be varied by the controller itself, taking over for the boilers' onboard firing rate controllers.

So why would it be better to control multiple boilers, serving a common hot water system, by a single controller, in lieu of simply allowing them to operate via their factory installed/wired operating controls? There are actually several reasons, most of which relate to operator maintenance issues. The main advantage of having multiple boilers controlled from a single controller is that there is only one setpoint. Without a boiler sequencer, for instance, two boilers serving a common hot water system would have to be set up so that one boiler (lead boiler) operates to maintain one setpoint, while the other boiler (lag boiler) operates to maintain a slightly lower setpoint. Thus, if the lead boiler cannot maintain its setpoint, the lag boiler will operate as well.

This method of "staggering" the setpoints of the boilers' operating controls is, at least in theory, quite acceptable. It's simple staged control of multiple boilers. However, in practice, this method of control leaves a lot to be desired. First, the operating controls of each boiler need to be set up appropriately, and thus not fiddled with. If the operating controls of the boilers are not staggered, and each set to maintain the same hot water temperature setpoint, then the boilers will tend to cycle rather randomly, each coming on and off, in an unpredictable manner, to maintain their individual setpoints. From an operation and maintenance standpoint, this is not a very desirable mode of operation. However, if the setpoints are staggered, then the lead boiler will predictably be the first to come on, and the last to turn off.

Now, in order to perform any type of "lead boiler alternation," the boilers' operating setpoints would need to be physically changed. In other words, the operating control of the lead boiler would need to turned down, and conversely, the operating control of the lag boiler would need to be turned up. In essence, the boilers' setpoints need to be swapped, for lead/lag alternation. This is a manual procedure required to be performed by the maintenance personnel, on a regular basis, if lead/lag alternation is desired. Otherwise, the boiler that is chosen to be the lead boiler, with its higher setpoint, will always be the first to fire and the last to turn off. With this type of setup, the lead boiler pretty much operates for the entire heating season, as the lag boiler sits idle for the milder periods of time, coming into operation only when the system heating load exceeds the lead boiler's available capacity.

With a boiler sequencer controlling the firing cycles of the boilers, the requirement for messing with each individual boiler's on-board operating controls goes away, as the hot water setpoint is set via the sequencer. The sequencer can operate the boilers as required to maintain this single adjustable setpoint. Automatic (and manual) lead boiler alternation can also be implemented, depending on the capability of the sequencer.

Application of a Central Controller

Applying a boiler sequencer to multiple single-stage boilers is straightforward. The number of stages that the sequencer is required to be capable of controlling is simply the number of boilers. This is staged control in its simplest form, with each boiler serving as a stage of heat. For multiple two-stage (low/high fire) boilers, the number of sequencing stages must be double the number of boilers. In this application, the sequencer will step through its stages of control, as a function of hot water temperature and setpoint, thus stepping through the low and high fire stage of each boiler, and moving on to the next boiler, as the hot water temperature deviates from setpoint. Alternatively, the sequencer can step through the low fire stages of each boiler, before moving on to the high fire stages, thus assuring more equal run times between the boilers.

In applying a boiler sequencer to boilers with modulating burners, it is important to take over the duties of both the operating control and the proportional firing rate controller of each boiler. In other words the sequencer should have an output to replace the operating control, and an output to replace the firing rate controller, per boiler. By taking over these functions with the sequencer, an array of different, specialized control schemes can be realized. With a smart controller, optimal "staged proportional control" can be established, with the appropriate algorithms. For instance, with a smart sequencer, a subsequent stage of boiler operation can be brought on before the previous stage modulates to full fire. This allows both boilers to operate at more efficient operating points, rather than have one boiler operating at full fire, and the other operating at its minimum firing rate. Short-cycling of stages can also be minimized with this type of sequencer, by instituting the proper control algorithm.

When a boiler sequencer is utilized to control the operation of multiple boilers, the boilers' on-board firing rate controllers and/or operating controls should remain wired. This is so that the boilers can be operated by their own controls, should a problem arise with the sequencer. With firing rate controllers, a switch should be provided (sometimes part of the factory wiring package), that can allow the boilers' firing rates to be controlled by either the boiler sequencer or by the boilers' on-board controllers. With operating controls, they can remain in the limit circuit, in series with the boiler sequencer's operating stages, and be set high enough so that they are always calling for operation, with the added function of serving as a secondary "automatic reset" high limit temperature controller.

Boiler sequencers are often specified and are quite functional, if properly applied. The store-bought version is readily available, in both the staged and the proportional varieties. Of course, the control systems designer with "higher end" capabilities may choose to design his own boiler sequencer, with electrical/electronic or digital controls, and save the installation some money at the same time, by designing in only those features of which are needed or specified for the particular project.

When a hot water system is specified to be designed with both boiler and pump sequencing controls, the store-bought option becomes a bit more difficult to apply. The designer may opt for a store-bought version of both required sequencers; the boiler and the pump sequencer. Yet, the nature of these two stand-alone controllers may hinder the practical aspects of hot water system temperature control. It would be more beneficial, to overall system operation, and for maintenance reasons, to integrate the control of the boilers and the pumps into a single control system. The most feasible solution to this challenge is to control all aspects of hot water system operation by a single digital controller, one having enough inputs and outputs to accommodate all of the required temperature control points. Once installed and wired, the designer/programmer of the control system need only to configure the controller in the appropriate manner, to arrive at the desired end result of a fully integrated hot water temperature control system.

Cautions in Applying Central Control

Not all multiple boiler applications warrant a central boiler sequencer. In fact, certain applications inhibit the proper functional operation of a (store-bought) boiler controller. Consider a two-boiler hot water system, with two single-stage boilers, and a three-way control valve. The three-way valve performs the main temperature control of the hot water system, typically hot water reset control based on outside air temperature. The boiler plant operates to maintain a constant "supply header setpoint," as sensed prior to the control valve. The valve operates to maintain system, or "loop" setpoint, by mixing various amounts of boiler supply water with amounts of bypassed boiler return water. As the loop hot water temperature rises toward and through setpoint, the valve bypasses more and more water around the boilers. When the loop temperature gets far enough above setpoint, the three-way valve is positioned to bypass the full amount of water around the boiler plant. Thus, in this instance, no water flows through the boilers.

When the system is significantly loaded, the temperature control performed by the three-way valve will be such that the valve is positioned to always allow some water to flow through the boilers. However, when the system is under light load conditions, the valve will most often be in a position of full bypass. This is especially true with reset control, if the outside air temperature is high enough to cause the loop setpoint to be substantially lower than the "supply header setpoint." As an example, consider the boilers to be operating to maintain a 180-degree supply header setpoint, and the outside air temperature to be warm enough such that the loop setpoint is 130 degrees. The water returning from the system must mix with the 180-degree boiler water to maintain the 130-degree loop setpoint. Given this scenario, the valve will likely operate in such a way as to "inject" a small amount of boiler water into the loop, overshoot loop setpoint, and subsequently revert back to the full bypass position, until the loop temperature once again falls back toward setpoint. In these conditions, the valve will spend the majority of the time in the full bypass position, with no water flowing through the boiler plant.

A boiler sequencer will not always work well in this application. Remember that the hot water temperature sensor wired back to the sequencer must be installed in the hot water supply piping, common to both boilers. For the controller to even stand a chance of working properly, the sensor must be located before the three-way valve. When the system is significantly loaded, there will always be some flow through the boiler plant, and thus across the hot water sensor. With flow through the boilers and across the sensor, the sensor can accurately measure the boiler common supply water temperature, and therefore the sequencer can adequately perform boiler temperature control.

During periods of "no flow" through the boiler plant, the temperature sensor cannot accurately sense the temperature of the boiler water. The water temperature at the sensor will tend to drop, thus initiating a call for boiler operation. At least one of the two boilers will fire. With no flow from the boilers to the sensing point, the water temperature at the sensor will likely remain below setpoint, thus continuing the call for boiler operation. Meanwhile, the temperature at the boilers (and at the boilers' on-board controls and limits) continues to rise, eventually rising to the setpoint of the (operating) boiler's high limit controller. If the high limit controller is an automatic reset device, then the boiler proceeds to "cycle" on this controller. If the high limit controller is a manual reset device, then the boiler is shut down completely.

With this type of application, it may be better to simply allow the boilers to operate via their own controls. The main disadvantage of this is that there are multiple setpoints (as discussed previously). There are a couple of feasible methods, however, to make a central boiler controller work in this scenario.

The first method would require that the controller know when there is little or no flow through the boiler plant. This can be accomplished via a flow switch installed in the common header, or even via an end switch on the control valve, set to open when the valve is in a position of substantial bypass.

The boilers' operating controls need to be left in place, and set to maintain a low limit temperature setpoint, say around 150 degrees. The output stages of the boiler controller need to be wired in parallel with the boilers' operating controls. Now, when there is any amount of flow through the boilers, the boiler controller is allowed to operate the boiler plant, to maintain setpoint (180 degrees). This of course accomplished via the output stages of the controller. Since the water temperature at each boiler is in excess of the low limit setpoints (by quite a bit!), the boilers' operating controls are open, and remain that way, for as long as the boiler controller is in control of the plant.

If flow through the boiler plant drops off to the "critical" point, as determined by either the flow switch or by the three-way valve's end switch (choose your method), then the boiler controller is disabled, and control is relegated to the boilers' on-board operating controls, which are set at 150 degrees.

With this method of control, the dilemma of overheating the boiler plant is resolved. With no flow through the plant, the boilers simply operate to maintain their on-board low limit setpoints. Once flow is restored through the plant, the boiler controller again takes over, and controls the boilers to the central operating setpoint of 180 degrees.

This scheme of control may briefly beg the question "Why not simply disable the boilers during periods of no flow?" Of course, we remember from earlier, allowing a boiler to cool down and fire up repeatedly is tough on the poor guy. At least by implementing this control scheme, we limit how far the boiler plant can cool down to, so that recovery is not as dramatic when ample flow through the plant is re-established.

There are some cautions and concerns with this method of control. First, boiler control cannot be disabled by simply turning off the central controller, for the boilers will continue to operate via their own controls. There needs to be an additional "enable/disable" interlock between the central controller and the boiler plant. This can be handled by installing a relay (or two) whose coil is energized when the boiler controller is on, and de-energized when the boiler controller is off. The normally open contacts of the relay(s) can be wired into the limit circuits of the boilers, or even directly in series with the boilers' main on-off switches.

Another concern is that the above description of control focuses on single-stage boilers; boilers with single-stage operating controls. For boilers with two stages of control, this application becomes a bit more challenging, for there are twice as many stages to wire in. Not to say that it won't work or can't be done. Just a little more complicated, that's all.

It should be noted that this type of application is not *readily* adaptable to boilers with modulating burners and proportional firing rate controllers. With staged control, the preceding descriptions are quite feasible, yet with proportional control, this method of boiler plant control is not practicable. However, it is still possible to apply a central boiler controller to such a system, at least for the function of lead/lag control. The boiler controller can be a simple two-stage temperature controller (one stage per boiler), with some additional controls components to provide manual or automatic alternation of the lead boiler. Here's how this works.

Each boiler is set to maintain the same setpoint (180 degrees, for instance). A temperature sensor is installed in the supply piping, common to both boilers and before the three-way valve. The sensor is wired back to the controller. The controller only enables and disables the boilers, with the boilers' on-board controls performing temperature control. The setpoint of the controller can be set at 185 degrees for the first stage, with an "interstage differential" of ten degrees, so that the second stage is 175 degrees.

With any amount of flow through the boilers, the temperature sensed at the controller's sensor will be an accurate measurement of boiler plant temperature. Upon initial startup, the hot water temperature will be below the controller's first and second stage setpoints, and both boilers will be allowed to fire. Both boilers operate (at full capacity), and the water temperature rises toward the boilers' setpoints (each at 180 degrees). When the temperature reaches 175 degrees, the lag boiler is disabled. The lead boiler continues to operate. Under lighter loads, the lead boiler will reach setpoint, and turn off. The lead boiler will then proceed to "cycle" on its operating control, to maintain its 180-degree setpoint. The lead boiler will remain enabled, as long as the water temperature sensed by the controller's sensor does not exceed 185 degrees.

So the lag boiler remains disabled, as long as the water temperature sensed by the controller's sensor does not drop below 175 degrees. As the system load increases, the lead boiler tends to remain in continuous operation, and it's firing rate increases toward maximum. As the load approaches and exceeds the lead boiler's capacity, the hot water temperature decreases toward 175 degrees. Once it drops below 175, the lag boiler is allowed to operate, and proceeds to fire via its own controls. This boiler's firing rate approaches maximum, resulting in both boilers operating at or near full capacity. With both boilers in full operation, the capacity of the boiler plant should be such that the temperature of the water sensed by the controller's sensor will rise back up above 175 degrees (plus differential). When this occurs, the lag boiler once again is disabled. Thus, under heavy load conditions, the lead boiler is in constant operation, and the lag boiler "cycles" on the boiler controller.

Under light load conditions, when there is no flow through the boiler plant, the temperature at the controller's sensor will likely tend to drop, thus possibly calling for the operation of both boilers. The boilers will both be allowed to operate, via their on-board controls. In this scenario, the boilers will each "cycle" on their own operating controls. Since the boiler controller is simply enabling and disabling boiler operation, and not actually controlling the boilers, the water temperature will not be allowed to reach the setpoint of the boilers' high limit temperature controllers.

In systems with three-way temperature control valves as discussed herein, a manufactured boiler sequencing controller may not work properly, installed right out of the box. With such a system, specifying central boiler control is tricky, and designing the controls for such a system is challenging at best. With boilers having modulating burners, application of a store-bought controller, with full modulating capabilities, may be downright prohibitive in such an application. Perhaps the best that can be done in this situation (within the realm of practicality) is to provide a simple controller, set up for the sole purpose of manual and/or automatic lead/lag boiler alternation, as discussed. So what do you do if the engineer specifies a boiler sequencer for a system having multiple proportional type boilers, and a three-way control valve? Time to raise the red flag! Honestly, unless you can take full control of the boilers and the valve by a common microprocessor-based controller, and incorporate the required "smarts" into it, the scenario should really be brought to the light of the engineer, who should then decide which course of action to take.

Steam Boilers— Single and Multiple Boiler Systems

Ah, good old steam! Nothing like it! While utilizing steam as a means of temperature control may be more challenging than utilizing hot water, there are definite benefits to going with steam. First, the potential for freeze-up is much less than with hot water, which makes steam a favorable choice for make-up air systems. It's commonly utilized as a means of humidification, in which a boiler produces steam and distributes it to steam-utilizing humidifiers. Perhaps the most attractive benefit to steam is that it needs no pump! Really! The steam is basically "self-transmitting," going wherever the pipes lead it to (provided there's enough pressure). Of course, getting the condensate back to the boiler is another thing. Normally though, this is done with a relatively small pump, or even by gravity.

Single Boiler Systems

When a single steam boiler serves a system, there is no reason not to use the boiler's factory installed operating controls. Externally controlling a single steam boiler offers little benefit, if any. Reset control is not an option; you cannot apply outside air reset control to a steam boiler. In other words, you cannot reset the steam pressure downward as the outside air temperature increases. A boiler shall maintain a constant operating pressure, and shall not be varied or reset. The only reason to take over control of the firing rate of the boiler, whether staged or proportional, is if the facility were equipped with a Building Automation System, and the customer and/or the engineer had some unconventional reason to have the BAS directly control the boiler. Otherwise, leave the boiler's operating controls alone, and simply impose "supervisory control" on the boiler.

Okay, got it. Don't mess with the boiler's on-board controller(s). Simply allow the single boiler to operate the way it was designed to, and only, if anything, impose supervisory control. One problem. What the heck is supervisory control? Fancy term, huh? And one that's been mentioned several times thus far, without any explanation as to its meaning. Supervisory control simply means that you are telling something what it can and can't do, and when. For example, telling a boiler that it can operate, and telling it when it can't. The functions accomplished by the boiler are done with the boiler's integral control systems, and not by some external system. The external system only "supervises" boiler operation, enabling it and disabling it as per some governing set of rules. Way, way back in Chapter 2, we brought up the term "hierarchy of control." This in essence is what is meant by that term. Since boilers are packaged equipment, when a networked Building Automation System is implemented for temperature control, often enough a boiler will be left alone by the control system, with the exception of being enabled and disabled, and possibly being monitored in a couple of different ways. This is especially true with single boiler systems, for with multiple boiler systems, often the networked digital control system will be taken advantage of, by having it perform the duties of boiler sequencing.

Okay, back to steam boilers. What else is required for proper control of a packaged steam boiler? The typical steam boiler will be equipped with a combination pump controller/low water cutoff. This control is typically shipped with the boiler, for field installation by the pipefitters. Control-wise, it simply monitors the water level in the boiler, and energizes a field mounted pump if the level gets too low. The pump in question is normally part of a package referred to as a boiler feed unit. This unit also consists of a tank that accepts condensate returning from the system, and accepts a make-up water connection as well. When the feed pump is called upon to operate by the pump controller, it pumps water from the feed tank to the boiler. If the water in the feed tank ever drops too low, it is simply made up via the make-up water connection.

The boiler feed unit must be provided with power. In addition, there needs to be a control interlock between the feed unit and the pump controller. The pump controller serves another purpose, as implied by its full name. It also serves as a low water cutoff. If the water in the steam boiler falls to a critical level, the low water cutoff, whose operating contacts are wired into the limit circuit, will shut down boiler operation, until the water level in the boiler rises back up to an acceptable level.

Not much more to talk about here, in regard to single boiler steam systems. Oh yeah. Condensate return. How does the steam, after it's utilized and turned back to water, get back to the boiler? In smaller systems, this may be accomplished simply by gravity, where the condensate line is pitched all the way back to the boiler. In larger systems, however, or in systems in which gravity return simply cannot be implemented, a condensate pump is needed, to pump condensate from the steam-utilizing equipment back to the boiler. Normally located remotely from the boiler, this is a packaged piece of equipment consisting of the pump, a holding/receiving tank for the condensate, and a level control to operate the pump. The pump must be provided with power. However, there is no requirement for an interlock between the boiler and the condensate pump package. The pump simply starts when the condensate tank becomes full, and stops when the tank is emptied out.

Systems with Multiple Boilers

Steam systems served by multiple boilers (Figure 17-4) are generally recommended to be controlled by a single sequencer, perhaps more so than as is the case with hot water boilers. Perhaps the critical nature of steam boiler operation lends itself to be more rightfully controlled by a single system. It's definitely more applicable, compared to hot water boiler systems, simply because of the properties of the controlled medium. Whereas hot water temperature is dependent upon flow and sensing location, steam pressure is generally not as dependent upon these variables. Furthermore, there is no three-way valve to throw a monkey wrench into the proper operation of the sequencer. It's true that the steam pressure sensor should at least be in close proximity to the boilers, in the common header. The sensor monitors steam pressure at this point, transmits the pressure signal to the sequencer, and the sequencer does its thing to maintain pressure setpoint, as set via the sequencer. There's not too much of an opportunity for misapplication here, when equipping a steam boiler plant with a boiler sequencer.

When two steam boilers serve the same system, each boiler will have a combination pump controller/low water cutoff. The pump controller portion of these devices will typically need to be wired back to a "duplex" boiler feed unit. This is essentially the same thing as the feed unit we discussed earlier, except that there are two pumps instead of one, with each pump controlled by either boiler.

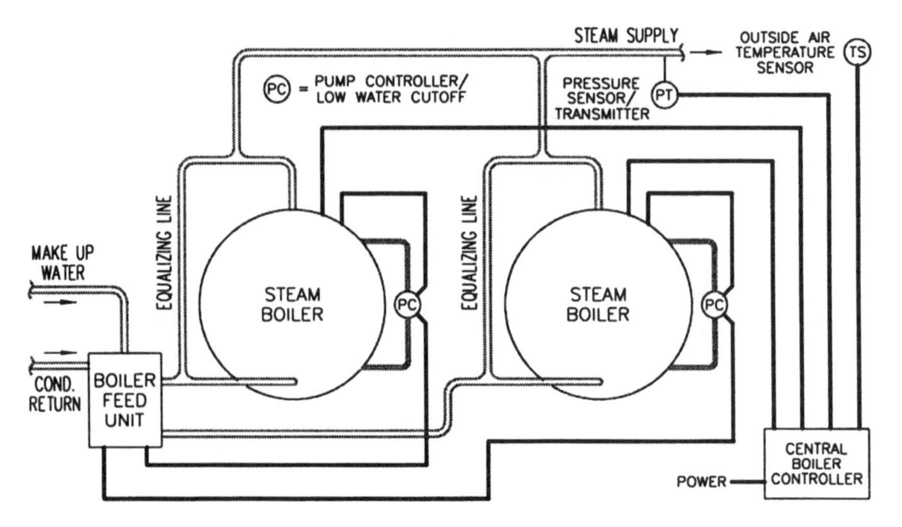

Figure 17-4. Two boilers serving a common steam system, controlled by an outboard central boiler controller.

Combustion Air

The requirement of combustion air for indoor gas-fired heating appliances such as boilers and furnaces is one that is code-mandated, but more importantly, one that is mandated by the operational nature of the gas-fired equipment. For combustion to take place, there must be three components present. The first two, fuel and ignition, are provided via the equipment. The third, which is oxygen, must be introduced into the space housing the equipment at a rate dictated by the size and capacity of the gas-fired equipment. An insufficient amount of combustion air will result in reduced combustion, in which the flame is actually choked down due to the lack of oxygen. The resulting situation is undesirable, for two reasons: reduced heating capacity, and perhaps more importantly, compromised operational safety!

To meet the requirements for combustion air in a typical boiler room, a hole (or two) will be cut through the outside wall of the room, and terminated with a louver and normally closed motorized damper. Operation is preferably such that the damper is closed when there is no equipment being fired, commanded open when there is a call for boiler operation, and proven open before actually allowing boiler operation.

Another means of handling combustion air requirements, one that's normally reserved for the larger scale boiler plants, is to have a "forced air" ventilation system. This is typically in the form of a 100 percent outside air unit (make-up air unit). Operation of the unit, again preferably, is such that the unit is off when no equipment is being fired, commanded to run upon a call for boiler operation, and proven to be running before combustion is allowed to take place.

With either method of providing combustion air, we see that the "preference" is to confirm combustion air before allowing boiler operation. With the simple combustion air damper, this is done via an end switch on the damper actuator, set to close when the damper has stroked substantially open. With the forced air system, this will typically be done by confirming airflow, via an air proving switch, or by confirming fan operation, via a current sensing switch.

When a mechanical system, be it hot water or steam, is served by a single boiler, interlocking the boiler and combustion air is a relatively straightforward task. The trick is to intercept the boiler's operating control, and use it as a means to energize the combustion air damper (or the make-up air unit). Wait a minute, we're not done yet. We then need to get the "means of proof" into the boiler's limit circuit, after the operating control. Now, when the operating control calls for boiler operation, the combustion air system is placed into operation, but the boiler doesn't fire right away, because its limit circuit is not quite made. Once the operation of the combustion air system is confirmed, the proving switch makes, and the boiler's limit circuit is completed. And away we go!

If, for whatever reason, the means of providing combustion air fails, the boiler is prohibited from operating. This is normally considered "the lesser of two evils." In other words, a better scenario than allowing the boiler to fire without sufficient combustion air. At least it will prompt a maintenance inspection or a service call, and the source of the failure will be discovered and rectified.

When a mechanical system is served by more than one boiler, and there is only a single source of combustion air common to the boilers, interlocking boiler and combustion air is not so straightforward. To demonstrate the relative complexity of this scenario, consider a two-boiler system, with each boiler operating via its own operating control. The preference, remember, is to prove combustion air before allowing combustion to take place.

First, we need to determine whether either boiler is calling for operation. Therefore, we need to monitor the status of each boiler's operating control. This can be done with a relay at each boiler. The coil of the relay is energized when its respective boiler is calling for operation. The normally open contacts of each relay can be wired in parallel, and the combustion air source can be called for by this parallel connection of contacts. In other words, if either contact closes, the combustion air system will be placed into operation. Now, when combustion air is proven, the "proof" must be in the form of not one but two switches, with each switch wired into each boiler's limit circuit. Recapping, on a call for boiler operation by either boiler, a field mounted relay local to the calling boiler is energized. The contacts close, thus pressing the combustion air system into operation. Once combustion air is proven, two switches or contacts are made, one in each boiler's limit circuit. At this point, both boilers are enabled for operation. The boiler whose operating control is calling proceeds through its ignition process, and fires.

So you can see how you need to perform some logical gymnastics with relays and switches, in order to prove combustion air before allowing boiler operation, when two (or more) boilers serve the same system and are not centrally controlled. What is the scenario when they are controlled by a central boiler sequencer? The logic would become much easier, if combustion air control could be incorporated into the sequencer. Upon a drop in temperature (or pressure) below setpoint, the sequencer will request the operation of the lead boiler. But the sequencer, if smart enough, can also send out a request for combustion air (via an output contact), and wait for confirmation (via an input), before allowing lead boiler operation.

Though on paper this looks to be an attractive and relatively straightforward approach to combustion air control, in practice it leaves a bit to be desired, and (in the opinion of this writer) shouldn't be considered as an option. The concern is what takes place if and when the sequencer breaks down or is taken out of service for some reason. Be that the case, the boilers will need to temporarily be controlled by their own operating controls, at least until the sequencer is pressed back into operation. What about combustion air during this period? If the sequencer has command over the operation and confirmation of the combustion air system, then what happens when the sequencer is taken out of service? At the very least, combustion air won't be initiated automatically. It can be implemented manually, by a maintenance attendant throwing a switch to get the damper open or to get the air handler running. However, the maintenance personnel must be made aware of this manual procedure, in order for the boilers to get the required combustion air. In the case of the simple damper, this raises concerns about the damper being open when neither boiler is firing, and cold outside air migrating into the boiler room for no reason.

Combustion air is a matter of operational safety, and the implementation of it should be integrated into the boilers' control systems, and not solely by any higher tier system that can be taken out of service and yet still allow the boilers to operate. In other words, the call for combustion air should be initiated whether it's the sequencer or the boilers' own operating controls calling for boiler operation. Furthermore, confirmation of combustion air should be wired into the boilers' limit circuits, and not back to a sequencer whose role in the operation of the boilers is subject to manual intervention.

Combustion air, whether provided simply by a motorized damper or by a forced air fan system, is a requirement in virtually all gas-fired heating plant applications. The interlock between heating plant and combustion air system is an important one, and one that must be addressed on a job-to-job basis. While it's a simpler task to fire the boiler and simultaneously send out a request for combustion air, most designers and engineers prefer that proof of combustion air be received before allowing heating plant operation.

Chapter 18

CHILLERS

Chillers come in many different sizes, types, and configurations. In HVAC, the main function of all chillers is the same: to chill water for air conditioning purposes. While there are variances due to design criteria and system requirements, the "default" water temperature that chillers are depended upon to maintain is 45 degrees. How they do that, and what they need in terms of outboard controls requirements, is the subject of this chapter.

Regardless of size or configuration, all chillers have at least a couple of things in common. First and perhaps most obvious is the chilled water piping connections. Chilled water is produced, pumped from the chiller out to the chilled water utilizing equipment, and returned to the chiller, just a bit warmer in temperature. Of course, how much warmer depends upon the demand for cooling by the equipment utilizing the chilled water; the more of a demand, the warmer it returns.

All chillers (that we consider here) utilize the principles of refrigeration to perform their function. As such, all require the basic components of the refrigeration cycle, namely, compressor(s), evaporator, and condenser. How these components are "brought together" to operate as one varies among chiller applications. A packaged air cooled chiller is a single piece of equipment that contains all of the required refrigeration components within itself, and is located outdoors. Split system chillers are air cooled chillers in which part of the refrigeration cycle is outdoors, and part of it is indoors, with refrigeration piping running between the indoor and outdoor equipment. A water cooled chiller is a packaged piece of equipment that is located indoors, requiring condenser water piping connections to it in addition to the chilled water connections. Application dictates the type of configuration best suited for the given project, and is evaluated on a job-to-job basis. While it is not the intent of this chapter to delve into the design criteria governing the selection and design of chiller systems, the upcoming sections will at least touch upon some of the possible reasons of going with one type of system as opposed to another.

Perhaps the final item that all chillers have in common is the requirement that water be flowing before chiller operation is allowed. In other words, any chiller, no matter what the size, type, or configuration, must have flow through it in order for it to operate. A chiller must not be allowed to work unless water is flowing (okay, point taken!). This is conventionally accomplished via a flow switch interlock to the chiller, although a differential pressure switch installed across the chilled water mains is increasingly becoming the preferred method of confirming flow.

With packaged microprocessor-based chillers, it is possible to confirm chilled water flow without the use of an external flow switch. Cool! How's that done? Well, first consider that a chiller with an integral microprocessor-based control system has access to many of the chiller's operating parameters, including chilled water supply and return water temperatures. The "smart chiller" will know when it is commanding its compressor(s) to operate, and will expect to read a "delta T" between the (factory installed) supply and return water temperature sensors. This would indicate that flow is present, cooling is taking place, and all is well. Were there no flow, then the supply and return water sensors would essentially register the same or similar values, with no temperature drop from return to supply. It is in this manner that an intelligent chiller control system can verify the presence or absence of flow. Even with this said, a flow switch is still often specified, for redundancy, by engineers and designers who simply feel more comfortable with a "hardware interlock."

Air Cooled Chillers

Air cooled chillers exist outdoors, in their entirety! All components required for the refrigeration cycle are integral to a single piece of equipment that is normally located either on grade or up on a roof. Refer to Figure 18-1 for the physical depiction of an air cooled chiller (along with some of its possible external controls components). The chiller consists of an air cooled condenser, an evaporator, and a compressor or two (or more). The condenser is the standard type; finned coil and condenser fans. The evaporator is a "shell" of which houses a "tube bundle" of refrigeration piping. Chilled water is pumped through the shell, and in the process rejects heat to the refrigerant flowing through the tube

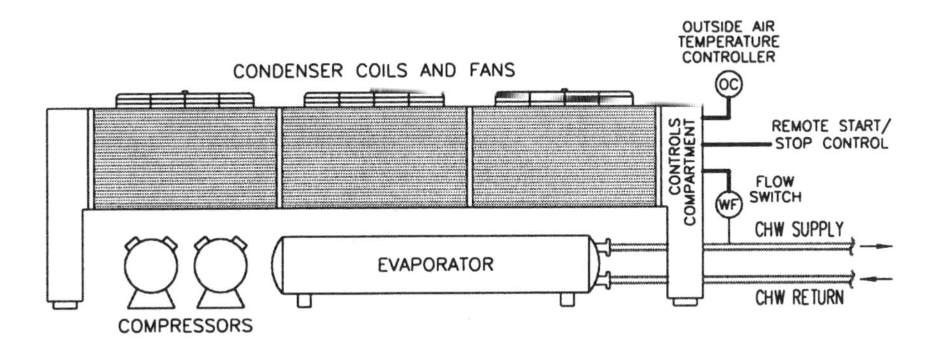

Figure 18-1. Air cooled chiller serving a chilled water system, operated via its integral, factory furnished control system. Flow switch wired into the control system prevents chiller operation unless flow is present. Outside air temperature controller and remote start/stop control interlock can provide additional means of chiller enable/disable.

bundle. The compressors, of course, enable the whole refrigeration process, by pumping refrigerant from the evaporator to the condenser, and so forth. For a concise description of the refrigeration cycle, please refer to the opening section of Chapter 8.

The chilled water is piped into the facility served, and distributed to the equipment. Depending upon the seasonal usage of the chiller and the climate, the chilled water mains that are exposed to the outdoors may require to be "heat traced." This generally takes the form of electrical resistance "tape" that is wrapped around the chilled water pipe. When given electrical power, the heat tape warms up, and prevents the chilled water pipes from freezing. Another means of preventing freeze conditions is to run the chiller with a solution of water and glycol (anti-freeze), which effectively lowers the freezing point of the medium. Adding glycol, however, also derates chiller efficiency. The third method of preventing freeze conditions is to simply drain the entire system during the winter months.

What are the reasons governing the decision to locate a chiller outdoors? The first, perhaps most common, is interior space restrictions. There may simply not be enough room to locate a rather large piece of equipment indoors. Another reason for going with an air cooled chiller is maintainability. With all components of the refrigeration cycle located within a single piece of equipment, maintenance and troubleshooting are simplified. Other reasons may have to do with first cost, operating costs, noise restrictions (compressors are loud, get 'em outside!), and simply user preference.

In a typical chilled water air conditioning application, a chiller will operate to maintain a leaving water temperature setpoint of 45 degrees. The 45-degree water is delivered to the cooling equipment, where heat is transferred from the spaces served by the equipment to the water. How much heat is transferred is a function of how much cooling is required in the zones served by the equipment. If no cooling is required for a particular zone, then no heat is rejected to the chilled water by the piece of equipment, and the water returns from the equipment at or close to the same temperature that it was delivered. This for example could be via a three-way valve in its fully bypassed position.

The more cooling required by the equipment, the more heat is rejected to the chilled water. Chillers are generally sized for a 10-degree delta T (maximum) through the chiller. If the chilled water temperature setpoint is 45 degrees, then when the water is returning to the chiller at 55 degrees, the chiller will be "fully loaded," meaning that it will be operating at its full capacity. This scenario occurs when all equipment utilizing chilled water for cooling purposes is calling for maximum cooling. The heat rejected from each piece of equipment to the chilled water is in the area of 10 degrees or so, such that the temperature of the full volume of water returning to the chiller is 10 degrees above the chilled water setpoint.

Operational Considerations

In the chapter on built up air handling units, we touched upon the airside and waterside equations for heat transfer. It is appropriate at this time to re-introduce the waterside member of that pair of equations. The equation is as follows:

$$Btuh = 500 \times gpm \times delta\ T$$

where gpm is the flow rate through the chiller, and delta T is the temperature difference of the water entering and leaving the chiller. Of course, by knowing any two of the three variables in the above formula, the third can be solved for. This equation, and its airside counterpart, are invaluable tools in the design of mechanical systems, and in the performance evaluation of such systems. The control system designer can gather quite a bit of insight into the design of a mechanical system, and acquire information as to how the system was designed to operate. With these tools, a designer can obtain a better feel for the mechanical system, and thus will be better equipped at providing the appropriate control system.

With that said, let's talk about how a packaged air cooled chiller is designed to operate. Given that the maximum delta T through a chiller is typically 10 degrees, the chiller might be equipped with more than just a couple of stages of control. Precise maintenance of chilled water temperature setpoint is pretty critical. With only two stages of control, the chiller doesn't stand much of a chance at precise temperature control. Four stages of control isn't bad (2.5 degrees per stage), and 6 stages is even better (1.6 degrees per stage). How the staging is accomplished with the mechanical equipment varies with the equipment. There may be one compressor for each stage of control, or the compressors may be equipped with unloaders, which allow them to operate at partial loads. Six stages of control could feasibly be accomplished with only two compressors, if each compressor were equipped with two unloaders.

How are these stages of chiller operation controlled? Well, remember that the chiller is a packaged piece of equipment, meaning that most, if not all, of the controls components required for proper operation are integral to the equipment. All chillers will come with their own leaving water temperature control system, whether it be microprocessor-based, or electrical/electronic in nature. For single chiller systems, rarely is there a need to disable this factory furnished control system and provide a separate outboard system of control. So for our air cooled chiller, the only requirement for outboard controls is that of the chilled water pump interlock. This of course being done with a flow switch that is wired into the chiller control circuit, enabling chiller operation only if chilled water flow is proven. For added safety and redundancy, some chiller manufacturers will recommend that a pump starter auxiliary contact be used in series with the flow switch. This would provide assurance of chiller disablement in the unlikely scenario of the flow switch failing closed and the pump being off.

Remote enablement and disablement of the chiller can be performed, if the application calls for it. An outside air temperature controller can be wired into the chiller control circuit, to prevent chiller operation when the outside air temperature is below a certain point (typically 50 degrees). Of course, unless equipped with low ambient controls, the chiller shouldn't be allowed to operate anyway if it's cold outside. A more common approach is to control the chilled water pump(s) with the controller, and have the chiller follow suit via the flow interlock.

Remote start/stop of a chiller, as described above, is commonly done when the entire HVAC system is controlled by a Building Automation System. Enabling and disabling of the chiller is performed via a binary output off of some digital controller. The chiller's temperature control system is normally left alone, unless the chiller manufacturer provides a controller that can accept a "remote setpoint." This is a very useful feature. The chilled water temperature setpoint is set via the BAS, and sent to the chiller control system in the form of an analog signal. The chiller operates via its own leaving water temperature controller, but to maintain this remotely provided setpoint. This gives the BAS user the ability of dialing in the desired chilled water setpoint through his "front end," yet leaves the critical temperature control process up to the factory furnished chiller control system.

Split System Chillers

The term "split system" has been mentioned several times thus far. The term was first introduced in Chapter 2, in the subsection on chillers, and its use was broadened in the chapter on built up air handling units. Now we explore the meaning of the term as it applies to chiller systems.

As we discussed in the previous section, a packaged air cooled chiller will have all of its refrigeration components located outdoors, in a single piece of equipment. Technically, an air cooled chiller may have one or more of its components located indoors, with the condenser at least being a part of the outdoor equipment. A chiller system with refrigeration components located both indoors and outdoors is classified as a split system, with refrigeration piping running between the indoor and outdoor equipment.

In order to explore the two unique configurations of split system chillers, let's start with the packaged air cooled chiller. For the first configuration, let's move the evaporator indoors. The condenser and compressors remain outdoors as part of the outdoor equipment, and refrigeration piping is run from the outdoor equipment inside to the evaporator. This type of chiller configuration is normally referred to as an "air cooled chiller with remote indoor evaporator," and is illustrated in Figure 18-2. The outdoor equipment, that which is referred to as "the chiller," is really nothing more than an air cooled condensing unit. Pretty much (mechanically) identical to that which would be used with a split system air handler. Of course the evaporator in this case does not cool air (as does a DX coil), but water. Since the evaporator is located indoors, all chilled water piping is also located indoors.

So why do something like this, as opposed to having a single packaged outdoor unit? Well, the biggest reason is what was mentioned in the immediately preceding paragraph; you get all of the chilled water piping indoors. Why is this preferable? Because you don't have to take steps to provide against freeze-up of the chilled water piping. No glycol, no heat tracing, and no need to drain the system during the down season. If the requirement is for year-round operation of the chilled water system, then this makes even more sense, especially in colder climates.

With this type of chilled water system, the chiller's control system will be located at the outdoor equipment. As with the packaged air cooled chiller, the control system governs all facets of chiller operation. At the indoor equipment, there needs to be at least a couple of controls

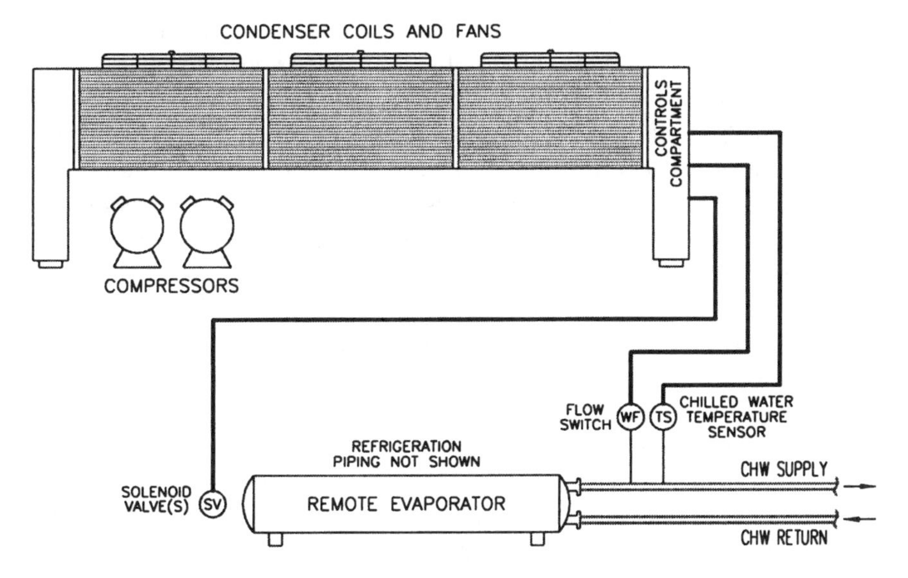

Figure 18-2. Chiller system consisting of an air cooled chiller and a remote indoor evaporator.

components. We had seen the need for refrigeration solenoid valves when we discussed split system air handlers. These valves were required to be installed local to the air handler's DX coil. The same applies here as well. In addition, the chilled water flow switch must be installed at the indoor equipment, in the chilled water piping leaving the evaporator. Finally, a chilled water supply temperature sensor needs to get into that chilled water piping as well (with a packaged air cooled unit, the sensor is local to the unit, and often factory mounted and wired).

The location of these components necessitates field wiring between the indoor and outdoor equipment. The chiller's control system is outside. In order for proper operation of the chiller, the control system must see, as inputs, flow confirmation (from the flow switch) and supply water temperature (from the sensor). In addition, the chiller's control system has command over the operation of the solenoid valve. The wiring between the indoor and outdoor equipment must account for these items.

For the second configuration of split system chillers, now picture the compressors coming indoors. With the compressors and evaporator indoors, all that's left outside is the condenser. Even the control system comes indoors! A chiller configuration of this type is often referred to as a "condenserless chiller with remote air cooled condenser," aptly depicted in Figure 18-3. It's interesting to note that the equipment referred to as "the chiller" is that which holds the compressors and the control system.

So why would this be considered, for a facility's chilled water system? First, as with the previous configuration discussed, the chilled water piping is indoors. That's a plus. In addition, the compressors are indoors, where they aren't subject to the wear and tear of "life in the outdoors." Serviceability is improved as well, for it's much easier to work on a compressor in a conditioned space, then it is on a rooftop when the temperature at the roofline is in excess of 100 degrees! Finally, with the control system indoors, maintainability is optimized; the maintenance personnel doesn't have to go to the outdoor equipment to make routine adjustments to the system.

Again chiller operation is governed by the chiller's control system, which is integral to the indoor unit. The chilled water supply temperature sensor is likely factory installed and wired, so you don't have to worry about that. The flow switch is to be installed right there at the indoor unit, in the chilled water supply piping, and must be wired back

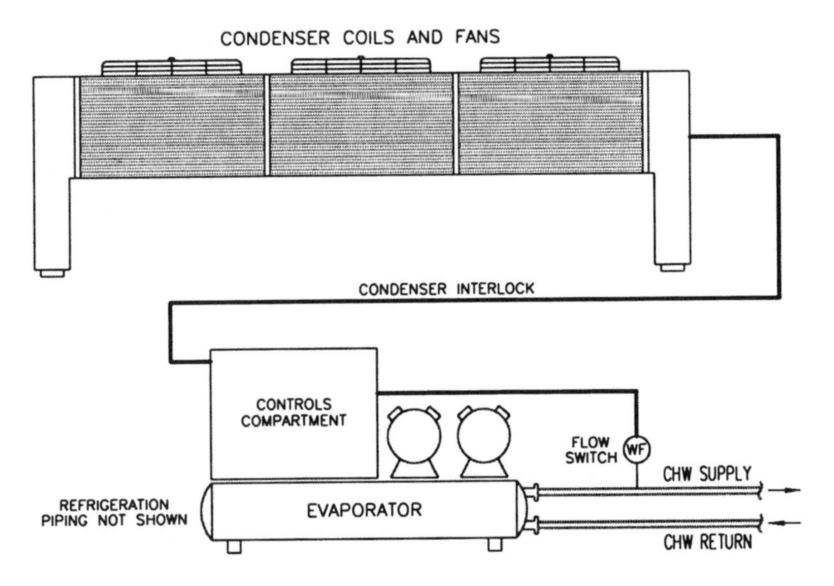

Figure 18-3. Chiller system consisting of an indoor chiller and a remote outdoor condenser.

to the chiller. Any solenoid valves required for chiller operation are undoubtedly factory piped and wired.

Field wiring must still be run between the indoor and outdoor equipment. The outdoor unit consists of nothing more than a coil and some condenser fans, yet the control system back at the indoor unit still must assume and maintain control over these fans.

This pretty much exhausts the topic of air cooled chillers. Hopefully this section has provided the reader with some insight into the different types of air cooled chillers, the reasons behind their applications, and how they operate and what they require in terms of external controls.

Water Cooled Chillers

A water cooled chiller (Figure 18-4) is a chiller that has all of its refrigeration components indoors, including the condenser. In fact, the major difference between air cooled and water cooled chillers is the type of condenser. Whereas with the air cooled chiller, the condenser is located outdoors in the form of a finned coil and fans, with a water cooled

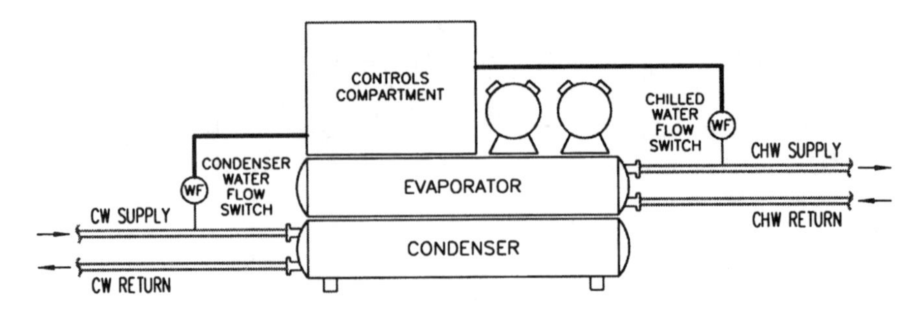

Figure 18-4. Water cooled chiller.

chiller, the condenser in indoors, an integral part of the chiller. The water cooled condenser takes the form of a shell and tube bundle, closely resembling the evaporator. In fact, upon first glance of a water cooled chiller, the casual observer probably can't tell the difference between the two barrels.

Heat rejection is performed by the condenser. Refrigerant runs through the shell, and is cooled by water that is pumped through the tube bundle. Incidentally, the evaporator of a water cooled chiller typically takes on the same form, with refrigerant in the shell and water in the tubes, which is opposite of how we described it for an air cooled chiller. Anyway, back to the condenser. The "condenser water" is carried off from the condenser to an outdoor piece of equipment called a "cooling tower." The cooling tower is primarily a big tank, open to the atmosphere. Condenser water enters the top of the tower, and is dispersed into droplets that fall to the bottom or sump of the tower. As the water droplets fall, they give off their heat to the ambient. The rejuvenated condenser water leaves at the bottom of the tower and is returned to the chiller to perform its function. The cooling tower also has a fan that will energize if the temperature of the condenser water leaving the tower is too warm. This is done by a local temperature controller at the tower, installed in the leaving condenser water piping. The purpose of the fan is to more rapidly reject heat from the water to the ambient.

Often a water cooled system such as this might call for a three-way valve up at the cooling tower. The purpose of this valve is to bypass condenser water around the tower, if the water returning to the chiller gets too cool. The valve is operated via a temperature controller installed in the condenser water piping returning to the chiller. This is simple proportional control; as the temperature of the water returning to the chiller drops to and through setpoint, the three-way valve modulates to allow more and more condenser water to bypass the cooling tower. The use of a three-way valve is more so a requirement in colder climates, and in year-round applications, where the cooling tower stands a chance of being "too good at its job."

One last thing about condenser water. From the last two paragraphs, we gather that the condenser water delivered from the cooling tower to the chiller can't be too warm or too cold. This is a requirement of the water cooled equipment (chiller) and its ability to efficiently and adequately perform its job. The temperature of the water delivered to the chiller's evaporator doesn't need to be "dialed in" to a precise setpoint. It just needs to be within a suitable range. The cooling tower's fan keeps it from exceeding the upper range, and the three-way valve prevents it from falling below the lower range. Condenser water temperatures are generally in the range of 70 to 90 degrees.

Water cooled chiller systems are costly, in terms of capital investment, as compared to air cooled systems. With the addition of a cooling tower and a pair of condenser pumps to the mix of components, the first cost of the equipment and piping labor is substantial, and the required maintenance is greater than it is with an air cooled system. Nevertheless, water cooled chiller systems are frequently installed, for many reasons. A good reason is that a condenser water system can serve more than one piece of HVAC equipment, and an overall HVAC system design for a building may require the use of other pieces of water cooled equipment, perhaps for computer room cooling applications. In this scenario, the condenser water system serves miscellaneous water cooled equipment in addition to the chiller. Other reasons for installing water cooled chiller systems have to do with overall system design and feasibility, physical and location concerns and limitations, efficiency, etc.

As with packaged air cooled chillers, water cooled chillers require minimal outboard controls components. The factory control system performs all facets of chiller operation, and the leaving water temperature sensor is likely factory installed and wired. Like with the air cooled chillers that we discussed in the previous sections, a chilled water flow switch, or similar interlock, needs to be wired into the chiller control system, so as to prevent chiller operation upon the absence of chilled water flow. Well, with a water cooled chiller, the same thing has to be done with, you guessed it, the condenser water side! The chiller must not operate unless there is condenser water flow to and through its condenser. Were there no flow, the chiller would have no means of rejecting heat, refrigeration pressure would build, and the chiller would trip out on its factory installed refrigerant pressure switch.

Remote start/stop can be performed as well, just as it can with any of the air cooled chiller systems discussed previously.

Multiple Chiller Systems

Like when more than one boiler serves a common hot water system, multiple chillers serving a common chilled water system should be controlled from a central controller (Figure 18-5), for all of the same reasons offered back in the chapter on boilers. As with boilers, chiller sequencers can be purchased for this task, or can be designed using electronic or digital controls. Chiller sequencing is pretty critical, more so than boiler sequencing. The role of the central controller is an important one in multiple chiller systems, and care should be taken in the design of the controller and of its functions, should the designer choose not to opt for the store-bought version. A couple reasons for this are now discussed.

The Critical Nature of Chiller Control

Maintaining precise chilled water temperature setpoint is important. With hot water systems, it's not uncommon to control hot water temperature to within a range of 10 degrees or more. Consider a two-boiler system, each with two stages of control. The first stage of control would be to maintain setpoint, say, 184 degrees. The second stage might be offset from the first stage by 4 degrees, the third stage by 8 degrees, and the fourth stage by 12 degrees. Thus, when fully loaded, the boiler system actually controls to the fourth stage setpoint, which is in the area of 172 degrees (184 – 12). With chilled water temperature control, the "band of control" must be much narrower, for a couple of reasons.

When a hot water system serves a hot water coil in an air handling unit, the temperature of the air passing through the coil is not expected to approach the temperature of the water flowing through the coil, at least not too closely. On the other hand, when a chilled water system serves a chilled water coil in an air handler, the temperature of the air passing through the coil is expected to approach the temperature of the water flowing through the coil, pretty closely. There's not much room for

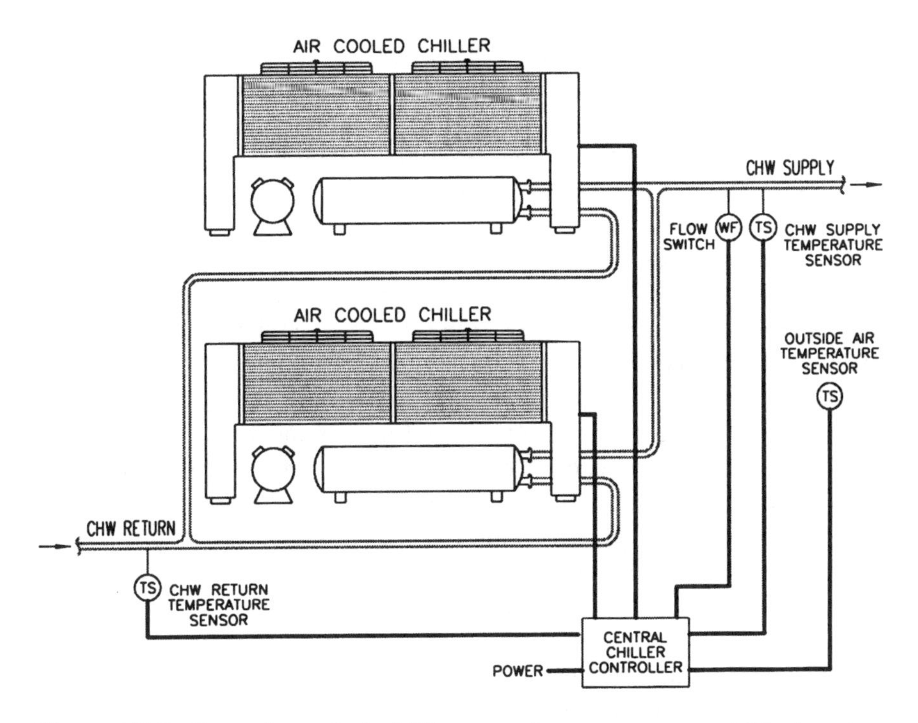

Figure 18-5. Two air cooled chillers serving a common chilled water system, controlled by an outboard central chiller controller.

play here. If the temperature of the chilled water is on the high end, the temperature of the air served by the chilled water coil might not be where it needs to be, in order to provide adequate comfort cooling and dehumidification. The chilled water temperature typically needs to be maintained at 45 degrees, and not much higher than that.

If the chilled water temperature is on the low end, other problems can ensue. First, operating chillers at temperatures lower than what they are designed to operate at results in reduced operating efficiencies. Second, and perhaps more importantly in terms of air handler operation, has to do with the requisite freezestat mounted to the leaving side of the chilled water coil. The purpose of the freezestat is to shut down air handler operation if the temperature sensed by the freezestat is below its setting, thus providing coil freeze protection. Freezestats are typically set at a "conservative" 40 degrees, to provide a good safety margin between setpoint and freezing point. If the water traveling through the chilled water coil approaches this temperature, this could result in "nuisance trips" of the freezestat, prompting the service personnel to lower the setpoint of the freezestat. Of course, doing that compromises the effectiveness of the device.

So the chilled water setpoint should be maintained as not too high and not too low. What is being asked for of the chiller is precise control to maintain setpoint. Yet prevention of excessive compressor cycling and wear and tear makes precise temperature setpoint control a challenge at best. This is one reason why it is more appropriate to let the factory installed chiller controller do its thing on single chiller systems, rather than trying to take over control of the chiller by some outboard means. In other words, leave the task of chiller control to the experts; the chiller manufacturers fit their chillers with control systems that suit their chillers very well, and you can't necessarily do any better with an outboard controller.

But what about these multiple chiller systems? For chilled water systems utilizing multiple chillers, there is added complexity in terms of staging and capacity loading. This is especially true with larger systems, systems with non-conventional piping configurations, and primary-secondary systems. For a system with any degree of added complexity, a microprocessor-based control system will most surely be specified, along with a proven, efficient control algorithm to be used for the process. For our discussions, we will consider the smaller, less complex systems, systems that are more traditional in nature, such as the common "two chillers in parallel" system that directly serves the chilled water utilizing equipment (single loop systems).

Central Control of Multiple Chillers

With two chillers serving the same chilled water system, the goal is to achieve and maintain a constant leaving water temperature setpoint, common to both chillers. Several methods are commonly employed for this task. The first is to simply stagger the setpoints of each chiller's factory controller. Set the lead chiller to maintain 45 degrees, and set the lag chiller to maintain 47 degrees. So that if the lead chiller can't maintain setpoint, the lag chiller will operate as well. Disadvantages of this method of control have been outlined in the corresponding section on multiple boiler control, and won't be restated here. Please refer to that section for a refresher.

Another method is to take "supervisory control" over the chillers. Set each chiller to maintain 45 degrees, yet allow only one chiller to operate if the load is light. How is this done? A temperature controller is required, along with some other components, depending on how fancy you want to get with things. The goal is to measure supply water temperature with the temperature controller, and enable the lag chiller if the temperature rises above some point. This type of control is crude at best, and does allow for some potential problems. An improvement to this scheme is to add a return water temperature sensor, and perhaps accomplish the scheme with a digital controller. Some intuitiveness can be incorporated into the scheme of control. For instance, determine when to enable the lag chiller by looking at supply water temperature, and determine when to disable it by looking at return water temperature. Build in some timing cycles, and you end up with mildly acceptable temperature control.

You don't necessarily want to take control over the chillers' stages with an outboard controller, as you would readily do with boilers. Not that you can't, but you must be careful. Remember that the chiller manufacturer equips his chiller with a very specialized temperature control system, perhaps even microprocessor-based. To break in to the chiller's factory control system for the purpose of taking over the staging duties is something that must be done with caution. With lower end chillers that do not have microprocessor-based control systems, this may not be that difficult, especially if the factory furnished controller is nothing more than an electrical or electronic sequencer. Decommission the sequencer aboard each chiller, and control the stages of chiller operation with a common outboard controller. Utilize a digital controller, and you can build in other features, such as automatic chiller lead/lag alternation, timing cycles, etc.

When chillers are equipped with microprocessor-based control systems, relegating the task of temperature control from the factory controller to an outboard controller is not a recommended practice, especially as per the manufacturer. Yet with microprocessor-based chillers, a special, extremely useful feature is often provided as a standard offering, and that's the ability of the on-board control system to accept an external setpoint signal. With this feature, multiple chillers serving a common chilled water system can very easily be controlled by a single controller, especially if the outboard controller is a digital controller. The controller can perform all of the typical functions of a central controller, such as lead/lag alternation, startup and shutdown based on external decisions, etc. In addition, the chilled water temperature setpoint can be established via the controller, and can be sent to the chillers. Actually, two setpoints need to be established, one to be sent to the lead chiller, and one to be sent to the lag chiller. The lead chiller receives actual chilled water temperature setpoint, and the lag chiller receives a setpoint somewhat higher, so that if the lead chiller can't handle the load, the lag chiller will begin to stage. Accomplishing multiple chiller operation with this method is extremely efficient, and really the best of both worlds. You get the centralized control that's preferred with multiple chiller systems, and you still allow the chillers' controllers to "do the dirty work" of compressor staging, timing cycles, etc.

An operational concern exists, when only one chiller is required to be in operation. The water flowing through the idle chiller is not conditioned, yet still mixes with water passing through the operating chiller. The result is that the leaving water temperature sensor, installed common to both chillers, will register the temperature not of the operating chiller's leaving water, but of the mixture of chilled and unchilled water. The control algorithm must take this scenario into account, by calculating what the lead chiller's setpoint should be, in order for the mixture of chilled and unchilled water to be at "system" setpoint. In other words, the calculation must result in a value sent to the lead chiller of less than the actual desired chilled water temperature setpoint. For instance, if the desired setpoint is 45 degrees, then the setpoint of the lead chiller must be less than 45 degrees, the value based on what the temperature of the water is passing through the idle chiller. The higher the temperature of water through the idle chiller, the lower the setpoint that is sent to the operating chiller, so that the result is a common leaving water temperature of 45 degrees.

A digital controller has the ability to perform such a calculation, and the designer should have the skills to build this into the operation of the controller. Another way to approach this is to install two leaving water temperature sensors, one per chiller. However, this approach doesn't relieve the programmer of performing a calculation within the control algorithm. A calculation must still be performed, the nature of it being a little different than what was discussed above.

The above discussions on chiller control hopefully have given the reader some idea of the critical nature of chilled water temperature control. Tight operating bands, equipment cycling concerns, and control methods all contribute to the complexity of multiple chiller control. This section also hopefully has served to show, however, that control of multiple chillers from a central controller is not only feasible, but also can be quite precise and efficient, especially if the chillers have factory control systems that lend themselves to being interfaced with an outboard controller. This is especially true of chillers that can accept external setpoint signals. With the continuing evolution of microprocessor-based controls and control systems, this should become the trend among chiller manufacturers, with chillers having this capability becoming the norm rather than the exception.

Refrigerant Monitoring/Ventilation Systems

We can't leave this chapter without discussing the requirement for and the applications of refrigerant monitoring and ventilation systems. In the event that any of the refrigeration components of a chilled water system are indoors, there may be a requirement to monitor the indoor equipment for refrigerant leaks. This seems to be more so a concern with the larger chiller systems, and not so much with the smaller systems. The requirement for such a monitoring system may be code related, specified by an engineer, or just good design practice on the part of the installing contractor (providing they can sell it!).

The concept is simple. Monitor the refrigerant level in the room that houses the chiller equipment, and upon detection of an excessive level, open a fresh air damper, turn on an exhaust fan, and enunciate the condition, with an audible or visual alarm (or both!).

Refer to Figure 18-6 for a complete monitoring/ventilation system. At the heart of the system is the refrigerant monitor, which is normally a stand-alone electronic device that can be configured to sense levels of refrigerant and affect the state of end devices based on the sensed levels. Refrigerant sensors must be installed in the space and near the refrigeration equipment, in a spot where refrigerant is likely to accumulate due to a leak. This typically means down low to the ground, driven by the fact that refrigerants are heavier than air, and they will tend to displace the air down at the floor level. The requirement for multiple sensors is dictated by the square foot area of the room, and is generally specified in the refrigerant monitor's supporting literature.

The monitor will have inputs to accept sensors, and will also have dry contact outputs, to drive end devices. The number of outputs varies, through typically there will be more than a single output, with each

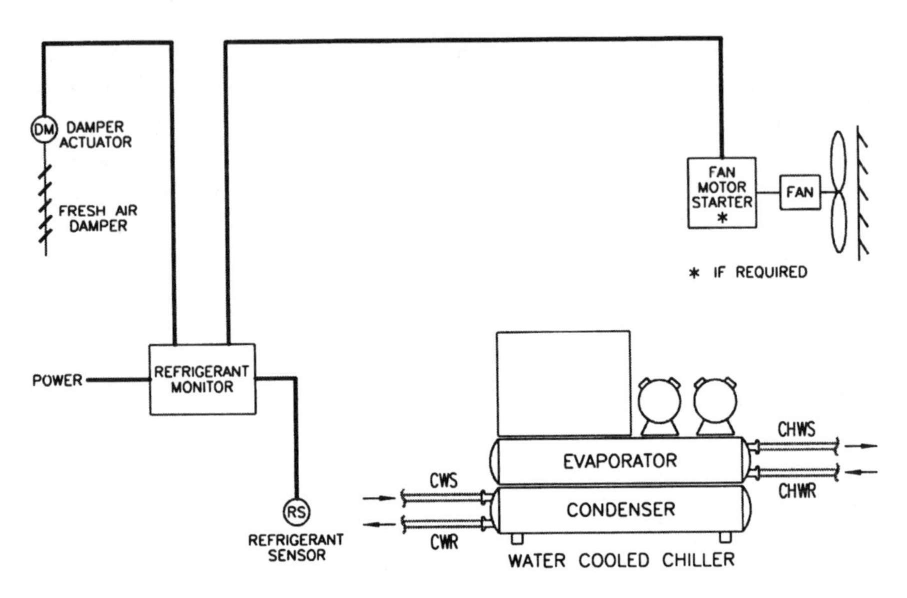

Figure 18-6. Refrigerant monitoring system consisting of a refrigerant monitor/sensor, an exhaust fan, and a motorized fresh air intake damper.

output being able to be configured to trip at a different user-defined level of refrigerant concentration. This allows the device to be used as a multi-level enunciator/controller. Perhaps upon reaching the first level of refrigerant concentration, the ventilation system is pressed into operation, and if the second level is reached, a horn is sounded and a light is lit.

The ventilation portion of the system consists of an intake damper and an exhaust fan. The damper is typically located in a wall opposite the fan. The fan is usually small enough to be single-phase, yet could be three-phase as well, depending upon the size of the space served, and upon the size of the refrigeration equipment. Operation is simple two-position control. Upon an increase of refrigerant concentration above the user-specified alarm level, the refrigerant monitor's output contact closes, energizing the damper and the fan at the same time. The notion of proving the damper open (via an end switch) before allowing fan operation is not valid in this application. Whether or not that damper strokes open, that fan better operate! We're talking harmful vapors here, and there are no prerequisites for that exhaust fan to run, other than the monitor telling it to. If the damper doesn't open, the room becomes negatively pressurized, which in this scenario is definitely the lesser of the two evils.

As far as the location of the monitor goes, common sense would dictate to have it outside of the room, so that it can be manipulated upon an alarm condition, without having to enter the room. Actually, the location of the monitor may not be all that important. Accepted practice is to locate the monitor within the room, and enunciate alarm conditions with a remote horn and/or light outside of the room. Another feature that may be required is to have a "horn silence" push-button, either local to the monitor or outside of the room as well.

Last, you may want to equip the monitoring/ventilation system with an external override switch. This switch allows the user to bypass the monitor's function and get the damper open and fan running by the flip of a switch. The switch will typically be located remotely, outside the room. It can be mounted local to the monitor, though the monitor itself may have the capability of being overridden through its own functions.

As stated above in the first part of this section, not every chiller application merits a monitoring/ventilation system. Obviously packaged air cooled chiller systems don't need it; the entire refrigeration cycle is outdoors! Other scenarios may lessen or "optionalize" the need for such a system, perhaps due to chiller size, room size, other ventilation, etc. Moreover, refrigerant monitors are store-bought items, and are relatively expensive equipment. Still, the use of such devices is often enough specified, and to say the least, it's good practice.

Chapter 19

HEAT EXCHANGERS

You might argue, after reading this chapter, that heat exchangers are primarily waterside equipment, and should be categorized as such, rather than falling under the catchall of Miscellaneous Systems and Equipment. To nip that argument in the bud, let's go way back to Chapter 2, where we initially define the term "heat exchanger." In that chapter, we defined a heat exchanger as a device that transfers heat from one medium to another. This general definition implies that the medium is not limited to water, but also can be air. This is true, and we iterated that in the same paragraph by qualifying the gas-fired heating section of a rooftop unit as a heat exchanger. Any coil, for that matter, can be considered a heat exchanger, with heat being exchanged between the medium flowing through the coil and the air passing over it.

The heat exchangers that we cover in this chapter will primarily be those that could be categorized as waterside equipment. We omit the air-to-air type heat exchangers that are normally part of packaged equipment. We also omit the air-to-water type heat exchangers that take the form of a heating or cooling coil, and focus on those types of heat exchangers that are component to a built up mechanical piping system.

Construction of such a heat exchanger will take one of two forms, as we also pointed out in Chapter 2. With either type of heat exchanger, there will be essentially four connections, two on the "primary side," and two on the "secondary side." The primary side is normally considered the "serving side" of the process, and the secondary side is considered the "side being served" by the process. For instance, with a steam-to-hot water heat exchanger, steam flows through the primary side, and water flows through the secondary side. The function of this type of heat exchanger is to give heat from the primary to the secondary side; to heat water with steam. Thus, the steam is *the server*, and the water is *the served*. As another example, consider a chilled water-to-chilled water heat exchanger. Pre-chilled water flows through the primary side, to chill the water flowing through the secondary side. The pre-chilled water is *the server*, and the secondary chilled water is *the served*.

It is these two types of heat exchangers (steam-to-water and water-to-water) that we will basically deal with in this chapter. Each section will briefly attempt to uncover the reasons for using such types of equipment, and will spend the rest of the time illustrating the methods of controlling the equipment. Rest assured, this is easy reading! Aside from a few good rules of thumb, there isn't a whole lot of complexity to this topic, and applying controls is as simple as abiding by the rules!

Steam-to-Water Heat Exchangers

Heat exchangers that convert steam to hot water are typically applied to projects where there is a need for both steam and hot water. An example of this would be an industrial facility, in which a steam boiler plant produces high pressure steam for process purposes. For HVAC purposes, a steam line is pulled off the boiler plant, reduced in pressure, and fed to a heat exchanger. The HVAC equipment utilizes the hot water for heating purposes. Another common application for steam-to-hot water heat exchangers is retrofit work, when an existing steam boiler plant is considered for the purpose of serving additional, new, hot water HVAC equipment.

When a heat exchanger is used to convert steam to hot water, it will take the form of "shell and tube" construction. The shell is basically a tank, of which the interior houses a network of piping. Steam enters the tank at the top, and exits as condensate from the bottom (via a steam trap). Water is pumped through the inside network of piping. As the water passes though the network, it is heated up by the steam. The steam gives up some of its heat to the water, and turns to condensate. Refer to Figure 19-1 for a visual of the typical shell and tube heat exchanger and of how it is generally equipped for such an application.

The rate at which the steam gives up heat to the water passing through the heat exchanger is regulated by a steam control valve, installed in the steam line prior to the shell. A temperature sensor is installed in the hot water piping, on the leaving side of the heat exchanger. The sensor conveys its information regarding hot water temperature to

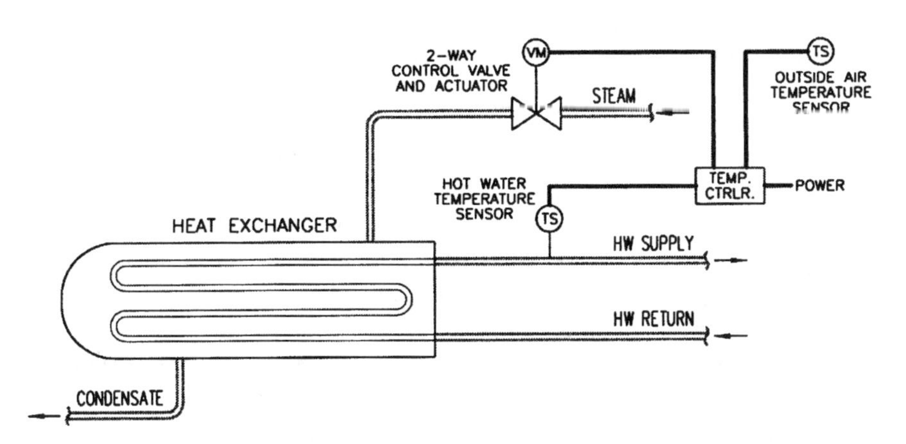

Figure 19-1. Shell and tube heat exchanger (steam-to-hot water). Temperature control is performed by an electronic or microprocessor-based controller operating a proportional steam control valve. Outside air sensor is required if the controller is to perform reset control.

a controller, which in turn modulates the proportional steam control valve to maintain hot water temperature setpoint, as set via the controller. Simple proportional control as we've come to know and love by this time!

General Control Considerations

The same concerns that apply to other steam control applications apply here as well. If the steam capacity is relatively small, only one control valve need be applied. However, if large enough, the steam capacity may dictate the use of two valves, sized for 1/3 and 2/3 capacity. The rule of thumb, again, for 1/3, 2/3 control is that if you require a steam control (globe) valve of larger than 2", then go with two valves.

When we talked about using steam as a means of controlling discharge air temperature, we saw that it was easier to let the steam run wild through the steam coil, and modulate face and bypass dampers to control to setpoint. It was pointed out that, simply because of the shear magnitude of the difference in steam temperature and desired discharge air temperature, is it extremely difficult to precisely hone in on and maintain setpoint. There tends to be a degree of constant hunting when this type of control is applied. The steam valve opens just a tad, in an effort to bring the discharge air temperature up to setpoint, and, because of the large temperature difference between the two mediums, a bunch of Btus are transferred from the coil to the air. The discharge air temperature setpoint is overshot, the valve closes off, the temperature drops back down, the valve opens, etc., etc. It was concluded that, if the temperature of the heating medium was closer to the medium being heated, the chances for more stable and accurate control were greater. For the same amount that the control valve opens, there would be less Btus (per hour) transferred, and therefore less of a change in the controlled medium. This leads to better controllability and smoother operation.

The point of the previous paragraph is to show how controlling a steam heat exchanger with a control valve is a better application than controlling a steam coil with a control valve. It's all about this difference in temperature between the controlling and the controlled mediums. When a heat exchanger is used to control hot water temperature setpoint by regulating steam flow through it, the difference in temperatures between the two mediums is not that great. With a typical hot water temperature setpoint being around 180 degrees, modulating a steam valve to maintain this setpoint would seem to be a pretty good match. Especially when you compare it with trying to maintain a 70-degree air temperature at the leaving side of a steam coil. While this should give you a good feeling about the application, there is a word of caution to be thrown out, and that is when you apply reset control. Remember that with reset control, the hot water temperature setpoint is reset downwards as the outside air temperature rises. Depending upon your reset schedule, you could begin to approach differences in temperatures between controlled and controlling mediums of 100 degrees or so, given that the temperature of the pressurized steam is in excess of 220 degrees.

So applying reset control in a steam-to-hot water heat exchanger application is more critical than applying straight hot water temperature control. Not to dissuade you from applying reset control. By all means, go ahead and do it, if your application calls for it! Just be aware up front that valve selection and rule of thumb implementation are a little bit more critical for the application, compared with straight hot water temperature control. Hey, if there are problems down the road with instability during milder periods, just go back and edit your reset schedule! Or do away with it entirely (application permitting)! Know up front that, with most reset controllers (at least any one that's electronic or microprocessor-based), you're not locked into the reset schedule, and it can be revised at any time.

Which way should the steam valve(s) spring return to, upon failure or removal of power? There could be two schools of thought here on this subject. The first being, have it fail open, so that the hot water equipment gets hot water even in the event of a heat exchanger failure. The other school of thought is, better have it spring return closed, for upon failure, we wouldn't want to run the risk of actually producing steam on the secondary side! So which is it? Spring open or spring closed? When it comes right down to it, which is the lesser of two evils for the given application? Is it more important to run the risk of producing steam on the secondary side and keep the people served by the hot water equipment happy? Or is occupant comfort "second fiddle" to equipment and system protection? Almost unanimously, you will find that the preference is to have the valve spring shut upon a failure, to protect the hot water side of the system. Furthermore, this will most definitely prompt a maintenance or service call, and the problem can be found and rectified quickly.

Up 'til now we have spoken generically about the application and control of the topic. Let's now touch upon the technical aspects of this type of control, specifically, how to do it! For this we look at two methods. First is the classic electronic method, with a simple electronic temperature controller. Then we will touch upon the use of a digital controller to perform the task, and why it might even be considered.

Standard Electronic Control

For simple single valve control of a steam-to-hot water heat exchanger using an electronic temperature controller, there's really nothing to it. Install the hot water temperature sensor in the hot water piping on the leaving side of the heat exchanger, and wire it back to the controller. Wire the proportional output of the controller out to the control valve, and wire power to the valve and the controller. You're good to go. Power it up, establish setpoint, and watch it go to work. You might have to play around with the throttling range a little bit, to achieve stable control, but that's typical with any proportional control application. You should be able to achieve and maintain accurate temperature control of the hot water, as long as you obeyed the rules of thumb and sized your valve correctly. If unstable control results and cannot be rectified, then you may be forced to consider going the two-valve route. Of course at this time the system is already completed and in operation, which would mean having to isolate the steam supply to the converter in order to remove the single valve and install two smaller valves.

Suppose that the above "nightmare" came true. Now you're challenged with the task of selecting two control valves, and operating them from the same temperature controller. Easier said than done. The temperature controller will send out only one control signal, which can be sent to both valve actuators. For two-valve control, it is normally required that the smaller of the two valves modulates open first, before the larger begins to modulate. From this requirement, you might gather that you need control valves with actuators that respond to signals within different ranges. For simplicity, consider that the standard proportional actuator is constructed to accept a 0 to 10 control signal. If the temperature controller produces a 0 to 10 signal as the temperature of the hot water decreases from above setpoint to below setpoint, this signal applied to a single valve actuator would result in the valve traveling from fully closed to fully open. For two valve control, the smaller of the two valves would need to be able accept a signal from 0 to 5, to go from fully closed to fully open. The larger of the two would need to be able to accept a signal from 5 to 10, to go from fully closed to fully open. Thus, as the hot water temperature falls to and through setpoint, the smaller valve would open first, reaching its fully open position at a control signal of 5 (at around setpoint), and the larger valve would open subsequently, reaching its fully open position at a control signal of 10. For a refresher on proportional control and signal concepts, please review the appropriate section in Chapter 3.

The last paragraph stated that the "standard" proportional actuator is constructed to operate on a control signal range of 0 to 10. Are there actuators out there that aren't constructed to this standard, that are available for the above application? Yes, there are! Actually, many actuator manufacturers will offer a "rangeable" actuator, whose operating range can be customized for a specific application. This fits the bill, and our task is complete. Size the two valves, order them with rangeable actuators, and get 'em installed and wired. Done deal? Not yet. The valve actuators now have to be "field calibrated," in order to have them operate in the ranges required. Not that this is any big deal. The commissioning technician should have no problem completing this task, yet he must be aware that such a task exists for him. Furthermore, the end user should be instructed to keep his hands off the valve actuators, for fear of messing up the calibration of the ranges.

Digital Control

Controlling a steam-to-hot water heat exchanger with two valves and an electronic temperature controller has its good points, as well as its drawbacks. It's a cheap, stand-alone method of getting it done. At the same time, there is a bit of room for error and improper setup. If the luxury of DDC is available and convenient, this is an ideal application for a digital controller. The controller accepts a signal from a hot water temperature sensor, and also from an outside air sensor, if reset control is desired or if there is a need to disable control altogether based on outside air temperature. The controller will use not one but two proportional outputs, one for each valve. The valve actuators can be of the standard variety. No rangeable actuators needed, and no monkeying around with calibrating operating ranges. And with microprocessor-based control, you have PID at your disposal, should you require "fine-tuning" of the control loops.

The controller algorithm is to be prepared as follows: The hot water setpoint is established via the controller. The controller modulates the valves in sequence, as a function of temperature and setpoint. The first valve to modulate is the smaller valve. As the hot water temperature falls toward setpoint, the small valve modulates from fully closed to fully open. The controller will begin to modulate the larger valve only after the smaller valve has modulated fully open. This can be done by setting up two distinct proportional control loops within the same controller. The first loop, that which controls the smaller valve, operates to maintain setpoint. The second loop, that which controls the larger valve, operates to maintain a value somewhat less than the main setpoint. For instance, the controller can be set up so that the smaller valve is modulated to maintain 180 degrees, and the larger valve is modulated to maintain, say, 175 degrees. Therefore, if modulation of the smaller valve is not sufficient to maintain setpoint, the hot water temperature falls toward 175 degrees, and the second valve is called upon to modulate. Control of two proportional end devices in sequence can be referred to as "staged proportional control," as mentioned in previous chapters. At any rate, a digital programmable controller can readily accomplish such a process, and is a good fit for this type of application. A slick temperature control engineer/programmer may even be able to build in fancier control algorithms, and optimize the control process throughout the entire range of heating loads. For example, force open the larger valve for loads greater than 70 percent, and modulate the smaller valve to maintain setpoint.

Water-to-Water Heat Exchangers

Water-to-water heat exchangers are typically used in applications that require isolation between the primary and secondary sides. They are generally of plate and frame construction. As an example, consider a facility whose HVAC systems are served by a condenser water system. The condenser water system pumps water to the water cooled HVAC equipment, which rejects its heat to the water. The water is then carried to a cooling tower, where it rejects heat to the outdoors, and then returns to the equipment for another go-around. The cooling tower is exposed to the elements, and the condenser water has the potential for picking up some debris. While this small amount of debris may be acceptable for the typical water cooled HVAC equipment serving the facility, certain water cooled equipment, such as those of which serve computer rooms and telecommunication rooms, are more critical in their roles, and are thus considered more important than the other equipment. They are therefore treated with "kid gloves." Normal condenser water isn't good enough for them. They require debris-free condenser water, and nothing less. Spoiled equipment, they are!

A heat exchanger can be used here, to fulfill this requirement. Normal "tower water" is pumped through the primary, and isolated condenser water is pumped through the secondary and to this equipment. The specialized water cooled equipment rejects its heat to the secondary loop, and the heat exchanger serves to transfer this heat to the primary loop, and out to the cooling tower. The secondary loop, which serves the equipment, is a closed circuit, and therefore the water stays nice and clean for the equipment that it serves!

Another application for this type of heat exchanger, one that may be found in the big cities, is when chilled water is offered as a "utility." In this application, a utility provider produces chilled water at a central plant, and distributes it in underground mains, offering it as any other utility. A facility "subscribing" to the utility would be required to have a water-to-water heat exchanger installed. The primary hooks up to the utility, and the secondary hooks up to the facility's chilled water distribution system. You can imagine some of the reasons behind this requirement, not the least of which is the fact that the utility doesn't want to be burdened with the task of pumping to and through the facilities' cooling equipment.

A third reason for utilizing a water-to-water heat exchanger is to serve the purpose of a "pressure break" in a high rise application. Consider a thirty story high rise, served by a boiler plant in the basement level. The plant serves unitary heating equipment on all thirty floors, and a pumping/piping system must support this. As piping systems grow in height, the fittings, service valves, control valves, etc., lower to the ground must be capable of withstanding the "static pressure" bestowed upon them by the weight of the water sitting in the pipes above (when the pump is off). The higher the water column, the heavier the water, and the more static pressure exerted on the piping components. There is a practical limit to how much static pressure a valve can withstand. If the height of the building results in a piping system whose static pressure potentials exceed those limits, then a heat exchanger can be installed, perhaps at the halfway point or in this example, the fifteenth floor. The first fifteen floors are served by the main system. At the fifteenth floor, the piping from the main system connects to the primary side of a water-to-water heat exchanger. On the secondary side, a separate pumping/piping system is connected, that serves floors sixteen through thirty. Instead of the valves and other piping components on the basement level seeing a static pressure attributable to thirty floors of piping above it, they only see fifteen floors of static pressure, for the heat exchanger breaks the entire pumping/piping system into two smaller systems.

The goal of any of the above-mentioned types of heat exchangers is to have the temperature on the secondary side approach the primary side temperature as close as possible. Is there a requirement for any type of control here? Not likely. What would a control valve on the primary side do for such a system? Not much. A temperature controller on the secondary side operating a control valve on the primary side would probably have the valve in a fully open position at all times. We talked about the difference in temperatures between controlling and controlled mediums to be a factor in the accuracy of maintaining the controlled medium. The larger the difference, the more challenging the task. On the other side of that coin, the smaller the difference, the simpler the task. With the applications that have been discussed in this section, the difference in temperatures in controlling and controlled mediums is, wishfully, zero. Thus, logic would have it that the ease of control in this type of application would approach infinity. Yes, it's true. This application needs no control, so why are we even wasting our time here talking about it? Well, it's good to know before you start sizing control valves and applying a control scheme to such a system.

Perhaps the only reason for providing a control valve and controller to a heat exchanger of this ilk is when it's serving as isolation between a chilled water utility and a facility's chilled water distribution system. The utility may provide chilled water at a "lower than normal" temperature, so a control valve on the primary side can regulate the temperature on the secondary side to normal operating temperature. This, coupled with concerns of unstable chilled water temperature on the primary side, might prompt an engineer (or the utility) to specify a control valve.

Chapter 20

HUMIDIFIERS

In HVAC applications demanding precise environmental control of the spaces served, humidification and dehumidification are often specified. We've discussed the process of dehumidification back in the chapter on Common Control Schemes. To implement humidification in any HVAC process, moisture must be added to the air stream of the main air handling equipment serving the space in question. This is accomplished via a duct mounted humidifier.

Regardless of the particular type of humidifier, all humidifiers have at least one thing in common. The **dispersion tube** provides the means of introducing moisture, mainly in the form of steam or mist, into the air stream. The dispersion tube is inserted into the ductwork, just a bit downstream of the main supply fan. This apparatus normally takes the form of a tube whose length is relative to the width of the discharge air duct that it is being installed in. In larger systems, there may be not one, but several tubes manifolded together. A dispersion tube has small holes uniformly spaced across the entire length of the tube, and is positioned in the duct so that the holes either face the air stream or face upward. Thus, when the fan is running and airflow is present, the mist travels out of the holes and into the wind, which turns it the opposite way and takes it to its final destination; the space served! The size of the holes, and the required orientation of the tube(s) in the duct, both help to assure that the mist is evaporated into the supply air and not collected on the bottom surface of the duct. Safety devices also provide protection from this, as we will see further on in this chapter.

The dispersion tube assembly connects to its steam source either by hard pipe or by flexible hosing. The two general types of humidifiers seen in HVAC applications are those which utilize steam that is produced not by the humidifier, but by some other means (steam boiler), and those which produce steam as part of the humidifier. The following sections are structured to offer insight to both types of humidifiers, and how they operate and are typically controlled. Without further hesitation then, let's open up our discussions with the topic of those types of humidifiers that don't make their own steam.

Steam-utilizing Humidifiers

These types of humidifiers are designed to utilize "pre-made steam" for humidification purposes. The humidifier consists of the dispersion tube, a specialized steam control valve/apparatus, and a steam trap. Steam is introduced into the air stream by regulating the control valve. This kind of humidifier, along with its required external controls, is illustrated in Figure 20-1.

The simplest of humidification applications are those requiring two-position control of the steam valve. Space humidification via a single zone constant volume air handler is one that would fit this bill. A two-position humidity controller, or humidistat, resides in the space served, normally adjacent to the temperature controller or thermostat controlling the air handler. Space humidity setpoint is established, and the humidistat operates the control valve in order to maintain setpoint. If the space humidity level drops below the setpoint of the humidistat, the valve opens fully to allow steam to flow through the dispersion tube and into the air stream. Once the humidity level in the space rises back above setpoint (plus differential), the valve closes fully.

More critical applications may call for more precise control: proportional control. The steam valve's actuator is a proportional device, and so

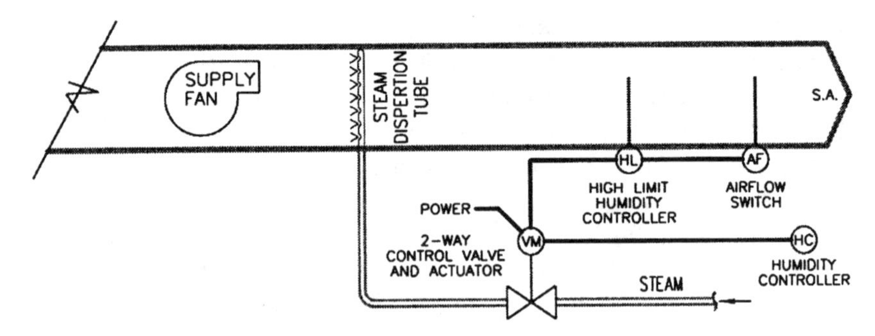

Figure 20-1. Duct mounted steam-utilizing humidifier, shown with controller and safeties.

is the space humidity controller. Just as we steer away from using the term "thermostat" when describing a proportional temperature controller, so we also do with the term humidistat. For proportional control of humidification, let's just call it a space humidity controller.

An application calling for this level of precision would be based on a few different criteria. Size is definitely one of them. Whereas you would more likely find two-position humidification on a small, single zone air handler, you would probably find proportional control specified for a larger system. The larger the air handler, the more likely it's serving not one zone, but multiple zones. From these premises, one could conclude that, for multi-zone air handlers, such as those serving VAV and reheat systems, proportional control of humidification would be the norm rather than the exception.

For air handlers employed to serve multiple zones, temperature control is normally via discharge air. There is no space temperature controller playing any part in the normal operation of the unit. What about the humidity controller? Where should it be located? It theoretically can be located in the general space served by the air handler, yet location of the humidity controller is critical, especially for VAV systems. And determining the best location is tricky at best. Practical experience dictates that the humidity controller take the form of a duct mounted controller, and be located in the return air duct, before it meets up with the outside air intake, or even in an exhaust air duct (application permitting).

For air handlers bringing in 100 percent outside air, humidification, if specified, is typically implemented as a means of conditioning the outside air. As the outside air drops in temperature and in relative humidity, humidification is employed to maintain a suitable discharge air humidity level. The humidity controller in this instance is mounted in the discharge air duct. Space reset or override of discharge air humidity setpoint may be employed, as is done with temperature control of such a system, should the application call for it.

The only other application that might call for discharge air humidification control would be that of a process or testing application. A lab set up for product or equipment testing may be such an application. An air handler serves the space, conditioning and recirculating the air. Requirements may dictate a constant discharge air temperature and humidity, with tight tolerances. While this type of control may be seen in laboratory or testing applications, it is not suited for environmental comfort control, and typically not employed in normal HVAC applications.

Now that we've covered some of the typical types of humidity control, we turn our attention to the other controls components required for any duct humidification process: the safeties! Two safety devices normally accompany any duct humidification control system, and these are the air proving switch and the high limit humidistat.

In duct humidification systems, it is important that the duct humidifier be disabled when the upstream supply fan is not running, for the obvious reasons. The same method of airflow proving that's applied with electric reheat coils is typically applied with humidifiers as well. An air proving (pressure) switch, with the high port piped into the supply air duct and the low port left open, provides airflow confirmation by monitoring the pressure in the duct. If the fan is running and the pressure switch is made, then the humidifier is allowed to operate. More critical applications may dictate an additional means of fan proving as well, such as a current sensing switch or an auxiliary contact off the fan starter. This provides redundancy, for failure of this safety could result in a very wet duct.

A high limit duct humidistat must be installed in the supply air duct, at least ten feet or so downstream from the humidifier dispersion tube. The purpose of the high limit humidistat is to prevent saturation of the supply air. In most duct humidification processes, humidification is controlled not at the discharge, but at some other location (space or return air). The humidity controller, be it two-position or proportional, operates the control valve with no regard to the level of humidity being produced within the duct itself. If the point at which the humidity controller was located consistently required humidification (humidity level continuously below setpoint), then the duct humidity could reach the saturation point, the point at which the supply air can hold no additional moisture. This is especially true with two-position control of the steam valve. The high limit humidistat's function is to override the controller and close the valve, before the saturation point is reached. A typical setting for the high limit humidistat is 80% R.H.

The high limit duct humidistat is not a manual reset device. If the duct humidity reaches the setpoint of the high limit, the steam valve closes, and the humidity level immediately begins to drop. Once the humidity level drops back far enough below the high limit's setpoint, the humidifier is once again enabled, and the steam valve opens again. During a condition in which the main humidity controller is continuously calling for humidification, the humidifier could actually be cycling on the high limit. This is not necessarily an undesirable mode of operation, and further goes to show the importance of the high limit, its proper placement, and its proper adjustment.

Upon a trip of either safety, the air proving switch or the high limit humidistat, control power to the valve should be interrupted. For this, a spring return normally closed valve is recommended, so that the valve shuts upon removal of power from it. The control valve actuator is supplied by the manufacturer as part of the humidifier, and is to be specified to be spring return. The safety devices are often provided with the humidifier, either as standard equipment or at least as an option, for field installation.

STEAM-GENERATING HUMIDIFIERS

Humidifiers that produce their own steam and are not dependent on a separate steam source are the topic of this section. Steam-generating humidifiers (Figure 20-2) are packaged, microprocessor controlled equipment. They are typically electric, though gas-fired humidifiers are available as well. The humidifier consists of a cylinder or vessel that holds water that is to be turned to steam. A make-up water connection is required, to replenish the water that is turned to steam and delivered to the dispersion tube. Water to the humidifier is made up automatically, either by means of a float valve/switch, or by an electronic probe sensor. With either method, the level of water in the vessel is monitored, and when it

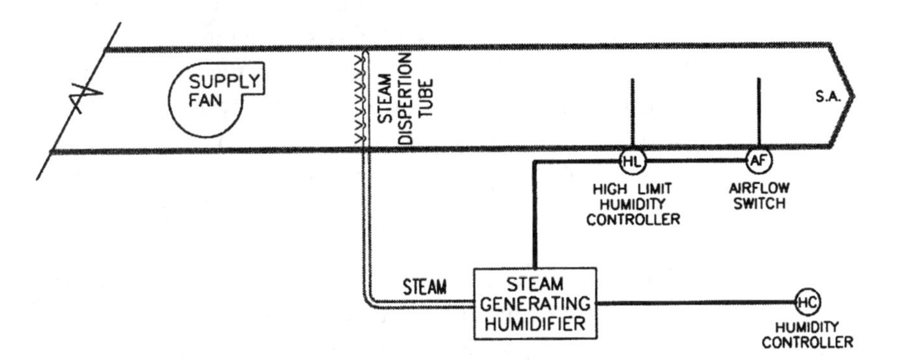

Figure 20-2. Duct mounted steam-generating humidifier, shown with controller and safeties.

gets low, make-up water is allowed to flow into the vessel, to bring the water back up to operating level. Both methods are factory furnished as "part of the package."

As far as humidity control goes, the same rules apply for steam-generating humidifiers that do for steam-utilizing humidifiers. For simple two-position or two-state control, the humidistat calls for operation, and the steam generator is pressed into operation. Contactors are energized, which in turn allow power to electrodes that are submersed in the vessel. Upon boiling, the steam will flow to and through the dispersion tube. When the call for humidity ceases, the contactors are de-energized, and the boiling process ends.

With proportional control, the power to the electrodes is varied, in much the same way as power is varied to the electric resistance heating elements of a duct heater equipped with an SCR controller. The variation in power to the electrodes is a function of the deviation in humidity level from setpoint; the more humidity required, the more power delivered to the electrodes, and thus the greater the steam production rate.

A microprocessor-based control system supervises all facets of humidifier operation. Operating parameters and special functions are normally accessed via a user interface integral to the humidifier. Typically a keypad with a small LCD readout, the interface allows the user to access or implement special features, such as drain/flush/fill cycles, timing cycles, etc.

Well, since I mentioned the term, and there really isn't much more to say about the topic of this section, I might as well touch upon "drain/flush/fill." As a steam-generating humidifier logs hours of operation, sediment tends to form - minerals boiled out of the water—at the bottom of the vessel. Every so often, this sediment needs to be removed from the vessel. This is done with a drain/flush/fill cycle. As per a user-defined interval, the humidifier shuts down and enters the cycle. As water is drained from the vessel, make-up water is allowed into the vessel. This cycle can last anywhere from a couple of minutes to ten minutes or more. When the cycle is completed, the vessel is allowed to fill back up, and normal operation of the humidifier is resumed.

The drain/flush/fill cycle is standard operating procedure of any packaged steam-generating humidifier. It makes good sense, from the manufacturer's standpoint as well as from the end user's standpoint. Suffice it to say that it's in the best interests of all parties. That being the case, the manufacturer typically will not allow disablement of the cycle, which could be detrimental in an application requiring precise, continuous, humidity control. During such a cycle, humidification shuts down! A cycle lasting ten minutes or longer means no humidification for that amount of time. Can the environmental control process at hand live with this cycle? Or is it compromising the entire process and defeating the purpose? For all but the most critical of HVAC applications, this cycle is acceptable. If the application is so critical that the drain/flush/fill cycle is counteractive to the process, then other alternatives must be explored. Disabling the cycle, while feasible, likely will void the equipment warranty and shorten the life of the humidifier. A manual cycle can be instituted, and hence this becomes a maintenance issue.

Chapter 21

Unitary Heating Equipment

The term "unitary" refers to a single piece of self-contained equipment, generally small in physical size, and typically controlled and operated to provide single point comfort control. A perfect example of this is the typical gas-fired unit heater sitting out in the warehouse of your office building. The heater hangs from above, and a thermostat that sits on a nearby column is wired up to it. Several unitary devices can theoretically serve the same space, as is normally the case with warehouses, in which not one but several unit heaters are installed, spread out to provide uniform temperature control.

This chapter deals with those types of unitary equipment that are strictly built for heating purposes. A fan coil unit serving a hotel room that has both hot and chilled water coils can be considered unitary. Yet for our discussions we reserve the term unitary for those types of self-contained equipment that are designed and manufactured solely for the purpose of heating.

Unit Heaters

Unit heaters are the classic means of providing heat in unfinished spaces, such as warehouses, electrical rooms, and mechanical equipment rooms. They can be powered by electricity, hot water, steam, or gas. These are forced air type heaters, which means that they consist of a fan that circulates air through the heater and distributes it to the space (refer to Figure 21-1 for an illustration of a steam unit heater). Typical operation dictates that, upon a call for heat from the operating thermostat, the heater energizes and the fan starts, and when the call for heat is satisfied, the heater de-energizes and the fan stops.

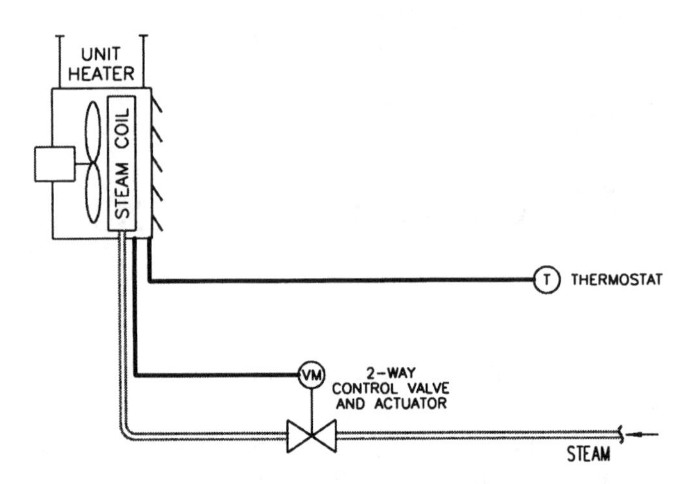

Figure 21-1. Steam unit heater controlled by a remote mounted thermostat. The control valve, though often specified, is not always required.

Unit heater thermostats are normally remote mounted, far enough from the heater so that the thermostat doesn't instantly feel the effects of the heating cycle, yet still close enough to be considered "in the zone" of temperature control. Unit heater manufacturers offer unit mounted thermostats as well, in which the thermostat is an integral piece of the equipment.

Electric unit heaters generate heat in much the same way as electric duct heaters: via electric resistance heating elements. Power is brought to the heater, for the fan and the heating elements, and is typically stepped down to 24 volts AC for control purposes. The control voltage is run through the thermostat, and back to the unit. Upon a call for heat by the thermostat, the control circuit is made, the fan and heating contactors are energized, and heat is produced and delivered. Electric unit heaters are applied in buildings that either don't have natural gas, or that the use of natural gas for unit heater purposes is not viable. Electric unit heaters are also seen in electrical equipment and switchgear rooms, where a concern exists in using any other type of unit heater. What's the concern? Well, using a hot water or steam unit heater could run the risk of wetting the electrical equipment, and using a gas-fired unit heater could run the risk of blowing the place up! The risks are attributed to springing a leak in the heating source. A hot water or steam leak could be detrimental to the equipment in the room, and a gas leak could be ignited by a spark generated by the equipment.

Hot water unit heaters consist minimally of a fan and a hot water coil. Power is brought to the heater, for the fan, and typically stepped down to 24 volts AC for control purposes, though if the fan is powered by 120 volts, a transformer may need not be there. The control voltage, therefore, would be 120 volts as well. The common application is to have the hot water "run wild" through the coil, and have the thermostat simply cycle the fan. With dual-temp systems or with hot water systems that operate seasonally, often an aquastat is attached to the pipe feeding the unit heater, that disables fan operation if the water in the pipe is not hot.

For applications in which there is a concern that the unit heater could give off too much heat in the form of radiation, in the case of when the thermostat is not calling and the fan is off, a two-way, two-position control valve may be required. This is normally not an up front concern, but if a problem arises, a control valve can be installed that disallows flow through the unit heater unless there is a call for heat. If using an aquastat and a two-way control valve, the aquastat must be located back at the supply mains, where hot water flows continuously. Remember that the aquastat basically disables the heater if the water that it senses is not hot enough. If installed right at the heater, and the valve is closed, the water in that segment of pipe will not flow, and the temperature of it will drop under the setting of the aquastat. The aquastat therefore disables heater operation, the valve stays closed, the temperature of the water at the aquastat stays cool, the heater remains disabled, the valves stays closed…

If there is the dual concern of overheating via radiation and of deadheading the main system pump when the control valve is closed, then a three-way valve may be utilized, so that flow bypasses the heater when there is no call for heat. In this case, the aquastat can be installed local to the heater, at the common port of the control valve, the port that continuously sees full flow. The requirement for a three-way valve in a unit heater application is a rare one. Yet there may be a need for it, and it's important to understand what that need is, in order to properly apply the valve and the associated controls.

Steam unit heaters (as in Figure 21-1) operate in the same manner as hot water unit heaters. Subtle differences exist between the two types of heaters. The first is that the potential for overheating via radiation is greater with steam than it is with hot water, so a control valve may be more frequently required. The control valve will be a two-way valve, and the aquastat, if used, will need to be located back at the steam mains. There is no pump, so there is no concern about deadheading. Furthermore, a three-way valve is not applicable in such a steam application; you cannot couple the steam supply main and the condensate return main with a control valve, at least not without a steam trap.

Gas-fired unit heaters are quite popular. They are more efficient than electric unit heaters, and unlike hot water or steam unit heaters, they don't need a boiler to support their task. Gas and electrical power are brought to the heater, and the power is stepped down to 24 volts AC, for operation of the gas valve. Operation of the typical gas-fired unit heater is similar to the operation of the gas heating cycle of a packaged rooftop unit. Upon a call for heat by the thermostat, ignition takes place and is confirmed. The heat exchanger heats up, and the fan starts. Once the call for heating is satisfied, the gas valve closes, and the heat exchanger cools down. The fan continues to run, to blow off any residual heat from the heat exchanger, and then shuts down.

For all types of unit heaters, there is one thing in common, and that is the thermostat. If the thermostat is specified to be an integral part of the heater, then there is no task required in this respect, for the controls designer. Unit heaters available with integral stats are typically those of the gas-fired and electric variety. If the thermostat is specified to be remote, then the unit heater supplier may provide a thermostat suitable for his equipment. If not, then a thermostat will have to be purchased separately. While the majority of unit heater applications utilize low voltage for control purposes, it is not uncommon to specify a line voltage type thermostat. These types of thermostats are generally more heavy-duty than their low voltage counterparts, and are typically of the snap-acting type. A rugged thermostat for a rough application.

Cabinet Unit Heaters and Wall Heaters

Cabinet unit heaters are packaged fan driven heaters that are installed in finished spaces, attempting to blend in with their surroundings. Typical applications include stairwells, corridors, and vestibules. Though sometimes specified to be controlled by a remote thermostat, the more common application is return air temperature control. Cabinet unit heaters are physically shaped like a cabinet (Figure 21-2). On one end of the cabinet is the supply opening, and on the other end is the return.

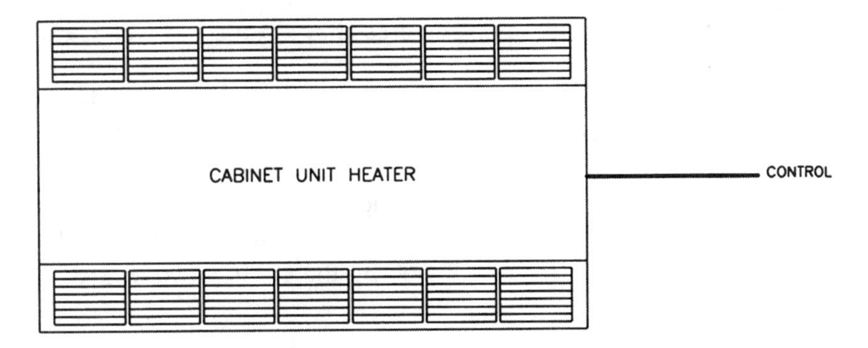

Figure 21-2. Ceiling or wall mounted (electric) cabinet unit heater. Temperature control can be integral or remote.

These openings may or may not be ducted, depending upon the mechanical layout of the space. Regardless, the heater operates to recirculate air in the space served, and is typically equipped with a factory installed return air temperature controller.

Electric cabinet unit heaters operate identically to electric unit heaters, if controlled by a remote thermostat. The thermostat calls for heat, and the heating elements and fan come to life. When the call for heat is over, the heater shuts down. If controlled via return air, the fan may be required to run continuously, with the controller cycling the heat. Running the fan continuously is not a bad idea to begin with, for either method of control, space or return air.

Hot water and steam cabinet unit heaters can be controlled in a number of ways. The first is similar to how a hot water or steam unit heater is typically controlled: let the hot water or steam flow continuously and cycle the fan. Of course with this method, return air control may not be the right application. If there is no call for heat and the fan is off, then there is no airflow across the temperature controller. The controller, mounted in the return air compartment and close to the heating coil, stays warm and cozy, while the space cools down to the point of needing heat. For this to work, the temperature controller really needs to be in the space served.

The more common method of control is to operate a control valve upon a call for heat, with the fan either running continuously or cycling upon calls. Control valve packages are generally offered by cabinet unit heater manufacturers, and can be specified as two-way or three-way, two-position or proportional, depending upon the application. In terms of aquastats, the same rules apply here as they do for hot water and steam unit heaters.

Wall heaters are nothing more than what the name states. These things are electric fan driven heaters, normally recessed in a wall. They are completely self-contained, with an on-board thermostat/temperature adjustment. The control systems designer need not concern himself with this type of equipment. In fact, often this is an item simply furnished by the mechanical contractor, for installation by the electrical contractor.

Baseboard Heaters

Baseboard heating and radiation has been around for eons. Perhaps you've been in an older building lately, and noticed hot water or steam radiators in the individual rooms of the building. The traditional description of a radiator puts the picture in your head of a tubular device, standing a few feet high, with perhaps a decorative radiator cover placed over it.

Over the years, radiation equipment has evolved and has become more efficient and less obtrusive. Radiators have no fan, and basically rely on the principles of radiation (and convection as well) to provide heat for a space. They are typically at ground level, stretching across the inside perimeter of an outside wall.

The classic description of a baseboard heater is that of a ground level cast iron radiator, in which steam or hot water flows through the entire length of it, and in the process gives off heat. Fin-tube radiation, though not really considered baseboard (at least not in the traditional sense), is installed the same way and serves the same purpose. A section of fin-tube consists of a pipe and strips of metal, or fins, that are spaced closely together in such a way as to conduct heat from the pipe and transfer it to the surrounding air. The pipe and fins are hidden by a decorative cover running the length of the heater. For the purpose of this chapter, we will refer to all types of perimeter ground level heating appliances as baseboard.

Electric baseboard heating is typically applied in buildings that don't have a source for hot water or steam, simply by design. The electric heat is normally operated via integral temperature controls, though can just as easily be specified to operate by a wall mounted thermostat. Each section of electric baseboard will be rated at a specific power draw.

Multiple sections of heat may be served by a single circuit breaker, as long as the combined power draw doesn't exceed the breaker's current rating. The individual sections of electric heat, or even groups of sections, can be directly controlled by a thermostat, provided that the thermostat is rated as an electric heat thermostat, and can handle the maximum current draw. Electric heat is resistive in nature, and special thermostats are manufactured to directly handle the power to electric baseboard heaters. Be aware that this is not low voltage here; the thermostat is directly handling the power that is delivered to the heater from the circuit breaker. Kinda scary when you think of it: the thermostat that is on the wall controlling your baseboard heater has line voltage running to it. All the more reason to use that thermostat specifically designed for the application. And of course that wiring to the thermostat from the heater will be in conduit!

Electric baseboard that is constructed as larger sections may have a control contactor and maybe a transformer. A thermostat controls the contactor, and not the heating element itself. When the section's thermostat calls for operation, the contactor is energized and power is allowed to the electric heating element. With this type of setup, an electric heat thermostat is not required.

Zones of electric baseboard may also be controlled not by individual thermostats, but by an interlock to a terminal unit serving the same zone. This is often the case when VAV boxes are employed to serve perimeter zones for cooling, and baseboard is employed to serve the same zones for heating. Rather than having two thermostats or two points of temperature control in each zone, with the potential of having the cooling and heating equipment "fighting" each other, temperature control is integrated into one device. The VAV box's zone sensor is wired up to its respective VAV box controller. The controller has the capability of operating auxiliary heating equipment, whether it be a coil at the box itself, or a section of remote baseboard. So the controller takes command of the electric baseboard serving that zone, thus eliminating the need for a separate temperature controller for the baseboard. Upon a drop in temperature below the setpoint of the VAV box temperature controller, the damper positions at minimum, and the section of electric baseboard for that zone is energized. The interlock between VAV box controller and baseboard mandates that the baseboard have a control contactor local to its section. Otherwise, one must be installed, to handle the power to the baseboard, since the VAV box controller's heating outputs, which are relay contacts or electronic two-state outputs, are rated for low voltage only. The contactor can be installed local to the baseboard, though it may be more beneficial to install the contactor back at the power source: at the circuit breaker panel. For multiple VAV/baseboard combinations, this makes sense; a panel can be built to house the required contactors for all zones of baseboard. The panel can be installed adjacent to the circuit breaker panel (space permitting), and all VAV box controllers zoned with baseboard can wire back to this central location. Depending upon the particular project, this may or may not be the most cost-effective way to go. But the consensus is that it's definitely the cleanest way.

Hot water and steam baseboard heaters can operate in a couple of different ways (Figure 21-3). The simplest by far is to equip the section of baseboard with a self-contained thermostatic control valve. This device has no electrical requirements. A gas-filled bulb normally mounts to the cover of the heater. The expansion and contraction of the gas inside the bulb as a function of the surrounding temperature directly affects the position of the valve. The colder it is around the bulb, the more the valve is open. The self-contained valve has an adjustment to set the desired temperature setpoint.

One step up from the self-contained thermostatic control valve is the motorized control valve. An electric control valve may be required if there is a desire to remotely control the section of baseboard, either via a wall mounted temperature controller, or via some other means. If controlled by a wall mounted controller, proportional control can be employed, though generally this is overkill for such an application. If zoned with a VAV box, the box's controller will typically have the capability of providing both two-position and proportional control. Depending upon the application, an engineer may specify proportional control of the base-

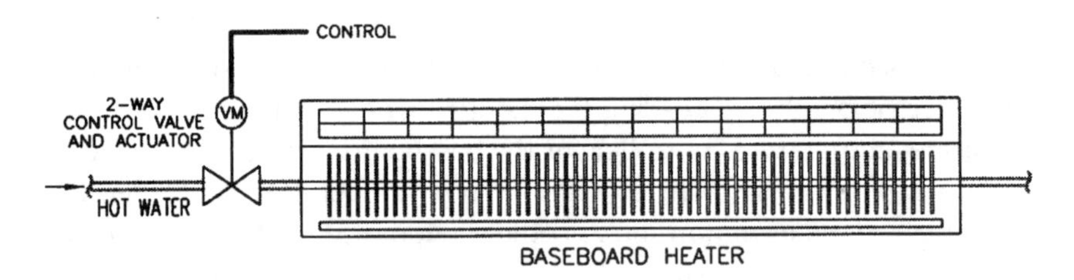

Figure 21-3. Hot water baseboard radiator. Hot water flow is controlled by either a self-contained thermostatic type control valve, or by a motorized control valve and some type of temperature controller.

board. Again, this is generally overkill for baseboard control, especially if the hot water system employs outside air reset control.

Last on the list of topics relating to baseboard control is a topic specific to hot water baseboard heating. Back in the chapter on boiler control, we discussed a type of zoning system in which individual zone pumps fed sections of baseboard. Each pump feeds a section (or sections) of baseboard and comprises a zone of temperature control. A thermostat in each zone operates its respective zone pump. Upon a call for heat, the pump is energized and hot water flows through the baseboard, and when the call for heat is satisfied, the pump turns off. This type of two-state control is similar in heating operation to that of two-position control of a baseboard control valve via a wall mounted thermostat: thermostat calls, water flows; thermostat satisfies, water ceases to flow. Operation is the same, the methods differ. As stated in the chapter on boilers, this type of "waterside zoning," via individual pumps (one per zone), is more a residential thing.

Chapter 22

Computer Room A/C Systems

Computer room style air conditioning units are packaged equipment designed for demanding applications. A room that houses a boatload of active computer equipment, or any other electronic equipment for that matter, is a candidate for such a unit. The electronic equipment, when in operation, generates quite a bit of heat. At the same time, this type of equipment is extremely sensitive to the environment surrounding it, and must be kept cool and happy. You might say that equipment of this type is its own worst enemy! It generates heat, but at the same time, too much heat can be detrimental to the equipment.

The types of units that are designed and manufactured as computer room "environmental control" units are primarily cooling units. Though heating isn't generally a concern with this type of application, the unit may have an electric reheat coil, to serve in a dehumidification cycle. The unit may also be equipped with an integral steam-generating humidifier.

As environmental control is of critical concern in such an application, a unit of this caliber will typically be equipped with an integral microprocessor-based control system. The system controls all facets of unit operation, in order to provide precise temperature and humidity control of the space served. The units normally will not bring in any outside air, and therefore recirculate 100 percent of the air delivered. Temperature and humidity sensors are factory installed in the return of the unit, and wired to the main controller. User interface is via a unit mounted or remote mounted display and keypad device. Temperature and humidity setpoints, as well as other parameters and monitoring points, are accessed through this device.

Redundancy is often a preference in these applications. To that end, multiple units will serve the same space. The units may each be full-sized, or they may be sized such that the total combined capacity substantially exceeds the full cooling load of the space. Staging of units, selection of the primary or lead unit, alternation, and automatic backup operation are all often performed by a separate microprocessor-based supervisory control system. Manufacturers of this type of equipment will likely also offer such a package, specifically designed with their units in mind.

Air Cooled Systems

An air cooled computer room A/C system (Figure 22-1) is classified as a split system, because the refrigeration cycle is not contained in one single piece of equipment, but is broken down into indoor and outdoor equipment. The indoor equipment contains both the evaporator (DX) coil and the compressor(s). The outdoor unit is nothing more than a condenser coil and condenser fan(s). When the outdoor unit does not house the compressors, then it is referred to not as an air cooled condensing unit, but as an air cooled condenser.

Why aren't the compressors outside at the outdoor unit? Isn't that where they belong? With this specialized equipment, the compressors are an integral part of the indoor unit, perhaps for a combination of reasons. In a computer room, noise level is generally not a huge concern. The very equipment that is generating the heat in the space and ne-

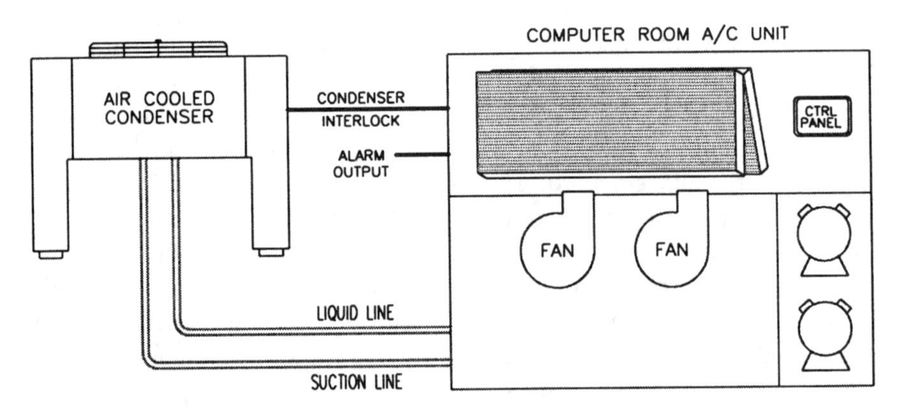

Figure 22-1. Computer room A/C "split system", consisting of an indoor microprocessor-based A/C unit, and a matching outdoor air cooled condenser.

cessitating the use of an A/C unit is very likely producing plenty of its own noise. The compressors of any A/C system are admittedly pretty loud when in operation. At least when you compare it with the noise level produced by the supply fan. It is for this reason that split system air handlers will normally have the compressors at the outdoor unit. Yet, if noise is not a concern, then it's pretty much a toss-up as to where the compressors are located, indoors or outdoors. With noise not being a concern, why not get the compressors indoors and out of the elements? Sounds like a good idea, for both operational and maintenance reasons. And with compressor operation being not just a comfort requirement, but a requirement critical to the capital investment of the equipment served, and to the functions that the equipment is counted on to continuously carry out, the compressor suddenly becomes a very important piece of equipment itself. Get it inside, where its chances of breaking down due to extreme ambient conditions are eliminated, and maintainability is simplified.

As with any split system, an interlock must exist between the indoor and outdoor units. When the indoor unit calls for compressor operation, the outdoor unit must be enabled for operation. When the interlock is made, the condenser fans are allowed to operate. Standard mode of operation for condenser fans is that they stage on as a function of ambient outdoor temperature, though other methods of condenser fan operation could apply as well.

So the indoor unit calls for cooling, the compressor starts, and an interlock calls for the air cooled condenser to operate. The condenser fans, if controlled as a function of ambient temperature, will stage accordingly. The greater the outdoor air temperature, the more fans are on. This is done by a factory furnished/wired multistage temperature controller, and individual fan contactors.

The interlock between indoor and outdoor units is an important one, and depending upon the separation between the two pieces of equipment, could be a difficult one to accomplish. The most obvious path between units is to follow the refrigeration piping with the interlock wiring. An argument could be made that, since the outdoor unit does not house the compressors, an interlock between indoor and outdoor units can be done without. Simply allow the condenser fans to do their thing, whether the compressor is in operation or not. Fact is, most matched systems are set up for the interlock; the condenser is looking for a contact closure. Sure, you can jumper out the unit so that the fans are continuously enabled for operation, but why would you want to allow the fans to operate when the indoor unit doesn't need them to operate? No, in all but the most unorthodox applications, the interlock is status quo.

Microprocessor-based A/C units generally will offer dry contact outputs that can be configured to close upon user definable alarm conditions. More generally, there may be a single "common" alarm output that can be configured to close upon any number of user definable alarm conditions. The manufacturer offers this as a means of communicating equipment failure or abnormal operation to a remote location. The alarm contacts, for instance, can be used to simply sound a horn or light a light in another room. Or they can feed a "dial-out" device, which in turn can initiate a telephone call to the pager of the individual in charge of maintenance. Or they can serve as an input to a digital controller that's part of a Building Automation System, where from there they can be utilized to perform any number of functions, from alarm enunciation to dial-out.

Tower Water Systems

Computer room A/C systems that are water cooled take on the form of one of two types of systems, the first of which we discuss in this section (as illustrated in Figure 22-2), and the second of which we tackle in the following section. Remember that equipment that is water cooled has all components of the refrigeration cycle integral to its package, with heat rejection taking the form of condenser water pumped to and from the equipment.

Tower water systems are systems in which often not one but many pieces of water cooled equipment are served by a common cooling tower. Such a system will consist of the cooling tower and typically a pair of main condenser water pumps. The main pumps operate continuously in order to deliver condenser water to all equipment served by the system. At any point in time, the equipment in need of condenser water will utilize it by rejecting heat to it, and the equipment not in need of condenser water will simply not reject any heat to it.

A typical condenser water system, as stated in the last paragraph, can serve many pieces of water cooled equipment. For instance, consider a facility served solely by water source heat pumps, which are basically packaged water cooled heating/cooling units. In addition to the heat pumps serving the purpose of occupant comfort control, a couple of

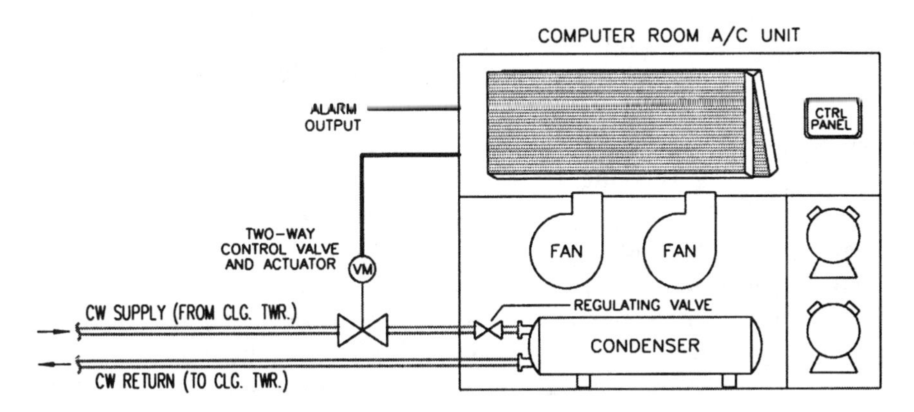

Figure 22-2. Water cooled computer room unit, factory equipped with a two-way water regulating valve. The external isolation control valve is field installed, and is optional.

computer room A/C units will typically tie in to the condenser water system as well, either directly, or via an isolation heat exchanger. For an explanation on the function of a heat exchanger in this type of application, please refer back to the chapter on Heat Exchangers, the section covering water-to-water heat exchangers.

With water cooled "computer room style" air conditioning equipment of any considerable size, a water regulating valve will be factory provided and piped, that controls the flow through the equipment's condenser bundle. When the piece of equipment is operating in a cooling mode, with its compressor running, the water valve will modulate to regulate the flow of water through the condenser as a function of refrigeration pressure. Simply put, the more heat rejection required, the more open the valve will be. This valve is mechanically actuated, directly operated by the pressure sensed in the refrigeration line. As such, the valve's actuator will have a pressure "tie-in" to the refrigeration line, and by way of a capillary between the tie-in and the mechanical valve actuator, the valve will modulate as a function of refrigeration "head pressure." The greater the pressure, the more the valve is open.

Generally the condenser water regulating valve furnished (as standard) with a water cooled computer room A/C unit will be two-way. Given the particular application, this may or may not be acceptable. Caution must be exercised here. The entire condenser water system must be evaluated. If the condenser water supply to the unit in question can be "choked off" without any possibility of deadheading the main condenser water system pump, then a two-way valve will suffice. Otherwise, a three-way valve must be specified. A manufacturer may offer, as an option, a two-way valve with a parallel "factory-piped" bypass line. The bypass line is configured such that if the two-way regulating valve closes off to the condenser, the bypass line will take on enough flow required to safely and continuously operate the main system pump without fear of deadheading.

Sometimes an engineer will specify a two-way, two-position control valve, configured so as to isolate the A/C unit from the condenser water loop when the unit is not in operation (refer again to Figure 22-2). This would appear to be a redundancy, for if the unit were not in operation, the absence of refrigeration pressure would have the water regulating valve in a fully closed position. Why the need for an additional control valve? Ask the engineer who specified it... perhaps he has some very legitimate issues and concerns that would lead him to specify this redundancy.

Anyway, this control valve would typically be wired into the control circuit of the unit so that when the unit calls for compressor operation, the valve is powered open, and when compressor operation ceases, the valve spring returns shut. The issue of main system pump deadheading is again a concern with this type of application, and the condenser water system should be evaluated. If found as acceptable in terms of system pump operation, then a two-way valve installed for isolation purposes is allowable. If the equipment is specified to have a three-way water regulating valve, or even a two-way valve with bypass, a two-way isolation valve would appear to be a "misapplication," for it would seem to defeat the purpose of the specified water regulating valve setup. In other words, why specify anything more (for the unit) than the standard two-way water regulating valve, if condenser water will be choked off to the unit anyway (by the isolation valve) when the unit is not calling for cooling? If both a three-way water regulating valve and a two-way isolation valve are specified for the same piece of equipment, the requirement for either valve must be re-evaluated.

Often enough, a computer room A/C unit will require a "booster pump," to help the condenser water get "to and through" the unit's condenser. In such an application, the pump will be wired so that it operates whenever the unit's compressor is in operation. Deadheading of the main system pump is normally not a concern with this scenario; the engineer has already deemed the requirement of the booster pump, surmising that the main pump wouldn't have enough head to handle the A/C unit, in addition to the other equipment served by the condenser water system. Furthermore, this type of application often takes the form of an "add-on," in which additional equipment is tied in to an existing condenser water system. The original design has likely taken into account any main system pump concerns, by way of the original equipment on the system. Adding more equipment won't have any adverse affects on condenser water pumping, except that the original system pump may not have the flow capacity to handle the additional equipment, hence the booster pump!

Last on the agenda here is the topic of interlock. Unlike the air cooled systems we talked about in the last section, computer room A/C equipment served by a common tower water system require no interlock to the heat-rejecting equipment. Generally the condenser water system serving this type of application is required to be in continuous operation, year-round. As such, the main condenser water system pump runs continuously, and the cooling tower is enabled and in its required mode of operation. Therefore, condenser water is readily available at all equipment, at all times. The individual pieces of water cooled equipment need not "call" for condenser water; the water is continuously at their disposal, and no interlock is required to make the call for it.

Glycol Cooled Systems

Glycol cooled A/C systems are those of which are generally categorized as "matched systems," where each piece of water cooled equipment has its own "matched" heat-rejecting unit. So there isn't a single cooling tower common to all A/C units, but a dedicated "fluid cooler" for each unit. These systems typically utilize a solution of water and glycol as the heat-rejecting medium. The glycol is simply anti-freeze; the higher the concentration of glycol, the lower the freezing point of the solution. It should be noted here that tower water systems can operate with glycol-based solutions as well. The term "glycol cooled system" is not necessarily restricted to the types of systems that we cover in this section. Standard terminology however, and the manufacturers' use of it, dictates that the term typically describes these types of matched systems. Figure 22-3 portrays such a system, showing indoor and outdoor units, and additional required components.

The heat-rejecting unit is referred to as a drycooler or a fluid cooler, depending on the manufacturer. For our discussions, we will refer to the outdoor heat-rejecting unit of a matched system as a drycooler, simply as a matter of preference by the author.

When water cooled equipment is served by a cooling tower, the piping system is an "open loop" system, meaning that, at some point in the system, the fluid is exposed to atmospheric pressure. In the case of a condenser water system served by a cooling tower, this occurs at the tower. With the types of systems discussed in this section, the condenser water or glycol piping system is a "closed loop" system, meaning that the fluid is completely sealed throughout the entire system. The drycooler, in essence, is a "closed circuit" cooling tower. The drycooler consists of a finned coil and a series of fans, the physical appearance being very similar to that of an air cooled condenser. The glycol solution is pumped through the coil, where it transfers its heat to the ambient air surrounding it. Fan operation is staged to maintain a decent fluid supply temperature. By factory controls, the fans are staged on as the fluid temperature rises from setpoint; the warmer the fluid pumped from the drycooler, the more fans will be on. The fans simply help to reject heat from the coil by drawing ambient air through the coil.

How the glycol solution is pumped from the drycooler to the A/C unit and back is generally performed by another packaged piece of

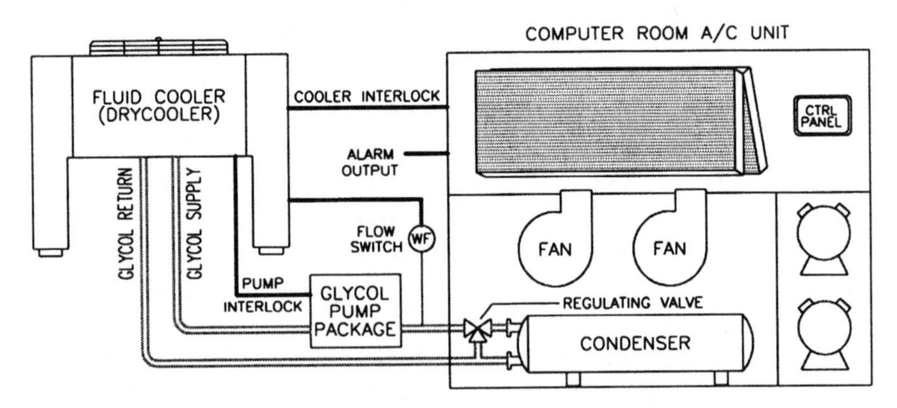

Figure 22-3. Computer room A/C system, consisting of a glycol cooled indoor unit (with a three-way water regulating valve), and a matching outdoor glycol cooler.

equipment, matched for the particular application. A "duplex" pump package will typically be offered by the manufacturer, to be used in conjunction with the drycooler and the A/C unit. The pump package consists of two full-sized pumps, of which actually derive their power feeds from the drycooler electrical system. In other words, the power feed for the drycooler is sized for both the drycooler and the pump package, and the drycooler feeds the pumps via cooler mounted contactors. It is in this manner that the drycooler control circuit has command of the pump package.

Why does the drycooler need control over the pump package? To answer that, let's consider how the system operates as a whole. Upon a need for cooling at the computer room A/C unit, the compressor starts and a request is issued by the A/C unit, in the form of an interlock, for drycooler operation. The drycooler is enabled, and its fans are engaged as required. At the same time, the drycooler commands one of the two pumps to run, by energizing the appropriate contactor in the drycooler electrical compartment. When the call for cooling subsides, the interlock is broken, the drycooler is disabled, and the pump turns off.

A flow switch must be installed in the glycol piping, downstream of the pump package. The flow switch wires back to the drycooler control circuit. Now the drycooler not only has control over the two pumps, but also knows of pump status. If the drycooler calls for pump operation, and flow is not confirmed in a timely manner, the drycooler control circuit will de-energize the contactor of the "primary" pump, and energize the contactor of the "backup" pump. Hey, this looks familiar! A primary/backup pumping system, from Chapter 16, with automatic backup upon failure of the primary pump! Indeed it's true; the drycooler itself is the pump controller. The drycooler may even have the capability of primary pump alternation, and of issuing an alarm upon failure of the primary pump, by means of a dry contact.

While this is the more commonly applied means of tackling pump operation, it is not unthinkable to have a separately furnished/installed pump that's not a "package" specifically matched to the equipment. A single pump, or pair of pumps, can be purchased separately, and piped accordingly. Care must be exercised in the pump selection, especially if the intentions are to control the pumping operation from the drycooler control system. Generally, the safest and simplest route is to just have the provider of the A/C unit-drycooler system to select and furnish the duplex pump package. In this application, criticality merits the package.

The water regulating valve for such a system will be a three-way valve, or at least a two-way valve with bypass. A simple two-way valve will not work in this application, and the manufacturer knows this. As such, he will furnish his A/C unit with the appropriate factory piping package. Consider why a two-way valve doesn't fly in this setup: The A/C unit calls for cooling, the interlock to the drycooler is made, the drycooler is enabled for operation, and the glycol pump starts. As the refrigeration cycle proceeds, the water regulating valve operates to maintain proper refrigeration pressure in the lines. The valve can very realistically modulate down to a point of allowing very little flow through the A/C unit's condenser, especially if the temperature of the glycol solution is lower than normal. Of course, what does this do to the pump? Well, the pump will tend to deadhead, which could cause the pump motor to overamp, or draw too much electrical current. The overload circuitry, whether in the pump motors themselves or back at the drycooler electrical compartment, catches the condition, and shuts down the pump. Flow is lost, and the backup pump is placed into operation. The backup pump operates under the same circumstances, and it too shuts down on overload. Now back at the A/C unit, where there is no flow to the condenser, the refrigeration pressure builds, to a point of exceeding the high pressure safety cutout, which is a manual reset device. At this point, the only thing left in working condition (not that it matters anyway) is the drycooler, which is disabled because the interlock has been broken upon the failure of the A/C unit.

Rule of thumb: For any glycol cooled computer room A/C system, one that consists of an indoor unit and a matched outdoor unit, the water regulating valve must be a three-way valve, or there must be a factory piped bypass that can allow glycol to flow through it when demand is low. The system provider knows this rule, and will furnish the proper equipment, as it is in his best interests. Still, for as simple of a rule it is to remember, it's worth knowing it.

The final topic of discussion for this section and for this chapter is the topic of "water detection." We can't leave this chapter without at least touching upon this all-too-important topic. What is water detection as it applies to these types of A/C units, and why is it important? How does this fit in with the subject of controls, and why is it brought up here?

Okay, okay. With any type of cooling device, be it a chilled water coil in a built up air handler or a DX coil in a self-contained computer room A/C unit, condensation tends to form as air is cooled down by the coil and moisture is wrung out of the air. This condensation collects in a drain pan, and is carried off by means of condensation piping, to a drainage location. If for whatever reason the condensation piping clogs, the drain pan will begin to fill up. Once full, the water will of course proceed to spill over the rim of the pan, and out onto the floor of the computer room. This is a big, big no-no for this type of application, especially if the A/C unit is not floor mounted but hung from above. In either case, water damage to computer room equipment is probable, and the losses incurred from such damage are often irrecoverable.

A water detection switch can be installed in the drain pan to monitor it for excessive water level. If the level in the pan reaches a dangerous level, the switch will break, and hence disable the operation of the A/C unit. These types of switches are often provided as factory equipment, either standard or as an option, though can also be furnished "after the fact" and field installed. These switches are not of the manual reset type, meaning that, if the water level drops back down to a safe level, they will automatically reset and allow operation of the unit. A slow condensation drain will result in cycling of the unit on the water detection switch. A complete clog will result in complete shutdown of the unit.

What about the sensitive electronic equipment in the space? If the A/C unit is disabled, the temperature in the space will tend to rise, and damage to the equipment could result. To put it bluntly, the "lesser of the two evils" is to disable the unit, rather than have the computer room equipment exposed to potential water damage. This here is one of the very reasons that redundancy is often built into computer room cooling systems. If an A/C unit fails or is disabled for any reason, there should be another unit just waiting to come on and save the day! At the very least, the failure should be immediately enunciated, so that the proper actions can be taken by the personnel in charge of maintenance.

Chapter 23

WATER COOLED SYSTEMS

We've already covered some water cooled systems and equipment in previous chapters, so this chapter will serve to "round out" that topic. In Chapter 18, we discussed water cooled chiller systems, and in Chapter 22, we talked a little bit about water cooled air conditioning equipment. The purpose of this chapter is to discuss not so much the equipment utilizing condenser water, but the condenser water system itself. We do devote a section to water cooled equipment, though short in comparison to the following section on condenser water systems. It is important to understand the operation of water cooled equipment, and its dependency on condenser water, before we can explore the many facets of condenser water systems.

This chapter revisits many concepts that have been covered thus far throughout the entire book, and kind of "pulls them together" into one topic. The focus of this chapter is to convey the understanding of what is meant by such terms as "water cooled" and "condenser water," and how controls play a major part in the successful operation of water cooled equipment and of the condenser water system that serves the equipment. For all practical purposes, this chapter can be considered more of a "bonus" chapter, and as many of the topics have been previously covered, it doesn't necessarily introduce many new concepts. Nevertheless, it will serve to demonstrate how all of these individual controls concepts and subsystems have to come together and operate as one.

WATER COOLED EQUIPMENT

To begin this chapter, we introduce a new term: "water source heat pump." Actually we brushed upon this term in the last chapter, with no explanation. The term is used to describe a packaged, single zone, water cooled heating and air conditioning unit. Comparable in size to a small fan coil unit, and typically residing in the space above the ceiling, the unit (minimally) consists of a supply fan, a refrigeration system/cycle, and ductwork to and from the space served. A thermostat in the space controls the unit in typical single zone fashion.

For any given unit, the fan runs continuously or intermittently, as determined by the fan switch on the thermostat. On a call for cooling, the unit's compressor starts, and the air passing through the refrigeration coil is cooled. Heat is rejected to the condenser water system, by way of the unit's condenser.

Now, what happens when there is a call for heating? The unit may be equipped with electric heat. More likely, the unit will utilize its refrigeration cycle to produce heat at the coil. Huh? How's that done? Well, the trick is to "reverse" the cycle, so that the evaporator (coil) becomes the condenser, and vice versa. This is done by means of a four-way reversing valve. Upon a call for heating by the thermostat, the reversing valve is energized along with the compressor. The refrigeration cycle works in the opposite direction, and heat is pulled from the condenser water system and transferred to the supply air via the coil. This is the basic concept of a heat pump; the term referring to the idea that heat can be pumped or transferred bidirectionally.

Small single zone units such as these will typically not require the use of condenser water regulating valves for proper operation. You will recall from the previous chapter the need for water regulating valves on computer room style A/C units. These "application critical" systems command precise control of condenser water right at the unit. Yet for the run-of-the-mill unitary heat pump, a condenser water regulating valve is more of an option than it is a requirement. For these types of unitary water cooled equipment, as long as the condenser water temperature can be maintained within an acceptable range, then the packaged heat pump need not be equipped with a water regulating valve. Of course this puts the onus on the condenser water system, yet maintaining acceptable condenser water temperature shouldn't be too demanding of a task, as we shall explore further on.

A final operating concern with unitary water cooled equipment, as with any water cooled equipment, is that the loss of condenser water flow to the unit will most definitely terminate unit operation. As pointed out in the last chapter, water cooled equipment will be equipped with a refrigeration pressure switch, which trips upon high pressure. Upon a call for cooling, the compressor starts, and if there is no means of rejecting heat (because there is no condenser water flowing through the condenser), then the refrigeration head pressure rises, to the point of tripping the high pressure safety cutout, which is a manual reset device. Unit operation is interrupted, and a service call is placed.

Water cooled equipment ranges from small, unitary equipment such as heat pumps, to midrange equipment such as computer room air conditioning units, to large systemic equipment such as chillers. A condenser water system can serve it all. The equipment of course is at the mercy of the condenser water system. A condenser water system that is down bodes ill for all equipment served by it. The importance of the functionality and reliability of the condenser water system is demonstrated above, and now we explore the components of a typical condenser water system, and how to make that system work!

Condenser Water Systems

The role of the condenser water system is an important one, with regard to the equipment served by it. As expressed in the previous section, no condenser water, no air conditioning! In simple form, the condenser water system itself isn't all that complex of a system. Yet the dependency of the water cooled air conditioning equipment on the condenser water system makes it a very crucial one, in terms of operation, maintenance, efficiency, and reliability. We start this section by listing the fundamental components that make up a simple condenser water system, and discussing how they are tied together. Further on, we will expand on the simple system, and explore different configurations and equipment variations. Finally, we summarize the requirement for controls. The latter part of this section will discuss many of the concepts covered already in previous chapters, but will serve to bring these concepts together as they are applied to a single mechanical system.

In its simplest form, a condenser water system consists of a main pump (usually two) and a cooling tower. The physical construction and operation of the cooling tower was explained in the chapter on chillers. To reiterate, a cooling tower is a large tank of which condenser water is pumped to the top of. By gravity, the water falls into the tank and is dispersed into droplets, thereby giving up its heat to the outdoors. The droplets collect at the bottom of the tank, referred to as the sump. The cooled down condenser water leaves the sump and is pumped back to and through the water cooled equipment. Cooling towers will typically have a fan or two (or more), physically situated so as to help release the heat from the droplets of water falling to the sump. The fan(s) will be controlled by a local temperature controller at the tower, whose sensor is installed in the condenser water piping leaving the tower. If the leaving condenser water is too warm, the fan(s) will be energized in an attempt to help reject the heat from the water. Finally, a cooling tower, if expected to be in operation year-round, may have a sump heater, if in a climate that would require it. The heater is an electric resistance immersion heater, controlled locally, to maintain the sump at a safe temperature above freezing.

A piece of equipment that we should talk about now, similar in function to a cooling tower, is what's known as a "closed circuit cooler." To the untrained eye, the physical appearance of this type of equipment resembles that of a cooling tower, at least from the exterior. Condenser water is pumped to the cooler. Within the cooler is a coil. The water is pumped through the coil, and gives up its heat to the ambient. A couple of on-board components help this process: a pump and a fan. The sump of the cooler is filled with water. Upon a rise in temperature above the desired condenser water temperature setpoint, the pump energizes to pull water from the sump and flow it over the closed circuit coil. Upon a further rise in temperature, the fan energizes to flow ambient air through the coil. The fan may be equipped with two speeds, so that the temperature control process at the cooler can actually be three stages: pump, fan on low speed, and fan on high speed. With all that's apparently involved here to reject heat from the closed circuit cooler, why use such a piece of equipment to begin with, and not just a standard cooling tower? Well, the cooler of course maintains the condenser water system as a closed loop system, thus providing isolation between the water and the elements. Without such a piece of equipment, the specifying engineer may dictate the use of a cooling tower in conjunction with an isolation heat exchanger. We'll expand on that concept further in our discussions.

The requirement for a three-way bypass control valve up at the tower (or closed circuit cooler) depends upon the climate. As you will recall from the prior discussions on water cooled chillers (see Chillers chapter), the purpose of such a valve is to bypass varying amounts of water around the tower, in the event that the condenser water traveling back to the water source equipment gets too cold. In essence, the valve serves as a low limit. A temperature controller, with its sensing element in the condenser water pipe leaving the valve, modulates the valve to maintain this low limit. Condenser water used for cooling purposes is generally kept in a safe range, between the high and low limits. The high limit, typically around 90 degrees, is maintained by the cooling tower controls. The low limit, typically in the area of 65-70 degrees, is maintained by the three-way valve and controller.

So the simple system, as illustrated in Figure 23-1, consists of a pair of condenser water pumps, a cooling tower or closed circuit cooler, a three-way valve for low limit control, and condenser water piping to and from. A system of this simplicity is fine for water cooled air conditioning equipment. But what if the equipment is expected to heat as well? In other words, what if the unitary equipment has the ability to reverse its refrigeration cycle and produce heat? It would follow that the condenser water system would not only require the ability to absorb heat from the water source equipment, but also to add heat to the equipment. In such a system, if all equipment served by the system was in a heating mode at the same time, all available heat from the condenser water loop would be extracted, with none added. In this scenario, it's likely that it's cold outside, so the condenser water system doesn't stand a chance of picking up any heat from the outdoors. What happens here in this hypothetical scenario is that the condenser water loop cannot furnish the amount of

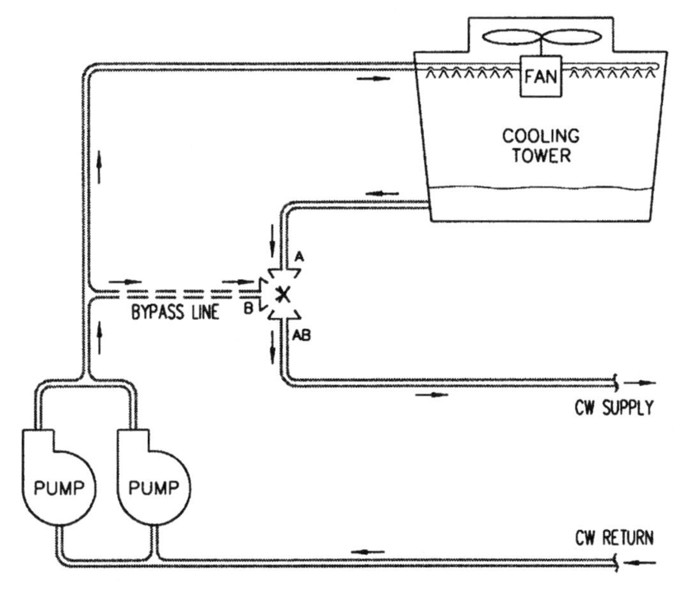

Figure 23-1. Condenser water system consisting of a pair of pumps, a cooling tower, and (as required) a three-way bypass control valve.

heat required by the water source equipment, and the equipment in turn cannot adequately perform the role of heating.

What's missing here? How about a boiler? Can we add heat to the loop with the addition of a boiler system? Certainly! Okay, but how's it done? First, you need a boiler and a pump or two. For this example, let's assume not one but two boilers, and a pair of hot water pumps. The boilers are piped in parallel, and the operating pump circulates water from the boiler plant. Heat is injected into the condenser water loop either directly, by means of a three-way valve, or indirectly, with the addition of an isolation heat exchanger (as in Figure 23-2). With either method, the boilers are controlled to maintain a hot water loop temperature setpoint, and the three-way valve is modulated to inject heat into the condenser water loop, to maintain the "low end" setpoint, which is typically in the area of 75-80 degrees.

It's important to note that at this point it's no longer terminologically correct to refer to this as a condenser water system. With the additional responsibility to "give" heat to the water source equipment, the loop pulls double duty, and not only serves to condense, but also to evaporate. This in reference to the refrigeration cycle integral to each and every piece of cooling/heating equipment on the system. When an individual piece of equipment is in a cooling mode, the water serves to condense the refrigerant, and when in a heating mode, the water serves to evaporate the refrigerant. For the remainder of this chapter, we will not make a distinction, and will continue to refer to the loop that serves the water source equipment as the condenser water loop. Just be weary of the use of the terms, and be understanding of the concepts involved, and we'll be fine.

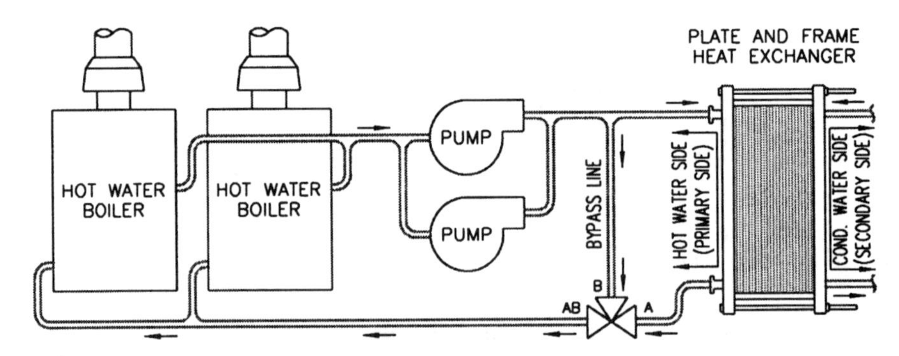

Figure 23-2. Diagram showing how a hot water piping/pumping system can tie in to and serve a condenser water system.

It is extremely interesting to explore the requirement for the boiler loop. As stated before in the hypothetical, or "worst case" scenario of all equipment on the condenser water loop calling for heat at the same time, a condenser water system with no boiler would not be able to satisfy the needs of the equipment. What about when some equipment is in a heating mode, and other equipment is in a cooling mode? The equipment in cooling is dumping heat into the loop, and the equipment in heating is pulling heat from the loop. Depending upon the balance, the boiler loop may be contributing very little to the system. In fact, in all but the coldest of days, there may be enough diversity to where the equipment calling for cooling may be dumping more than enough heat into the loop, to handle the equipment calling for heating. The term **diversity** refers to the fact that different parts of a building have different heating and cooling requirements at different times of the day. Southeastern exposures may require cooling in the morning, whereas western and northern exposures would require heat. Midday requirements might be for cooling for southern exposures, and for heating for northern exposures. And afternoon requirements could be for heating for the eastern and northern exposures, and for cooling for southwestern exposures. All this means is that the facility's HVAC system as a whole is rarely (if ever) loaded to its full capacity. Of course, engineers need to evaluate diversity, and still need to be conservative with their final load estimation. So the boiler loop, though often not theoretically essential, must nevertheless be included, for the worst case scenario.

Previously, we discussed the functions of cooling towers and closed circuit coolers. If a closed loop is required for the condenser water system, then a closed circuit cooler may be specified. Otherwise, another loop must be added to the system. Picture this. An isolation heat exchanger. On the primary side, a pair of pumps and the cooling tower. On the secondary side, the condenser water system. In essence, the condenser water system connects to the secondary side of the heat exchanger in exactly the same manner that it would connect to a cooling tower (or closed circuit cooler). Heat from the condenser water loop is rejected to the primary side of the heat exchanger, which carries it off to the open circuit cooling tower, by means of the "tower pumps." We'll call this particular loop the "tower loop." On the primary side is an open loop circuit. On the secondary side is a closed loop, isolated from the elements and kept nice and clean for the water source equipment served by it.

We've come to the point to where we've defined pretty much every piece of equipment that could possibly be component to a condenser water system. We've also defined three possible loops: condenser water loop, boiler or hot water loop, and tower water loop. For the purpose of attempting to pinpoint all of the possible required control schemes for such a system, let's take a look at each loop individually, and discuss equipment and controls resident to that loop.

Condenser Water Loop

Equipment:

- Cooling tower or closed circuit cooler
- Condenser water pumps

Main Controls:

- Tower controls
- Pump controller
- Three-way valve

The cooling tower or closed circuit cooler can generally come equipped with controls, at least to some extent. While it is possible to take control of the fans (and pump, if we're talking about a closed circuit cooler), it's often a good idea to let the manufacturer equip the unit with the appropriate controls. This goes for the electric sump heater as well, if so equipped.

The manufacturer can offer a control systems package, including a main system panel. The panel will house the controller for the fans, contactors and overload blocks, etc., and will have some supervisory devices as well (status lights, system switch, etc.). The panel is not necessarily weatherproof, and may need to be installed indoors.

If the tower is ordered without controls, a control system will need to be provided for it. The system will minimally consist of motor starters, a temperature controller, and a sensor. Being outdoor equipment, the motor starters may need to be weatherproof, and the temperature controller may either need to be housed in a weatherproof enclosure, or be located indoors.

A flow switch is a good idea, to enable and disable tower operation. If there is no condenser water flow, then disallow the operation of the fan (and pump).

An engineer may specify variable speed control of the tower fan. As such, the fan motor would be equipped with a VFD, controlled to maintain leaving condenser water temperature setpoint. As this is precise temperature control, it would seem to be "overkill" for a system that we described previously of typically operating to simply maintain a safe range of condenser water temperature. Nevertheless, the specifying engineer may have good reason to want this, and it is out there.

For condenser water systems, redundancy will be built into the pumping requirement, in the form of two pumps. If each is sized for half flow, then both will operate continuously. However, if each is sized for full flow, then an automatic primary/backup system is in order. A pump sequencer will be required, to provide automatic operation of the backup pump upon failure of the primary pump, and perhaps also to provide automatic alternation of the primary pump. A differential pressure switch or flow switch will be installed to monitor pump status, and will tie back to the pump sequencer.

The pump sequencer must not have a lengthy delay time between failure of the primary pump and operation of the backup pump. Delay between failure and backup is incorporated for the purpose of eliminating nuisance trips, and is acceptable for all pump sequencing applications. However, for condenser water applications, this delay must be on the short side. No flow conditions for water source equipment are detrimental to the operation of such equipment, and will not be tolerated by the equipment's limit controls. A unit calling for cooling and not getting condenser water flow will trip out on its own controls.

The three-way valve will be properly sized for proportional control. No 3-5 psi rule of thumb here, folks. This is a low pressure drop application, and the valve should be sized close to line size. A valve sized for a pressure drop falling within the 3-5 psi range will theoretically be undersized for the application. Better to "err" on the conservative side in the selection of the control valve, i.e., better to oversize the valve than to undersize it.

The valve will be controlled by a temperature controller that monitors leaving condenser water temperature, or in other words, the water that's delivered to the water source equipment. Local stand-alone control is acceptable, for this is typically a set-and-forget control scheme. For much of the time, the control valve is fully open to the tower. Only during colder periods will the valve begin to modulate. As it does, it bypasses varying amounts of water around the tower. The water going though the tower then mixes with the bypassed water, and is delivered to the equipment. The colder the water that is being sensed by the controller, the more water is bypassed around the tower. This valve should be spring return, and should fail open to the cooling tower.

Hot Water Loop

Equipment:

- Hot water boilers
- Hot water pumps
- Heat exchanger

Main Controls:

- Integral boiler controls
- External boiler controller
- Pump controller
- Three-way valve

The boilers will come with the full compliment of integral controls, which allow either boiler to operate completely stand-alone. However with two boilers, the use of a central boiler controller is warranted. A temperature sensor will be installed in the common supply header, which has full flow through it at all times. The sensor ties back to the controller, which operates the boilers in order to maintain loop setpoint (180 degrees typ.). A flow switch will also be installed and wired to the controller, for flow interlock.

Redundancy will likely be built into the pumping requirement, in the form of two pumps. A pump sequencer may or may not be specified or warranted. The importance of these pumps is not as great as the condenser water pumps, and manual backup may be sufficient. If a pump sequencer is warranted, then flow must be monitored, in the form of a differential pressure or flow switch, etc. It's good practice to enable and disable pump operation based on outside air temperature. If the outside air temperature is above 50 or 60 degrees, then the hot water loop will most definitely not be needed. The outside air temperature sensor/controller will tie back to the pump sequencer, and disable pump operation if the outside air temperature is above setpoint. This in turn disables boiler operation, per the flow switch interlock to the boiler controller.

The three-way valve will be properly sized for proportional control, and the 3-5 psi rule may or may not apply. If the valve is configured within the piping system to directly inject boiler loop water into the condenser loop, then the application is considered a low pressure drop application, and the valve will be sized at or close to line size. However, if the system design includes a heat exchanger, then the valve can be sized in accordance with the pressure drop through the heat exchanger, which may very well fall within our trusty rule of thumb.

The valve will be controlled by a temperature controller that monitors condenser water supply temperature. Local stand-alone control will suffice, though a boiler controller with additional smarts can provide the control for this valve. The valve will maintain the "low end" setpoint (75-80 degrees), by injecting hot water from the hot water loop directly into the condenser water loop. If an isolation heat exchanger is involved, then the valve modulates flow through the primary side of the heat exchanger, to maintain temperature setpoint on the secondary side of the heat exchanger. The valve should be spring return, and should fail closed to the condenser water loop, or to the heat exchanger.

Once a boiler is determined to be required in a condenser water system, the role of the cooling tower three-way control valve can be "downgraded" from proportional control to simple two-position control. The premise behind this is that, the valve will no longer be required to maintain a low limit condenser water setpoint. It is now the job of the hot water loop's three-way valve to do that. And since the system will be employed not only to extract heat from the equipment, but also to add heat, the low end setpoint will be in the range of 75-80 degrees, which is higher than what the cooling tower valve would be set to maintain. So the cooling tower valve is out of a job? Well, not completely. The valve can be employed for two-position duty, so that, when the system calls for heat injection (via the boiler loop), the valve can be positioned to fully bypass the cooling tower. Otherwise we might have the cooling tower working to remove the heat that we are adding to the loop, which would be counterproductive, and not very energy conscious. Caution is warranted, however. There must be a means of protecting the cooling tower and its associated piping from freezing during periods of no flow. This must be addressed, by adding glycol to the system, by heat tracing the outdoor piping and ensuring that the tower is equipped with a sump heater, or by other means.

Tower Water Loop

Equipment:

- Cooling tower
- Tower water pumps
- Isolation heat exchanger

Main Controls:

- Tower controls
- Pump controller

We've already discussed cooling tower controls, so we need not reiterate here. Note that the requirement for a tower water loop is driven by the desire for a closed system on the condenser water side. If the system is equipped with a closed circuit cooler in lieu of a cooling tower, then this loop need not be provided. If there is no requirement for a closed loop system, then a cooling tower alone will suffice, with no tower loop.

Now, if there is a requirement for a closed loop system, and this is not accomplished with a closed circuit cooler, then the tower loop will be required. The cooling tower leaves the condenser water loop, and "becomes one" with the tower water loop. Controls requirements are pretty much the same as if it were part of the condenser water loop.

As with the condenser loop, redundancy will be built into the pumping requirement of the tower loop, in the form of two pumps. This again is a prime candidate for a pump sequencer, no doubt about it. If the primary pump fails, the backup pump should automatically be pressed into operation, to provide uninterrupted service of the condenser water system.

The cooling tower three-way control valve can basically disappear with the addition of this loop. During periods of heat injection into the condenser water loop, the tower loop pump can simply be commanded to be off. One way to accomplish this (among other ways) is with an end switch off the hot water loop three-way valve. As the valve begins to modulate, the end switch would be set to open, and can be wired into the tower loop pump sequencer as an "external shutdown command." Now, with no flow on the primary side, no heat is exchanged from the secondary side. In other words, no heat is drawn from the condenser water loop to the tower water loop, this being the "preferred" operating mode during periods in which the hot water loop is dumping heat into the condenser water loop. Allowing the tower pump to run and the heat exchanger to pull heat from the loop would serve to help defeat the purpose of the operation of the hot water loop. So simply turn the tower pump off whenever the hot water loop is injecting heat into the condenser water loop. Again, the means of freeze protection must be addressed as applicable, for the tower water loop during periods of no flow.

Overall Control of the Condenser Water System

Figure 23-3 shows the hypothetical system that we have developed over the past few pages, made up of the three loops described (condenser water, hot water, and tower water loops). Again, the tower water loop is necessitated by the desire for a closed loop system, and the decision not to use a closed circuit cooler. By studying the illustration, we can analyze the three separate piping loops, assess the mechanical equipment making up each loop, and summarize the requirement for controls for each individual loop.

Wow! That's a lot of equipment making up a single mechanical system! And a lot of control loops as well! With all that's going on in this system, wouldn't it seem to make sense to attempt to integrate all facets of control into a single control system, instead of having numerous stand-alone controllers performing various functions? The simple answer to this question is yes, and the solution is to provide a distributed DDC system, consisting of digital controllers performing the various operations, all tied together via a network. This is the basic concept behind Building Automation Systems: utilize digital controllers and distribute the control tasks among these controllers. Yet instead of allowing them to operate in stand-alone fashion, tie them all together, and allow them to "work off of each other," to share information, to perform monitoring of critical functions and parameters, and to ultimately operate seamlessly together, as a single control system. Tie a front end (computer interface) to it, and you have a BAS, with the front end serving as the main user interface to the control system. No need to physically go around and check the operation of each controller on the network. All of the information that you need is right at your fingertips. The front end can provide trending of critical values, and enunciate failures as well, and can even initiate dial-outs upon system malfunctions.

The trend toward networked digital control systems is one driven by many factors. Originally, the primary driver was energy conservation. Nowadays, that is still very much a concern, however other benefits afforded by such a system are valid drivers as well, in the decision to go with networked digital control. This chapter demonstrates one of those additional drivers, and that is the ability of pulling many different control schemes of a single mechanical system together, to provide a single, coordinated control system.

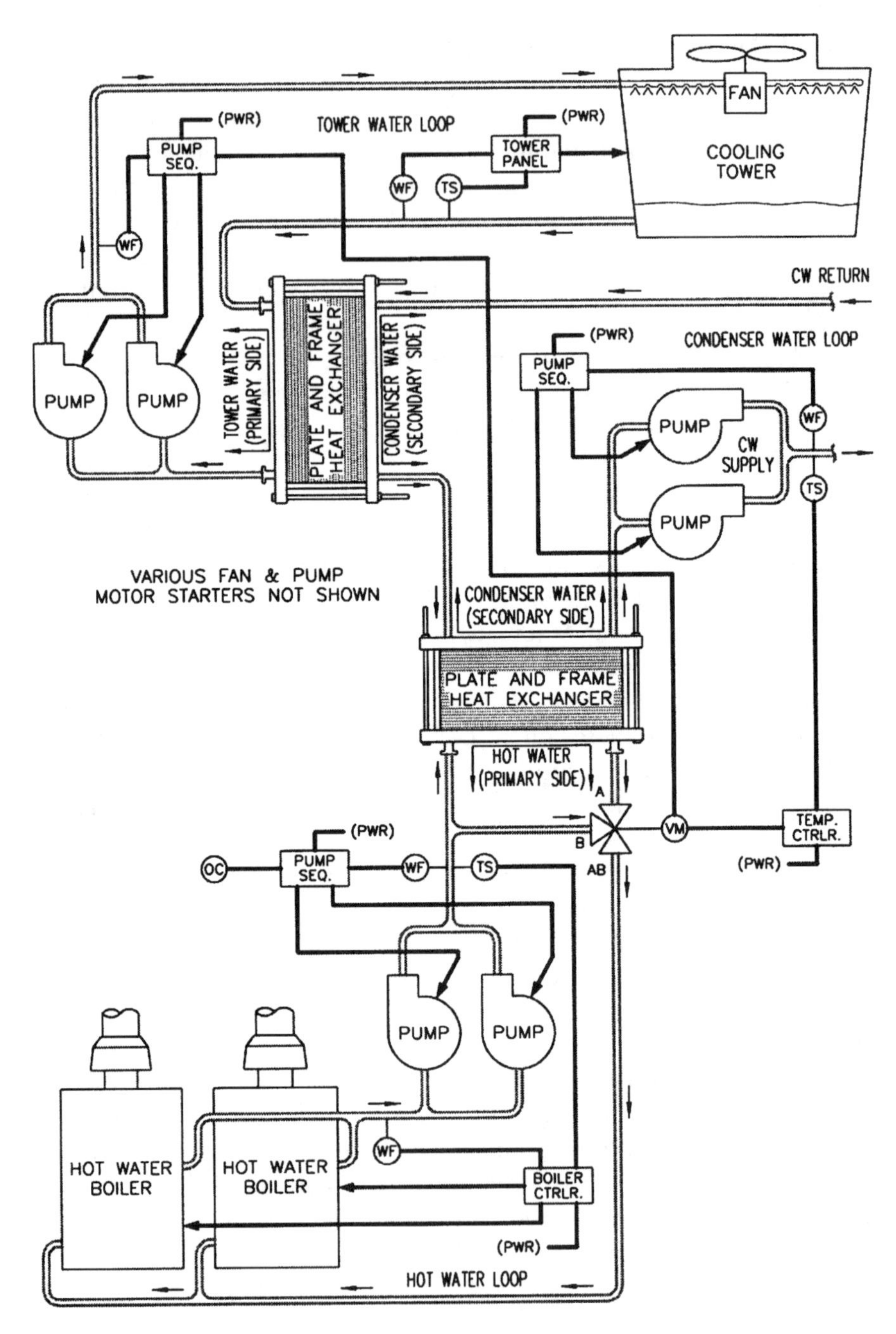

Figure 23-3. "Three-loop" system.

Chapter 24

CONCLUSION

This book has been written to serve a very distinct purpose. As the title suggests, it was written to illustrate the control of mechanical systems, using practical means and methods. Yet in order to serve that purpose, a good deal of the book was devoted to developing insight into the *design* of mechanical systems, in terms of how they are intended to operate and be controlled. A prerequisite knowledge and understanding of mechanical systems can be followed up with an education in controls and control systems. Hopefully this book has covered some ground in both of these areas.

The content of this book is meant to serve as a resource, a reference guide if you will, for future applications. There are some key issues, however, that the reader should come away with upon finishing this book. This short chapter touches upon those key issues that should remain with the reader, and tries to summarize the basic intents of the author.

Chapter 1 introduced the reader to the definition of an HVAC system, as being a mechanical system plus the control system required for its proper operation. The chapter also introduced the term "sequence of operation," and demonstrated the importance of it.

The distinction between packaged and built up equipment was a key issue put forth in Chapter 2, as was how a packaged piece of equipment can be a part of a built up "system." "Hierarchy of control" was the term introduced to describe how packaged equipment with factory controls can be integrated into a higher tier system of control.

In general, three methods of control exist for the operation of mechanical systems. They were introduced in Chapter 3 as two-position control, staged control, and proportional control. This chapter made special reference to proportional control, and included P+I and PID control, as well as floating control, as types of proportional control. The chapter also formally introduced the term "setpoint."

Chapter 4 defined the functions of sensors and controllers, and made the distinction between the two. In short, a sensor itself is not a controller, but can be an integral part of a controller. A controller's purpose is to gather information from a sensor, process it, compare it with a setpoint, and affect an action upon the "controlled variable," in preferential accordance with the setpoint. The chapter also touched upon DDC, networking, and Building Automation Systems, illustrating that technology plays a powerful role in HVAC control. Yet though technology will continue to simplify efforts and improve end results, the basic concepts of control will tend to remain.

Chapters 5 and 6 finished off the general topic of controls. Chapter 5 gave a detailed overview of end devices, their relation to the three methods of control, and the role that they play in an HVAC system. And Chapter 6 introduced the reader to some of the more common control schemes applied in this day and age, and set out to instill a familiarity with the terms describing these schemes.

The rest of the chapters really got into the bulk of what the book is all about. This is where the book becomes more of a reference manual, for the design and evaluation of HVAC control systems. However, there are key issues located throughout, some of which are reiterated here.

When it comes to our business, we seem to have our own language. The terminologies used throughout the industry are far from standardized, and nowhere is this more prevalent than with controls. Terms are used and abused, and one individual's or manufacturer's lingo may be quite different than another's. The key issue here is to be aware of the terminologies, and understand their use and their context. While it may be acceptable to talk in simple concepts and "technically" misuse common terms, it must be done so with each side understanding the general ideas behind the terms, so that misinterpretation is avoided and expectations are met.

As for "rules of thumb," we tend to sometimes overuse and even misuse these handy little tools. So much of what we do has been done so many times before, that there is generally no reason to "re-invent the wheel" every time we embark on a new project. Hence we use these rules of thumb to assist our efforts and accelerate our design processes. They are extremely useful, and should be among the contents of every designer's tool box. The key issue is that they are to be used with caution, with an understanding of where they originated and how they relate. In other words, know the concepts behind the rules.

Also among the control system engineer's tools should be a solid understanding of the formulas that govern many of the processes that constitute a mechanical system. Strong analytical skills are an invaluable attribute of any designer. To know how to manipulate the equations and formulas that characterize any mechanical system is a powerful skill to possess. It is the controls engineer's responsibility to not simply design a control system, but to also know the criteria used in the design of the mechanical system, so that he knows how to design his control system, in alignment with the mechanical engineer's intended method of operation for the mechanical system.

It must be known: controls can't work magic. A control system is only as good as the mechanical equipment that it is controlling, and cannot compensate for mechanical malfunctions and inadequacies. An example of the flip side of this "conventional wisdom" is the occupant who turns the thermostat controlling his rooftop unit down from 70 to 60 degrees, because it's 80 in the space. Of course, when it's 80 in the room and the thermostat is set at 70, turning it down any further will do no good. The thermostat has done all that it can do. This example demonstrates in simple fashion that not all environmental control problems can be blamed on the control system. Yet the controls are frequently the first to be blamed. This is seen more and more as digital controls propagate the industry, where the simple temperature control system is replaced with "black box" controls that people can't tell by looking at whether or not they're doing their job. For the vast majority of circumstances, the properly commissioned control system will be doing what it's supposed to be doing, and the failures will be mechanical in nature.

In line with the above train of thought, controls cannot compensate for a poorly designed mechanical system. And problems that surface upon start-up are not always attributable to the control system. The reliance on the control system should not be such that the controls will be able to overcome mechanical system shortcomings. Sure, a slick control systems engineer may be able to help out an ill-conceived mechanical system. But the control system in general will not be a panacea for all mechanical design problems, and should not be looked upon in that manner.

The importance of the control system should be evident, in what it brings to the HVAC system as a whole. It must be solid in concept and in design. And it must allow the well designed mechanical system to operate as intended. If properly applied and implemented, it can bring value to any HVAC system, even the ill-conceived system, to an extent.

Yet if poorly designed and inadequately commissioned, the control system can wreck even the most well thought out mechanical system.

The final chapter to this book (prior to this one) closed by touching upon the advantages of "distributed DDC." The technologies available in this day and age open up a whole new world of control system options and flexibility. Distributed control systems have become commonplace, and will continue to grow both in size and in number. As we as an industry move away from simple controls and control systems to integrated microprocessor-based systems and networks, we must hone our own skills and stay up with the trends and the technologies. Yet in the end, our installations must still appeal to the end user. A sophisticated control system must be made to "appear" simple, at least to the recipient of the system. That may be our biggest challenge for the future, because when all is said and done, the average Joe will still continue to hold on to the value of simplicity.

If there was any one particular purpose of writing this book, it would be not necessarily to teach the technical aspects of controls and mechanical systems, but perhaps more so to provide the reader with insight into the proper operation of these systems. And also to hone that skill, in hopes that the reader can perform his own evaluations and come to his own conclusions on how mechanical systems should operate and be controlled. To the individual that has gotten this far and completed the book, hopefully you have acquired at least a bit of that insight. And if you understand the key issues as well, then this book has succeeded in its goal.

Appendix 1

FORMULAS

Following is a list of formulas and equations in the order that they appear in the book, with a description of each:

$$P = (gpm/Cv)^2$$

Pressure (in psi), expressed as a function of flow and Cv (valve coefficient.

$$MAT = \{(RAT \times \%RA) + (OAT \times \%OA)\}/100$$

Mixed air temperature, expressed as a function of return air temperature and percentage, and outside air temperature and percentage.

$$\%OA = \{(MAT - RAT)/(OAT - RAT)\} \times 100$$

Percentage of outside air, expressed as a function of mixed air, return air, and outside air temperatures.

$$Btuh = 1.08 \times cfm \times delta\ T$$

Airside heat transfer equation

$$Btuh = 500 \times gpm \times delta\ T$$

Waterside heat transfer equation

$$Btuh = kW \times 3410$$

Conversion of kilowatts to Btuh

$$kW = Btuh/3410$$

Conversion of Btuh to kilowatts

$$psi = ft.\ hd./2.31$$

Conversion of “feet of head” to psi

Appendix 2

Control Symbols Abbreviations

Following is a list of the abbreviations used in the illustrations to symbolize controls devices and components:

AF Airflow Switch

AQ Aquastat

CD Control Device

DM Damper Actuator

DS Discharge Air (Temperature) Sensor

EC Economizer Controller or Enthalpy Controller

ES Enthalpy Sensor

FS Freezestat

HC Humidity Controller

HL High Limit (Temperature, Pressure, or Humidity) Controller

MS Mixed Air (Temperature) Sensor

OC Outside air (Temperature) Controller

OS Outside Air (Temperature) Sensor

PC Pressure Controller or Pump Controller

PT Pressure Transmitter

RS Return Air (Temperature) Sensor or Refrigerant Sensor

S Sensor

SD Smoke Detector

SS Space (Temperature) Sensor

SV Solenoid Valve

SW Switch

T Thermostat

TC Temperature Controller

TS Temperature Sensor

VM Valve Actuator

WF Water Flow Switch

ZS Zone (Temperature) Sensor

Appendix 3

Glossary

ctuator—An electrical device (motor) that can operate and control the position of a control valve or damper.

Alternation—For "primary/backup" equipment, a standard procedure by which the primary unit is taken out of service, and the backup unit is placed into service. For "lead/lag" equipment, the procedure of promoting the lag unit to "lead unit" and demoting the lag unit. For both scenarios, the procedure is typically performed on a daily or semi-weekly basis, to ensure even run-times and thus even wear on each piece of equipment.

Analog—Being able to assume any continuous value within a given range. The term is typically used to describe inputs and outputs on a digital controller that can possess any value (within a range) at any given instance. Analog inputs can accept signals from sensors and transmitters, and analog outputs can operate proportional end devices.

Application Specific (Controller)—Referring to a controller manufactured for a specific control purpose. Typically for a zone level application, such as a fan coil unit or a VAV box.

Automatic Reset—Referring to a controller (typically a safety device) that will reset itself after a trip, once the fault condition that tripped the device clears.

Auxiliary (Aux.) Contact(s)—A set of dry contacts furnished as an option or add-on accessory for a motor starter. The contacts change state when the starter is energized. The contacts can provide information about motor status, or can serve as an interlock or control point for a separate (related) control sequence.

BAS (Building Automation System)—A system of digital (DDC) controllers networked together and interfaced with via a front end.

Binary—Having two possible states. The term is typically used to describe inputs and outputs on a digital controller that possess one of two values at any given instance. Binary inputs can accept contact closures (open or closed) from external devices, and binary outputs can affect the state of two-state end devices (on/off, open/closed, etc.).

Btuh—British thermal units/hour

Built Up (Equipment)—Referring to mechanical equipment that comes from the manufacturer with no controls, and thus must be equipped with an engineered "built up" control system.

Canned—Referring to software (or firmware) that has been developed for a specific purpose or application, to where custom programming from scratch need not take place. For an application specific controller, this refers to the firmware that is burned into the controller's microprocessor chip, that simply needs to be configured to fit the given application.

cfm—cubic feet per minute

Changeover—Referring to the switching between two modes, performed either manually, as per a user switch, or automatically, as a function of some variable. An example of manual changeover is switching between the heating and cooling modes of a rooftop unit, via the system switch on the thermostat controlling it. An example of automatic changeover is switching between economizer enable and disable modes, as per an outside air enthalpy controller.

Contact(s)—A pair of electrically conductive plates or strips that, when in contact with each other, complete an electrical circuit by allowing current to flow between the two. Similar in context to an electric switch.

Control Band—See Throttling Range.

Control Point—In proportional control, the point at which the controller controls to at any given time. Control point is "load dependent." Theoretically, if the load on the control process is at "half capacity," then the controller controls to the desired setpoint, and thus the control point is the actual setpoint. For all other operating conditions, the controller will control to within the control band or throttling range, but to a point other than the actual setpoint. This is an inherent characteristic of proportional control.

Control Signal—In general terms, the output of a controller that is sent to an end device. The term is typically restricted to proportional control, wherein the signal is a varying, proportional control signal.

Control Valve—A mechanical device that, when fitted with an actuator, can control the passage of water or steam in a piping system.

Controlled Device—See End Device.

Controlled Variable—The variable that any control process is attempting to control to setpoint. For example, in controlling the hot water valve of a fan coil unit to maintain space temperature, the space temperature is the controlled variable. The hot water is the means of maintaining the controlled variable at setpoint, and is referred to as the "controlled medium."

Controller—A device that gathers information on a sensed variable, processes it, compares it with a setpoint, and acts upon the (controlled) variable in an effort to bring it closer to the desired value (setpoint).

CV—Constant Volume

Cv—Valve Coefficient, defined as the flow rate (in gpm) through a valve (when fully open) at a pressure drop across the valve of 1 psi.

Cycling—A term used in the context of two-position control to simply describe the manner in which an end device is controlled by the controller, as being turned on and off, on and off.

Daisy Chain—Typically referring to the manner in which digital controllers are connected together with a communication cable, to form a network. The cable runs from one controller to the next, connecting all controllers on the network in serial fashion.

Damper—A mechanical device that, when fitted with an actuator, can control the passage of air in a duct system.

DCV—Demand Controlled Ventilation. A control process in which the level of CO_2 in the space served by an air handling unit provides the basis for the amount of outside air to be brought in by the air handler's outside air intake. CO_2 is a human by-product, and can be used as an indicator of building occupancy. The higher the level of CO_2, the more occupants, and therefore the more outside air required.

DDC—Direct Digital Control. The term used to describe microprocessor-based control. Analog information (from sensors, transmitters, etc.) is converted to digital information, so that it can be processed and manipulated by a microprocessor. Likewise, digital results are converted to analog information, to control the operation of "real-world" end devices.

Deadband—A term used when referring to controllers that have both heating and cooling capabilities. Deadband is defined as a "no action" zone between heating and cooling setpoints. When the sensed temperature is in this range, no actions are taken upon the end device (or equipment) served by the controller.

Delta T—The difference between entering and leaving air temperatures of a coil at any given operating point. Also (for water coils), the difference between entering and leaving water temperatures of a coil at any given operating point.

Design Day—A day in which a mechanical system is theoretically at full load. Design days provide the basis for the capacity sizing of mechanical equipment and systems. Chicago design days are –10 degrees and 90 degrees.

Differential—A term used when referring to two-position control. Differential is defined as the amount of change in the controlled variable that must take place, once a two-position controller has taken action upon a controlled device, for the controller to relinquish its command upon the device. For example, how many degrees a space temperature must rise (above setpoint) once two-position heating is engaged, in order for the process to be disengaged.

Digital—A term with roots in digital electronics. The term stems from what is meant by the ability of a piece of electronic information (bit), to assume one of two states: either a "zero" (absence of voltage) or a "one" (presence of voltage). Real world information is converted (via analog to digital converters) to digital information (zeros and ones) before it can be processed by a computer. The term digital is often used interchangeably with the term binary, though it does have more meaning.

DIP Switch (Dual In-line Package)—A set of two-position switches packaged together in one device that is typically mounted on an electronic printed circuit board. The device is used to (help) configure the electronic board (controller) for the appropriate application.

Diversity—Referring to the notion that different parts of a building have different heating and cooling requirements at different times of the day. Following this notion, an HVAC system sized for the sum of the peak loads of every zone within a building would theoretically be oversized under all operating conditions.

Dry Bulb—The temperature of air that is measured by an ordinary thermometer, giving no information about humidity.

Dry Contact(s)—Contacts that do not source any voltage or current on their own.

DX—Direct Expansion

EAT, EWT—Entering Air Temperature, Leaving Air Temperature

Economizer—The section of an air handling unit, consisting (minimally) of outside and return air dampers, that is used to provide "free cooling" if outdoor air conditions permit. The dampers are controlled in unison (outside air damper opens as return air damper closes), to vary the amount of outside air brought in as required to meet cooling needs.

End Device—A device designed to accept the commands from a controller, and affect a change in the value of a controlled variable.

End Switch—A set of dry contacts furnished as an option or add-on accessory for a valve or damper actuator. The contacts change state at some point in the actuator's stroke, and is often adjustable. The contacts can provide information about valve or damper status, or can serve as an interlock or control point for a separate (related) control sequence.

Equipment Level (Controller)—Referring to a controller (usually digital) suited for equipment level control (e.g. built up air handler), and having enough inputs and outputs to accommodate equipment of such size.

Equipment Schedules—Tables that are found on the mechanical plans of a project that list out the mechanical equipment and detail the attributes of each piece of equipment.

Error (Proportional Control)—See Offset.

Face/Bypass Control—The process of controlling a set of dampers to regulate the amounts of air passing through and around an active coil, to achieve the appropriate mixture of both air streams in order to maintain temperature setpoint.

Firmware, Hardware, & Software—Terms describing typical components of DDC. Firmware is essentially software that is burned into a controller's microprocessor, that can't be altered. Hardware is the actual electrical and electronic components making up a controller. And software is what is used to configure and program a controller.

Floating Control—A method of control in which the controller issues commands to the end device as a means of keeping the controlled variable within a suitable range, or "null band." Specifically, the controller issues a "drive open" or "drive closed" command (whichever is appropriate) to the end device (valve or damper actuator) if the controlled variable is out of range, and issues no commands if the controlled variable is within range.

Free Cooling—See Economizer.

Front End—Simply the computer (laptop or desktop) connected to a Building Automation System that serves as a means of interface to the controllers on the network.

gpm—gallons per minute

Half-travel—In proportional control, the condition in which the controller is signaling the controlled device to be at half of its full operating range. Such a condition corresponds to the controlled variable, as sensed by the sensor and processed by the controller, being precisely at setpoint (given "steady state" conditions).

Hardware—See Firmware, Hardware, & Software.

Hardwire Interlock—An interlock made between two pieces of equipment or two related control schemes that is accomplished with physical wiring, as opposed to other "less direct" means, such as logical or software interlocks.

Hierarchy of Control—Referring to the concept that packaged equipment with factory furnished control systems can be integrated into higher tier systems of control.

High Limit (Low limit)—Referring to the enforcement of a limit (high or low) by a controller. A two-position controller can enforce a limit simply by shutting down the process if the measured variable passes the established limit. A proportional controller can enforce a limit by modulating a controlled device to maintain the measured variable within the constraints of the established limit.

Human Interface Device (or Module)—A device used to interface with a digital controller. In general, such a device can be anything from a portable service tool to a laptop or personal computer, though is conventionally a hand-held device with a readout and a keypad, used as a means to interface directly to a controller.

Hunting—An undesirable condition associated with proportional control in which the end device controlled by the proportional controller is in continuous motion, traveling back and forth within (or beyond) the control band, without being able to "dial in" on the control point.

HVAC—Heating, Ventilating, and Air Conditioning

HVAC System—As the author defines it, an HVAC system is a mechanical system plus the associated controls and control system required to operate it.

Hydronic—Referring to the utilization of water as a means to transfer heat in HVAC. Under this definition, both hot and chilled water systems would be considered hydronic systems. Steam heating systems would not.

IAQ—Indoor Air Quality

In. W.C.—Inches of water column

Input—Referring to a termination point on a controller that can accept either a contact closure (as from a two-position controller), or a signal (as from a sensor or transmitter). Inputs that accept contact closures are referred to as binary or digital inputs, and inputs that accept signals are referred to as analog inputs.

Interlock—Referring to a "connection," either physical or virtual, between two pieces of equipment or two separate but related subsystems, in which the operation of one affects or influences the operation of the other. An example is a make-up air unit interlocked to an exhaust fan: when the exhaust fan is placed into operation, the interlock forces the make-up air unit to operate as well.

Intermittent—Referring (typically) to the operation of the supply fan (of any air handler) that does not run continuously, but rather cycles on and off upon calls for heating and cooling.

Interstage Differential—In staged control, the difference between the setpoints of two adjacent stages of control. Closely related to offset.

Jumper Wire—A removable conductor mounted between two points on an electronic printed circuit board. As with a DIP switch, the device is used to (help) set the configuration of the board for the appropriate application. Removing a jumper would result in a different configuration than if the jumper were left in place.

kW—Kilowatt

Ladder Logic—A traditional method of control system design using relays and switches to accomplish some desired function. By connecting contacts of different relays in series and in parallel, in some meaningful manner, and controlling the coils of the relays as a function of some condition, "operational logic" can be performed. The result of this logic can be utilized to control end devices in their intended manner.

LAT, LWT—Leaving Air Temperature, Leaving Water Temperature

Lead/Lag—Referring to when two (or more) pieces of equipment are sized so that the sum of their capacities equals the maximum system requirements. The term implies "staging," where stage one is the lead, stage two is the lag, etc. Often confused with the term Primary/Backup.

Load—The heating or cooling requirement at any point in time of a space (or entire building) that is served by an HVAC system. A full load situation equates to the particular space requiring the full capacity of the heating or cooling system served by it. The system itself is said to be "fully loaded" in this scenario.

Low Limit—See High Limit.

Manual Reset—Referring to a controller (typically a safety device), that, upon a trip, must be intervened with manually (via an on-board push-button) in order to restore normal operation of the controller.

MAT, OAT, RAT—Mixed Air Temperature, Outside Air Temperature, Return Air Temperature

MCC—Motor Control Center

Mechanical Cooling—As the author defines it, any method of using energy to perform cooling (e.g. direct expansion or chilled water cooling). The opposite of "free cooling."

Mechanical System—Mechanical equipment and the means of connecting the equipment, to other equipment, and to the real world, in some meaningful, functional manner. In HVAC, mechanical systems are typically designed to perform heating, cooling, and ventilation of spaces requiring such types of environmental control.

Microprocessor-based—Referring to controllers and control systems that use digital technology as the basis for control. A digital controller will have, as the brain of the device, a microprocessor, which processes information from the controller's inputs, and coordinates output values in alignment with a predetermined mode of operation. The desired mode of operation must be set within the microprocessor, typically by programming or configuring the controller via an external interface device.

Minimum Position—The position at which the outside air damper of an air handling unit is set so as to provide the minimum allowable amount of outside air. During periods of economizer operation, the outside air damper is modulated open from this position. However, when economizer operation is disabled, the damper will not be allowed to close beyond the setting of this parameter.

Modulating Control—See Proportional Control.

Morning Warm-up—The process of engaging air handling unit heating, normally at full thrust, first thing in the morning, as a means to quickly bring the space temperature up to occupied comfort levels before going into normal occupied mode of operation. Typically associated with VAV and reheat systems.

Multizone—An abbreviated term meaning "serving multiple zones." A multizone system is a mechanical system that serves not one, but multiple zones of comfort control. Most often used in the context of air handling units.

Night Setback—The process of shutting down an air handling unit at night, having it only to come on if the temperature in the space drops below the unoccupied mode setpoint.

Normally Closed—When referring to a set of contacts, it means that the contacts will allow current to flow when in their "normal" position. Typically refers to the contacts of a relay or contactor, in which "normal" means that the coil of the relay is unpowered. When referring to a control valve or damper, it means that the device will not allow fluid or air to flow when in its "normal" position. In this case, "normal" means that the "spring return" actuator of the valve or damper is unpowered.

Normally Open—The opposite of Normally Closed.

Nuisance Trip—Referring to the tripping of a safety device not necessarily related to a fault condition that the device is responsible for monitoring, but perhaps more so related to the improper setting, placement, or application of such a device. The term pertains to a situation in which a manual reset safety device keeps tripping unnecessarily, causing the "nuisance" of having to keep resetting it.

Null Band—A term used when referring to floating control. Null band is defined as a "no action" zone centered about the setpoint of a floating controller. When the sensed variable (temperature, pressure) is in this range, no actions are taken upon the end device served by the controller.

Offset (Multistage Control)—In staged control, the differences between the main setpoint (first stage setpoint), and the setpoints of subsequent stages. Each subsequent stage of control (after the first stage) has a corresponding offset value associated with it.

Offset (Proportional Control)—In proportional control, the difference between the control point and the actual setpoint at any given time. Refer to Control Point.

Operator Interface—See Human Interface Device.

Output—Referring to a termination point on a controller that can control two-position end devices, or can control proportional (modulating) end devices. Outputs that control two-position end devices are referred to as binary or digital outputs, and outputs that control modulating end devices are referred to as analog outputs.

Override—The process of forcing an input or an output to a desired value. Most often associated with two-position controllers and end devices. An example is a push-button that will override an air handler into a "run mode" when the air handler is scheduled to be off.

P+I Control—Proportional Plus Integral Control. P+I attempts to minimize the offset that is inherent in straight proportional control, so that the control point is closer to the actual setpoint for any given load.

Packaged (Equipment)—Referring to mechanical equipment that comes from the manufacturer with a factory furnished control system, allowing the equipment to operate virtually "out of the crate."

Parallel (Series)—referring to the arrangement of electrical devices (or even equipment within a piping or duct system) with respect to one another. See Figure 4-9 for illustrations of these terms as they pertain to sensor wiring configurations.

Personality—A term used to describe the configuration of an application specific controller. The controller, either through software or via DIP switches and jumper wires, is configured to fit the application, and is thus given a "personality."

PID Control—Proportional Plus Integral Plus Derivative Control. PID is a modification to P+I control, in which the "rate of change" of the controlled variable is also taken into the calculation. Adding the derivative element to the control process allows the process to evaluate how quickly the controlled variable is moving toward or away from setpoint, so that additional corrective action can be taken.

Pilot Device—A device that allows supervisory control of a specific process, or monitoring of a specific condition. The term typically refers to selector switches and push-buttons (for control), and indicating lights (for monitoring).

Pilot Duty—Typically refers to "low energy" control using relays or switches whose contacts are not rated for large amounts of current. For example, relays whose contacts are rated for pilot duty are used in the implementation of ladder logic.

Plant (Boiler or Chiller)—A term typically used with boiler and chiller systems, describing larger scale systems in which more than one piece of equipment (boilers, chillers) serves a common piping distribution system.

Plenum Return—Refers to when plenum space is used as the return air path to an air handler. Typical for VAV systems, the air supplied into the spaces by the terminal units leaves the spaces via "lay-in" return grilles.

The air continues on, via the plenum space, back to the air handler, whose return air opening is open to the plenum.

Plenum Space—The space above a typical lay-in ceiling.

Points (Points of Control)—A term used to describe a control operation, whether it be a sensing action or a controlling action. A point can be anything from a temperature sensor, to an output of a proportional controller operating a control valve. With digital controllers, a tally of the used inputs and outputs can provide an overall "point count" of a control system, and in this respect can serve as a quantitative evaluation of the given system.

Poll/Broadcast—Referring to communication over a network of controllers, whereby a single "network controller" may poll a group of unit level controllers for zone level information, and also broadcast information and commands to the unit level controllers.

Potentiometer—An electronic device whose resistance varies manually as the device is adjusted. Often in the form of either a rotating knob or a slidebar, a potentiometer allows manual adjustability of whatever parameter it is set up for and designed to represent. An example is a setpoint adjustment of the minimum position of an outside air damper.

ppm—parts per million

Primary/Backup—Referring to when two pieces of equipment are each sized to handle a single application, with only one operating at a time, while the other serves as a backup, in case of failure of the primary. The term implies "system redundancy."

Proportional Control—A method of control in which the controller can command the end device (regulating a process) to be in any position within its entire range of operation. With proportional control, if the controlled variable is at setpoint, then the controller positions the end device to be at half of its range. As the controlled variable strays in either direction from setpoint, the end device is repositioned accordingly. There is a linear relationship between the "deviation" of the controlled variable from setpoint and the control signal sent from the proportional controller to the end device.

psi—pounds per square inch

Psychrometric Chart—An imposing scientific chart that graphically displays the properties of air and moisture over a wide range of conditions, and allows for the relationship of these properties to be plotted and determined.

Pump Curve—a graphical representation of a pump relating capacity (in gpm) to pressure (in feet of head). For any given pump, capacity drops off as pressure increases. This relationship is plotted graphically, and is depicted as a curve. A pump working against increasing system pressure and delivering less and less gpm is said to be "riding back on its curve."

Reheat—The process of heating air that has already been conditioned (cooled and/or dehumidified) by a separate process, in order to maintain the appropriate comfort level(s) in the space(s) served by the processes.

Relay Logic—See Ladder Logic.

Remote Setpoint—Referring to the ability of establishing the setpoint of a controller remotely, as opposed to having to do so locally at the controller.

Reset—The process of automatically adjusting a setpoint as a function of some changing condition. For example, "hot water temperature reset control based on outside air" describes the process of automatically raising and lowering the hot water temperature setpoint as the outdoor air temperature varies. More precisely, the colder it is outside, the higher the setpoint.

Safety Device—A two-position controller designed to break a control circuit and affect an equipment-wide shutdown upon sensing an undesirable condition. Such a device typically requires a manual reset to restore the protected system to normal operation.

Sensor—A device that accurately measures a variable such as temperature, pressure, or humidity, and offers a control signal that varies in proportion with the sensed variable.

Sequence of Operation—A written description detailing the intended operation of a mechanical system by the accompanying controls and control system. The description should describe in detail how each piece of equipment and each subsystem should operate, so that the system as a whole is properly functional.

Series—See Parallel.

Service Tool—A term used to describe a human interface device used by commissioning and service personnel to interface with digital controllers.

Set and Forget—Referring to a control process that need only be set up once, and subsequently be left alone and "forgotten about." Applies to control processes that do not require continual adjustment of operating parameters (setpoint, differential, etc.).

Setpoint—A preference. A setpoint is given to a controller, and the controller operates an end device in "preferential accordance" with the setpoint.

Software—See Firmware, Hardware, & Software.

Solid State—A term in electronics traditionally used to describe electronic devices having "no moving parts."

Split System—A refrigeration system where the components of the refrigeration cycle are not in one piece of equipment, but split up as indoor and outdoor equipment.

Spring Return—The ability of an actuator to return to a "normal" position upon removal of power from the actuator, by means of a wound up spring within the actuator.

Staged Control—A method of control that breaks up the process into steps or stages. The further the controlled variable is from setpoint, the more stages of control are activated by the controller. Two-position control is a form of staged control, with the number of stages equaling one. Staged control can be thought of as "incremental control" of a process.

Staged Proportional Control—referring to the process of staging two (or more) proportional control loops in sequence. For instance, as the controlled variable deviates from setpoint, the first proportional control loop will be engaged, and its respective end device is modulated. Once the end device has modulated fully, if the controlled variable continues to deviate, the second proportional control loop is engaged.

Stand-alone—Referring to equipment that does not rely upon any type of external support in order for it to operate in proper fashion. In DDC, the term pertains to digital controllers (serving equipment) that are not networked as part of a larger system. In this context, each controller operates its respective equipment in "stand-alone" fashion, with no external support or hierarchical system of control.

Stratification—Referring to air passing through a duct that is not uniform in temperature. Typical for the mixed air section of an air handling unit, in which the mixture of outside and return air has not yet had a chance of blending thoroughly.

Subzone—Literally, a zone within a zone. An area of comfort control that is located within a larger, more general area of comfort control. A subzone will typically have limited control over its comfort needs, generally relying, at least to some extent, on what comfort level the larger zone is being maintained at.

Supervisory Control—Referring to the concept of taking additional control of equipment and systems that are already equipped with some form of control. Allowing and disallowing something to operate, as per some governing set of rules. Enabling and disabling a chiller via a higher tier control system is an example of supervisory control. Related to Hierarchy of Control.

Thermistor—An electronic device whose resistance varies as a function of the temperature of the medium surrounding it. Specifically, as the temperature increases, the resistance of the device decreases. Widely used in HVAC for temperature sensing purposes.

Thermostat—As the author defines it, a space temperature controller designed to operate a unitary piece of equipment typically consisting of fan, heating, and cooling sections. Control of heating and cooling is staged, generally limited to two stages per mode.

Throttling Range—A term used when referring to proportional control. The range above and below setpoint, that it takes a proportional controller to output a control signal from zero to maximum, and thus advance a proportional end device throughout its entire range or operation. Alternately referred to as control band.

Transformer—An electrical device that converts a voltage to a lesser voltage (e.g. 120-volt × 24-volt transformer).

Transmitter—Basically a sensor whose signal representing the sensed condition is amplified and conditioned with active electronics. Whereas temperature sensing can be done with a simple passive device (thermistor), pressure and humidity sensing must generally be done with transmitters.

Trip—Referring to when a manual reset safety device senses an undesirable condition and breaks the control circuit, thus affecting a unit or system shutdown. Once tripped, the device remains tripped and the control circuit remains broken until the reset button is pressed (and the fault condition is cleared).

Tri-state Control—See Floating Control.

Two-position Control—A simple method of control in which the controller issues a command to begin a process if the controlled variable strays far enough from setpoint, and relinquishes its command once the controlled variable comes back to setpoint. Only two states of control over the process are achievable (all or nothing).

Unit Level (Controller)—Referring to a controller suited for unitary control (e.g. VAV box). Unit level controllers are typically "application specific."

Unitary—Referring to mechanical equipment small in physical size, typically serving only a single zone.

User Interface—See Human Interface Device.

VA—Volt-amps

VAV—Variable Air Volume

Ventilation Schedule—A table that is found on the mechanical plans of a project that lists out the ventilation (outside/exhaust air) requirements for each and every space served by the mechanical systems.

Zone—An area of temperature control.

Index